Poly(lactic acid) Science and Technology
Processing, Properties, Additives and Applications

RSC Polymer Chemistry Series

Editor-in-Chief:
Ben Zhong Tang, *The Hong Kong University of Science and Technology, Hong Kong, China*

Series Editors:
Alaa S. Abd-El-Aziz, *University of Prince Edward Island, Canada*
Stephen Craig, *Duke University, USA*
Jianhua Dong, *National Natural Science Foundation of China, China*
Toshio Masuda, *Fukui University of Technology, Japan*
Christoph Weder, *University of Fribourg, Switzerland*

Titles in the Series:
1: Renewable Resources for Functional Polymers and Biomaterials
2: Molecular Design and Applications of Photofunctional Polymers and Materials
3: Functional Polymers for Nanomedicine
4: Fundamentals of Controlled/Living Radical Polymerization
5: Healable Polymer Systems
6: Thiol-X Chemistries in Polymer and Materials Science
7: Natural Rubber Materials: Volume 1: Blends and IPNs
8: Natural Rubber Materials: Volume 2: Composites and Nanocomposites
9: Conjugated Polymers: A Practical Guide to Synthesis
10: Polymeric Materials with Antimicrobial Activity: From Synthesis to Applications
11: Phosphorus-Based Polymers: From Synthesis to Applications
12: Poly(lactic acid) Science and Technology: Processing, Properties, Additives and Applications

How to obtain future titles on publication:
A standing order plan is available for this series. A standing order will bring delivery of each new volume immediately on publication.

For further information please contact:
Book Sales Department, Royal Society of Chemistry, Thomas Graham House, Science Park, Milton Road, Cambridge, CB4 0WF, UK
Telephone: +44 (0)1223 420066, Fax: +44 (0)1223 420247
Email: booksales@rsc.org
Visit our website at www.rsc.org/books

Poly(lactic acid) Science and Technology
Processing, Properties, Additives and Applications

Edited by

Alfonso Jiménez
University of Alicante, Alicante, Spain
Email: alfjimenez@ua.es

Mercedes Peltzer
University of Alicante, Alicante, Spain
Email: mercedes.peltzer@ua.es

Roxana Ruseckaite
National University of Mar del Plata, Argentina
Email: roxana@fi.mdp.edu.ar

THE QUEEN'S AWARDS
FOR ENTERPRISE:
INTERNATIONAL TRADE
2013

RSC Polymer Chemistry Series No. 12

Print ISBN: 978-1-84973-879-8
PDF eISBN: 978-1-78262-480-6
ISSN: 2044-0790

A catalogue record for this book is available from the British Library

Published by The Royal Society of Chemistry,
Thomas Graham House, Science Park, Milton Road,
Cambridge CB4 0WF, UK

Registered Charity Number 207890

For further information see our web site at www.rsc.org

Preface

Introduction

Issues related to sustainability, petroleum shortage and fluctuating oil prices together with the enormous increase in the use of polymers in areas of great demand, such as packaging, have intensified the development of new cost-effective, greener alternatives. The challenge is to adopt some fairly new design criteria to produce materials with the necessary functionalities during use, but which undergo degradation under the stimulus of an environmental trigger after use. Bio-based and biodegradable polymers have emerged as renewable alternatives to depleting ones and have the distinctive ability of being entirely converted into biomass and harmless by-products through microbial activity in appropriate waste management infrastructure.[1] For these reasons there is great interest in the so-called green polymer materials to offer an answer to sustainable development of economically and ecologically attractive technologies, contributing to the preservation of fossil-based raw materials, reducing the volume of garbage through compostability in the natural cycle leading to climate protection and reduction of carbon dioxide footprint.[2]

Recently, the possibility of replacing petroleum-derived synthetic polymers with natural, abundant and low-cost biodegradable products has gained much interest in both academic and industrial fields.[3,4] For instance, the production of plastics in Europe reached 57 million tons in 2012, mostly divided between polyethylene, polypropylene, poly(vinyl chloride), polystyrene and poly(ethylene terephthalate) production.[5] These fossil-based plastics were consumed and discarded into the environment, generating 10.4 million tons of plastic waste, most of which ended up in landfills (Figure 1).

Therefore the development of "environmentally friendly" materials will result in huge benefits to the environment and will also contribute to reduced dependence on fossil fuels. Polymers produced from alternative

RSC Polymer Chemistry Series No. 12
Poly(lactic acid) Science and Technology: Processing, Properties, Additives and Applications
Edited by Alfonso Jiménez, Mercedes Peltzer and Roxana Ruseckaite
© The Royal Society of Chemistry 2015
Published by the Royal Society of Chemistry, www.rsc.org

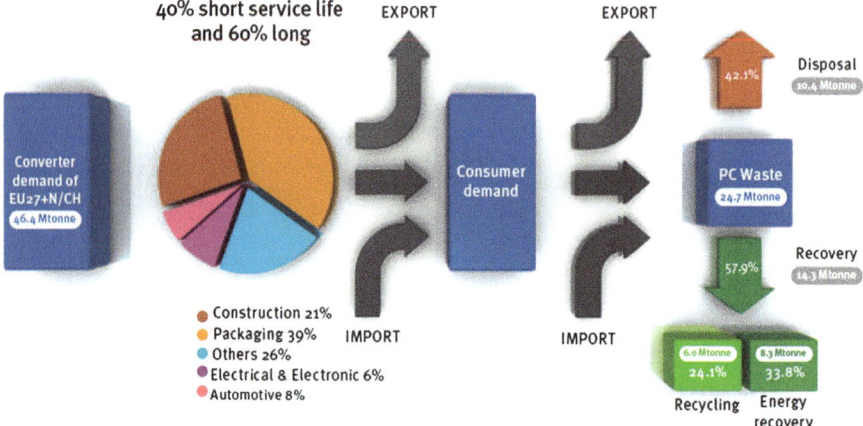

Figure 1   Plastics recovery in Europe for 2012.[5]

resources are a crucial issue, especially for short-life range applications, as they can be easily degraded by abiotic media or micro-organisms.[6,7] Nevertheless, the properties of these biomaterials are often behind those of common thermoplastics and some improvements are needed in order to make them fully operative for their industrial use.

The most optimistic forecasts have indicated that the potential of substitution of biopolymers to conventional synthetic ones can reach 15.4 million tons in the European Union (EU) (33% of the production of plastic at present).[5] The development of biopolymer matrices and their use in common applications is the subject of increasing interest by numerous research groups, since these materials are considered capable of substituting to certain synthetic thermoplastics. The reasons for this increase in the number of studies on these materials reside in the improving major concern on society with regards to environmental aspects and sustainability of the consumer goods, as well as to strict governmental regulations in the use of non-degradable thermoplastics.

General Issues on PLA Science and Technology

There are many kinds of bio-based and biodegradable polymers, among which one of the most promising is poly(lactic acid) (PLA), a biocompatible thermoplastic aliphatic polyester.[8] PLA is a linear thermoplastic polyester produced by the ring-opening polymerization of lactide. In general, commercial PLA grades are copolymers of poly(L-lactic acid) and poly(D,L-lactic acid), which are produced from L-lactides and D,L-lactides, respectively. The ratio of L-enantiomers to D,L-enantiomers is known to affect the properties of PLA,[8,9] *i.e.* if the materials are semicrystalline or amorphous. Until now, most of the efforts reported in order to improve the properties of PLA have been focused on the semicrystalline material.[8]

PLA is a most promising biopolymer, since it offers unique features of biodegradability as well as thermoplastic processability. PLA is used for many different applications, from packaging to agricultural products and disposable materials, as well as in medicine, surgery and pharmaceutical fields.[10,11]

Advanced industrial technologies of polymerization can be used in order to obtain pure PLA with high molecular weight, which is responsible for the improvement in the mechanical properties, with a better resistance to hydrolysis even under wet environment and adequate compostability.[2] There are some problems relative to its application, such as difficulty in controlling the hydrolysis rate, poor hydrophilicity as well as the high rigidity and poor control of its crystallinity. These drawbacks can be solved by the addition of nanofillers and/or by blending PLA with other biodegradable polymers.

The increasing research interest in PLA and PLA-based materials for multiple applications is represented by the large number of publications in scientific journals and monographs dealing with specific aspects in PLA science, technology and applications. A short survey of the recent literature reveals that the number of published articles related to PLA increased exponentially over the past decade, which is another clue to the worldwide rising interest in PLA research and innovation. Figure 2 shows some data taken from the Scopus® scientific database, where the increase in publications related to PLA research is clearly observed. The criterion for searching was the term "poly(lactic acid)" in the article's title or keywords.

Research in PLA involves a great amount of institutions and researchers around the world and reviews and monographs are necessary to extend knowledge in this important area in bio-based and sustainable materials. Some excellent works in reviewing the current research have been recently reported, most of them devoted to specific areas in the development of innovative PLA-based nanocomposites and applications.[12–17] A complete and

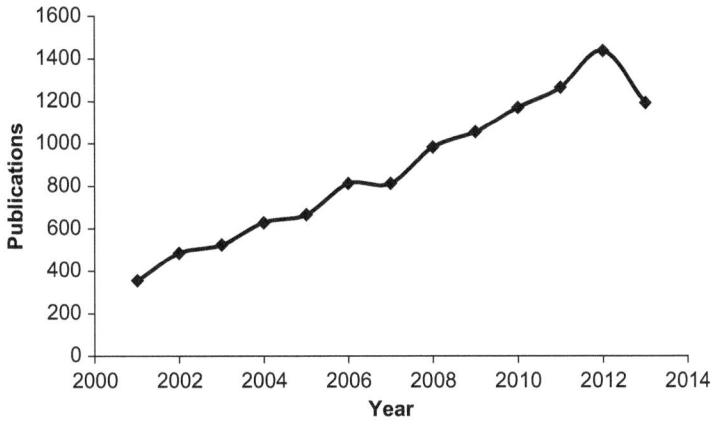

Figure 2 Number of publications related to PLA from 2001 to 2013 (source: Scopus®).

exhaustive monograph edited by Auras *et al.*[17] is particularly remarkable, where prominent authors in polymer science and technology highlight the most important features in PLA, from the origin to applications.

This volume will cover the main issues related to PLA science and technology and the goal will be to introduce the reader to PLA basics, in particular processing, properties and applications. The book aims to provide a broad survey of specific topics of PLA from the R&D stage to industrial developments and applications, including sustainability, legislative issues and the final market introduction of some selected items. In addition to highlighting cutting-edge technologies in PLA production, this book will address issues such as synthesis, plasticization, blending, compounding, bio/degradation and environmental traceability, future studies and perspectives, industrial uses and governmental policy issues.

With contributions from international experts, the editors' idea was to produce a book that will provide a unified perspective to these varied topics and is available for immediate use for a broad spectrum of potential readers, from students of chemistry, food technologies (particularly packaging) and materials engineering to researchers in the field of bio-based and bio-degradable polymers and composites. This is a hot topic and we, the editors, have carried out a selection and hopefully have chosen enough good reviews and unique works in the branch of PLA and its properties and final applications. These works are at the front edge of science and will be of interest for a large readership of scientists and students in the branches of chemistry and physics of polymers and biopolymers. We anticipate that you, the readers, will be happy with this selection.

This book is divided into six parts, where the most important issues in PLA science and technology are covered. The first part is devoted to PLA synthesis and polymerization, where different strategies for PLA synthesis from lactic acid will be described. The main physical properties of PLA will be also introduced, in particular those related to crystallization and stereo-complexation as major features in PLA structure. In the second part of this volume the use of innovative processing strategies, such as reactive extrusion and electrospinning, and the addition of plasticizers and elastomers to modify some of the key PLA properties, will be detailed. In the third section, two chapters will be devoted to innovation on nano-biocomposites based on PLA, where processing, properties and the main applications in packaging and biomedical sectors will be described. Different approaches and applications will be considered, in particular those with nanoclays and nano-celluloses to improve mechanical and barrier properties of PLA. These applications in key industrial sectors will be covered in Part 4, where the use of PLA in active packaging systems will be described. These new developments in food contact materials are very promising alternatives to the addition of synthetic additives to food formulations to increase food shelf-life. The use of bio-based and biodegradable polymers and natural additives will permit the design of fully environmentally friendly materials. Other important uses in the biomedical area, particularly but not only tissue engineering, will be also covered in Chapter 11. New approaches and different routes for degradation of PLA formulations in natural environments, particularly under hydrolysis, will be described in Part 5 as well as biodegradation in environmental conditions. Finally, industrial approaches for PLA formulations will be reviewed in Part 6, with a clear focus on the legislative issues regarding the use of bio-based polymers, particularly PLA, on key industrial sectors such as packaging, automotive, biomedical and others.

We are very confident that all these issues, covered by researchers with high expertise in PLA science and technology, will raise your interest and all comments and suggestions for further research work will be highly appreciated. We should acknowledge and thank all the authors for their hard work resulting in high-quality chapters as well as the Royal Society of Chemistry for their confidence in the preparation of this volume, in particular to Mrs Alice Toby-Brant for her infinite patience and hard work, essential for the conclusion of this book. We would like to thank our institutions, University of Alicante (Spain) and National University of Mar del Plata – National Research Council of Argentina, for allowing us to dedicate part of our effort and time to the completion of this edited book as well as to our colleagues, families and friends who supported us in this adventure.

Alfonso Jiménez
Mercedes A. Peltzer
Roxana A. Ruseckaite

References

1. R. Narayan, *Degradable Polymers and Materials ACS Symposium Series*, 2006, **18**, 282.
2. R. Auras, B. Harte and S. Selke, *Macromol. Biosci.*, 2004, **4**, 835.
3. A. Gandini, *Green Chem.*, 2011, **13**, 1061.
4. G. E. Luckachan and C. K. S. Pillai, *J. Polymer. Environ.*, 2011, **19**, 637.
5. www.plasticseurope.org. Accessed March 2014.
6. R. Chandra and R. Rustgi, *Progr. Polymer Sci.*, 1998, **23**, 1273.
7. A. Gandini, *Macromolecules*, 2008, **41**, 9491.
8. S. S. Ray and M. Okamoto, *Macromol. Rapid Comm.*, 2003, **24**, 815.
9. J. L. Feijoo, L. Cabedo, E. Giménez, J. M. Lagarón and J. J. Saura, *J. Mater. Sci.*, 2005, **40**, 1785.
10. K. Madhavan Nampoothiri, N. R. Nair and R. P. John, *Bioresources Technology*, 2010, **101**, 8493.
11. B. Gupta, N. Revagade and J. Hilborn, *Progr. Polymer Sci.*, 2007, **32**, 455.
12. T. Mukherjee and N. Kao, *J. Polymer. Environ.*, 2011, **19**, 714.
13. H. Zhou, J. G. Lawrence and S. B. Bhaduri, *Acta Biomater.*, 2012, **27**, 52.
14. P. K. Bajpai, I. Singh and J. Madaan, *J. Thermoplast. Compos. Mater.*, 2014, **27**, 52.

15. V. Piemonte (ed.), *Polylactic Acid: Synthesis, Properties and Applications*, Nova Publishers Inc., New York, USA, 2011, pp. 1–340. ISBN: 978-1-621-00348-9.

16. J. Ren (ed.), *Biodegradable Poly (Lactic Acid): Synthesis, Modification, Processing and Applications*, Springer, Heidelberg, Germany, 2011, pp. 1–297. ISBN: 978-3-642-17595-4.

17. R. Auras, L. T. Lim, S. E. M. Selke and H. Tsuji (ed.), *Poly(lactic acid). Synthesis, Structures, Properties, Processing, and Applications*, John Wiley & Sons, Hoboken, NJ, USA, 2010, pp. 1–499.

Contents

PLA Synthesis and Polymerization

Chapter 1 PLA Synthesis. From the Monomer to the Polymer   3
Kazunari Masutani and Yoshiharu Kimura

1.1	Introduction	3
1.2	Synthesis of Lactic Acids	4
	1.2.1 Stereoisomers of Lactic Acid	4
	1.2.2 Fermentation with Lactic Acid Bacteria	5
	1.2.3 Isolation and Purification of Lactic Acids	5
	1.2.4 Chemical Synthesis of Lactic Acids	7
1.3	Synthesis of Lactide Monomers	7
	1.3.1 Stereoisomers of Lactides	7
	1.3.2 Synthesis and Purification of Lactides	7
1.4	Polymerization of Lactide Monomers	8
	1.4.1 Structural Diversities of the Polylactides	8
	1.4.2 Thermodynamics for the Polymerization of D- and L-Lactides	9
	1.4.3 Metal Catalysts	11
	1.4.4 Cationic Catalysts	15
	1.4.5 Organic Catalysts	15
	1.4.6 Stereo-controlled Polymerization	18
	1.4.7 Stereo-block Copolymerization	21
	1.4.8 Copolymerization	22
1.5	Polycondensation of Lactic Acids	25
	1.5.1 Solution Polycondensation	27
	1.5.2 Melt/Solid Polycondensation	27
	1.5.3 Stereo-block Polycondensation	29
	References	31

RSC Polymer Chemistry Series No. 12
Poly(lactic acid) Science and Technology: Processing, Properties, Additives and Applications
Edited by Alfonso Jiménez, Mercedes Peltzer and Roxana Ruseckaite
© The Royal Society of Chemistry 2015
Published by the Royal Society of Chemistry, www.rsc.org

**Chapter 2 Polylactide Stereo-complex: From Principles to
           Applications**   37
Suming Li and Yanfei Hu

         2.1 Introduction   37
         2.2 Synthesis and Structure-Properties of PLA
              Stereo-complex   38
              2.2.1 Synthesis of PLA Homopolymers and
                     Stereo-copolymers   38
              2.2.2 PLA Stereo-complex by Co-crystallization   39
              2.2.3 Properties of PLA Stereo-complex   40
              2.2.4 Degradation of PLA Stereo-complex   44
         2.3 Biomedical Applications of PLA Stereo-complex   47
              2.3.1 PLA Stereo-complex Nanofibre Scaffolds   48
              2.3.2 PLA Stereo-complex Microparticles   49
              2.3.3 PLA Stereo-complex Micelles   50
              2.3.4 Stereo-complex Hydrogels   56
              2.3.5 Other Stereo-complex Systems   61
         References   61

Chapter 3 Crystallization of PLA-based Materials   66
A. J. Müller, M. Ávila, G. Saenz and J. Salazar

         3.1 Introduction   66
         3.2 Crystal Structure and Single Crystals   67
         3.3 Melting and Glass Transition Temperatures   68
         3.4 Superstructural Morphology   73
         3.5 Crystallization Kinetics   74
              3.5.1 Spherulite Growth Kinetics   74
              3.5.2 Application of the Lauritzen and Hoffman
                     Theory to PLLA   76
              3.5.3 Overall Crystallization Kinetics   80
         3.6 Block Copolymers Based on PLA   84
              3.6.1 PLLA–PEO Copolymers   84
              3.6.2 Copolymers *versus* Blends of PLLA and PCL   87
              3.6.3 PE-*b*-PLA Block Copolymers   89
         3.7 Conclusions and Outlook   92
         References   93

Processing, Characterization and Physical Properties of PLA

**Chapter 4 Reactive Extrusion of PLA-based Materials:
           from Synthesis to Reactive Melt-blending**   101
*Jean-Marie Raquez, Rindra Ramy-Ratiarison,
Marius Murariu and Philippe Dubois*

         4.1 Introduction   101

4.2   Polylactide (PLA)-based Materials: General Aspects
        and Reactive Extrusion (REX) Processing   102
4.3   REX Synthesis of PLA-based Materials *via*
        Ring-opening Polymerization   105
4.4   REX Coupling Reactions from PLA Precursors   107
        4.4.1   REX Isocyanate-based Coupling Reactions   108
        4.4.2   REX Epoxy-based Coupling Reactions   110
4.5   REX Free-radical Grafting Reactions of PLA Chains   113
4.6   REX Transesterification (Exchange) Reactions   117
4.7   Conclusions   118
Acknowledgements   119
References   119

Chapter 5   **Plasticization of Poly(lactide)**   **124**
             Alexandre Ruellan, Violette Ducruet and Sandra Domenek

5.1   Introduction   124
5.2   Principles of Plasticizing   125
5.3   Plasticizer Permanence, Migration and Interaction
        with Contact Media   129
5.4   PLA Plasticizers: Properties, Effects and
        Processability   134
        5.4.1   Plasticizer Impact on Mechanical Properties   134
        5.4.2   Plasticizers as Processing Aids   153
5.5   Long-term Stability of Plasticized PLA   155
        5.5.1   Physicochemical Studies of Long-term
                    Stability   155
        5.5.2   Chemical Stability and Degradation   159
5.6   PLA Plasticizers Derived from Biomass   160
5.7   Conclusion   163
Symbols and Abbreviations   164
Monomeric Plasticizers   164
Oligomeric Plasticizers   165
References   165

Chapter 6   **Electrospinning of PLA**   **171**
             Laura Peponi, Alicia Mújica-García and José M. Kenny

6.1   Production of PLA fibres by Electrospinning   171
        6.1.1   PLA Electrospun Nanofibres: Diameter,
                    Morphology and Orientation   175
        6.1.2   PLA Electrospun Nanofibres: Crystallinity   177
        6.1.3   PLA Electrospun Nanofibres: Mechanical
                    Properties   177

6.2   PLA Matrix Nanocomposite Electrospun Fibres   179
    6.2.1   PLA Electrospun Nanocomposite Fibres with
         Montmorillonites   180
    6.2.2   PLA Electrospun Nanocomposite Fibres with
         Halloysite Nanotubes   180
    6.2.3   PLA Electrospun Nanocomposite Fibres with
         Hydroxyapatite   181
    6.2.4   PLA Electrospun Nanocomposite Fibres with
         Carbon Nanotubes   182
    6.2.5   PLA Electrospun Nanocomposite Fibres with
         Graphene   183
6.3   Applications of Electrospun-PLA Fibres   183
    6.3.1   Tissue Engineering   184
    6.3.2   Wound Dressing   187
    6.3.3   Drug Delivery   188
6.4   Conclusions   191
References   191

Chapter 7   Modification of PLA by Blending with Elastomers   195
N. Bitinis, R. Verdejo and M. A. López-Manchado

7.1   Copolymerization   195
7.2   Blending with other Polymers   196
    7.2.1   Blending with Biodegradable Polymers   196
    7.2.2   Blending with Non-biodegradable
         Petroleum-based Polymers   197
    7.2.3   Blending with Elastomers   198
    7.2.4   Blending with Natural Rubber   203
7.3   Conclusions   208
References   209

PLA-based Nano-biocomposites

Chapter 8   Polylactide (PLA)/Clay Nano-biocomposites   215
Jose M. Lagarón and Luis Cabedo

8.1   Introduction   215
8.2   Nanoclays for PLA Nanocomposites   216
    8.2.1   PLA/Clay Nanocomposite Processing   219
    8.2.2   PLA/Clay Nanocomposites Properties   220
References   222

Chapter 9   PLA-nanocellulose Biocomposites   225
Qi Zhou and Lars A. Berglund

9.1   Introduction   225
9.2   Types of Nanocellulose   226

9.3 Surface Modification of Nanocellulose 228
9.4 Processing/Mixing Strategies 230
9.5 Properties 232
    9.5.1 Crystallization 232
    9.5.2 Thermal and Mechanical Properties 233
    9.5.3 Barrier Properties 238
9.6 Concluding Remarks 239
References 239

PLA Main Applications

Chapter 10 PLA and Active Packaging 245
    Ramón Catalá, Gracia López-Carballo,
    Pilar Hernández-Muñoz and Rafael Gavara

10.1 Introduction to Active Packaging 245
10.2 Antimicrobial Active Packaging 249
    10.2.1 Antimicrobial Agents 250
    10.2.2 Incorporation of the Agent into the
           Polymer Matrix 251
    10.2.3 Antimicrobial PLA-based Packaging
           Developments 251
10.3 Antioxidant Active Packaging 257
    10.3.1 Antioxidant Agents 258
    10.3.2 Incorporation of the Agent into the
           Polymer Matrix 258
    10.3.3 Other Packaging Applications 262
Acknowledgements 262
References 262

**Chapter 11 Biomaterials for Tissue Engineering Based on
           Nano-structured Poly(Lactic Acid)** 266
    Ilaria Armentano, Elena Fortunati, Samantha Mattioli,
    Nicoletta Rescignano and Josè Maria Kenny

11.1 Tissue Engineering 266
11.2 Stem Cells 267
    11.2.1 Embryonic Stem Cells 268
    11.2.2 Adult Stem Cell Sources 268
    11.2.3 Induced Pluripotent Stem Cells 268
    11.2.4 Differentiation Induction Factors in
           Tissue Engineering 269
11.3 Biomaterials and Nanotechnology 269
11.4 Nanostructured PLA 270
    11.4.1 Blends 271
    11.4.2 Nanoparticles and Nanoshells 274

      11.4.3   Nanocomposites   276
      11.4.4   Surface Modification   277
  11.5 Conclusions   279
Acknowledgements   279
References   279

Degradation and Biodegradation of PLA

Chapter 12   Abiotic-hydrolytic Degradation of Poly(lactic acid)   **289**
Kikku Fukushima and Giovanni Camino

  12.1 Introduction   289
  12.2 Molecular Hydrolytic Degradation Mechanisms of
       PLA   290
  12.3 Factors Controlling Hydrolytic Degradation of PLA   292
      12.3.1   Degradation Medium Conditions   292
      12.3.2   Structure and Properties of PLA-based
            Materials   295
  12.4 Possible Benefits of Hydrolytic Degradation of PLA   305
      12.4.1   Modifications in the Surface
            Hydrophilicity Level   306
      12.4.2   Structural Modification   306
  12.5 Conclusions   308
References   309

Industrial and Legislative Issues

Chapter 13   Industrial Uses of PLA   **317**
Stefano Fiori

  13.1 Introduction   317
  13.2 PLA Mechanical and Thermal Properties   318
  13.3 Stereochemistry   319
  13.4 Use of Additives   320
      13.4.1   Plasticizers   321
  13.5 PLA Processing   322
      13.5.1   Extrusion   322
      13.5.2   Film and Sheet Casting   322
      13.5.3   Stretch Blow Moulding   323
      13.5.4   Blown Film Extrusion   323
      13.5.5   Thermoforming   324
      13.5.6   Spinning and Electrospinning   325
      13.5.7   Foaming   325
  13.6 PLA Commercial Applications   326
      13.6.1   Food Contact Materials   326
      13.6.2   Medical and Biomedical Applications   328

	13.6.3	Agriculture	329
	13.6.4	Electrical Appliances	329
	13.6.5	Textiles	330
	13.6.6	Home Furnishing	331
	References		331

Chapter 14 Legislation Related to PLA   334
Mercedes A. Peltzer and Ana Beltrán-Sanahuja

14.1	Food Packaging and Legislation	334
14.2	Biodegradation and Compostability Legislation	337
14.3	Nanomaterials in Food Packaging Legislation	339
14.4	Life Cycle Assessment (LCA)	341
14.5	Conclusions	343
	References	344

Subject Index   347

PLA Synthesis and Polymerization

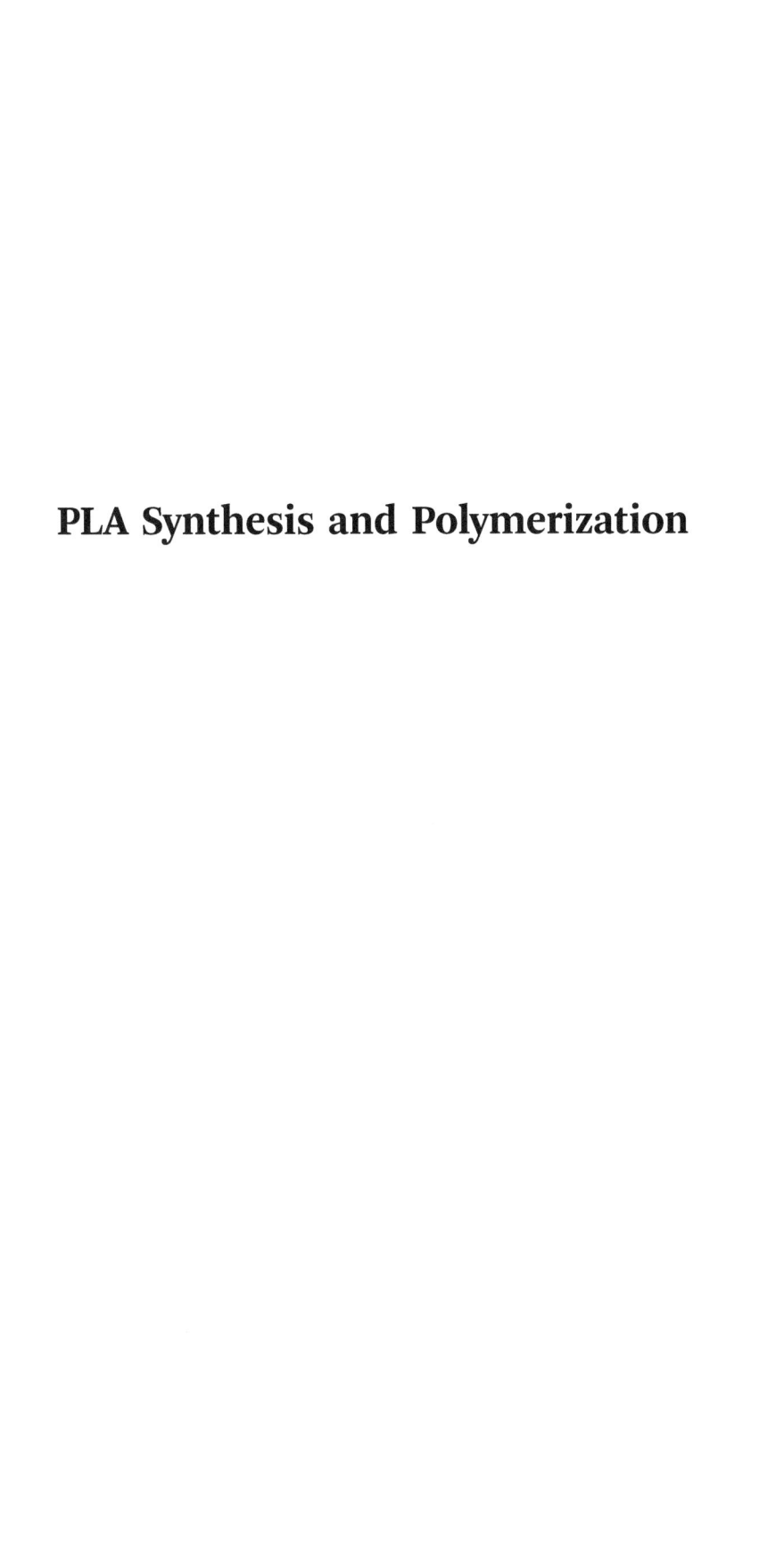

CHAPTER 1

PLA Synthesis. From the Monomer to the Polymer

KAZUNARI MASUTANI AND YOSHIHARU KIMURA*

Kyoto Institute of Technology, Japan
*Email: ykimura@kit.ac.jp

1.1 Introduction

It is recorded that Théophile-Jules Pelouze first synthesized poly(lactic acid) (PLA) by polycondensation of lactic acid in 1845.[1] In 1932, Wallace Hume Carothers *et al.* developed a method to polymerize lactide into PLA. This method was later patented by Du Pont in 1954.[2] Until the late 1970s, PLA and its copolymers were developed as biomedical materials based on their bioabsorbable and biocompatible nature and have been utilized in many therapeutic and pharmaceutical applications such as drug delivery systems (DDS),[3,4] protein encapsulation and delivery,[5,6] development of microspheres[7–10] and hydrogels[11] *etc*. Recently, the biomedical application of PLA has been extended to tissue engineering[12] including scaffold materials[13] as well as to biocompatible materials for sutures and prostheses[14] in which high- and low-molecular-weight PLAs are used, respectively. In the early 1990s, a breakthrough occurred in the production of PLA. Cargill Inc. succeeded in polymerizing high-molecular-weight poly(ʟ-lactic acid) (PLLA) by ring-opening polymerization (ROP) of ʟ-lactide in industrial scale and commercialized the PLLA polymer in the mid 1990s. Showing high mechanical properties in addition to a biodegradable nature, PLLA was thought to provide large opportunities to replace non-degradable oil-based polymers, such as poly(ethylene terephthalate) (PET) and polystyrene (PS).

RSC Polymer Chemistry Series No. 12
Poly(lactic acid) Science and Technology: Processing, Properties, Additives and Applications
Edited by Alfonso Jiménez, Mercedes Peltzer and Roxana Ruseckaite
© The Royal Society of Chemistry 2015
Published by the Royal Society of Chemistry, www.rsc.org

Since then, PLA has been utilized as biodegradable plastics for short-term use, such as rigid packaging containers, flexible packaging films, cold drink cups, cutlery, apparel and staple fibres, bottles, injection- and extrusion-moulds, coatings, and so on.[15] All of them can be degraded under industrial compositing conditions.[16] In the late 1990s, the bio-based nature of PLA was highlighted and its production as a bio-based polymer started. In this case, the newly developed polymers ought to have high-performances and long-life utilities that can compete with those of the ordinary engineering plastics. Various types of bio-based polymers are now under development, and several PLA types are also developed as promising alternatives to commercial commodities. In particular, PLLA polymers comprising high L-contents and stereo-complex PLA polymers showing high melting temperatures are now expected to be candidates for high-performance materials. The above historical view reveals the three specific features of PLA in terms of application, *i.e.* bio-absorbable, bio-degradable and bio-based.

Now, the synthesis of PLA polymers can be performed by direct polycondensation of lactic acid as well as by ring-opening polymerization of lactide (LA), a cyclic dimer of lactic acid. While the former method needs severe conditions to obtain a high-molecular-weight polymer (high temperature of 180–200 °C, low pressure as low as 5 mmHg and long reaction times),[17–19] the latter method can afford a high-molecular-weight PLA with narrow molecular weight distribution at relatively mild reaction conditions (low temperature of 130 °C and short reaction times).[20,21] Consequently, ROP of L-lactide is adopted in the ordinary industrial production of PLLA. On the other hand, since Ikada discovered the formation of stereo-complexes of PLLA and its enantiomer poly(D-lactic acid) (PDLA) in 1987, many trials have been done for its industrial production.[22] Manufacturing of D-lactic acid and improvement of the stereo-complexibility of the enantiomeric segments have been the big challenges in the trials thus far.[23,24] Synthesis of stereo-block polymers consisting of PLLA and PDLA macromolecular chains is a promising method for the preferential formation of stereo-complexes.[25–29] This chapter deals with the whole synthetic aspects of these PLA polymers and their starting monomers.

1.2 Synthesis of Lactic Acids

1.2.1 Stereoisomers of Lactic Acid

Lactic acid (2-hydroxypropanoic acid) is the simplest 2-hydroxycarboxylic acid with a chiral carbon atom and exists in two optically active stereo-isomers, namely L and D enantiomers (*S* and *R* in absolute configuration, respectively), as shown in Scheme 1.1. These L- and D-lactic acids are generally synthesized by fermentation using suitable micro-organisms. Racemic DL-lactic acid (*RS* configuration) consisting of the equimolar mixture of D- and L-lactic acids shows characteristics different from those of the

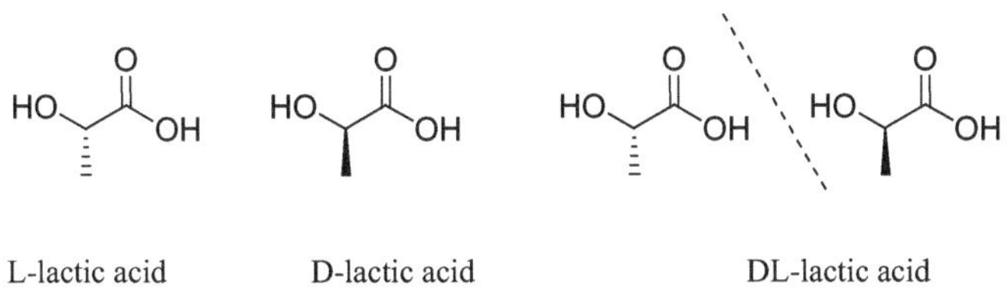

L-lactic acid   D-lactic acid   DL-lactic acid

Scheme 1.1   Structures of ʟ-, ᴅ- and ᴅʟ-lactic acids.

optically active ones. ᴅʟ-lactic acid is conveniently synthesized by chemical method rather than fermentation.

1.2.2   Fermentation with Lactic Acid Bacteria

Lactic acid fermentation is one of the bacterial reactions long utilized by mankind[30] along with alcoholic fermentation. The lactic acid bacteria are generally divided into several classes in terms of cell morphology, *i.e.* *Lactobacillus*, *Streptococcus*, *Pediococcus*, *Aerococcus*, *Leuconostoc* and *Coryne* species. They are also divided into various genera. Most of them produce ʟ-lactic acid while some produce ᴅ- or ᴅʟ-lactic acids. Table 1.1 compares which of ᴅ- or ʟ-lactic acid is produced by different bacteria. The species belonging to the same *Lactobacillus* genus produce either ʟ- or ᴅ-lactic acid preferentially. *Lactobacillus helvetics* and *Sporolactobacillus* produce ᴅʟ- and ᴅ-lactic acids, respectively. In the lactic acid formation, therefore, stereo-selectivity is much lower than in the amino acid formation where the ab-solute ʟ-selectivity is shown. Table 1.2 shows the mono- and di-saccharides assimilated by the lactic acid bacteria.[31] Each bacterium assimilates most mono-saccharides, but shows its own assimilation ability for di-saccharides. This difference in assimilation ability is important in the selection of bac-teria. Since the breakdown of cellulose and starch often produces di-saccharides, the species that can assimilate these di-saccharides must be used in the fermentation. In the ordinary lactic acid fermentation, the yields of ʟ- and ᴅ-lactic acids reach 85–90% and 70–80% based on carbon usage, respectively.[24]

1.2.3   Isolation and Purification of Lactic Acids

The fermenting liquor finally obtained in the above fermentation contains lactic acid together with various impurities such as un-reacted raw ma-terials, cells and culture media-derived saccharides, amino acids, carboxylic acids, proteins and inorganic salts. Therefore, the isolation and purification steps are needed for obtaining a highly pure product needed in the poly-mer's synthesis. In the usual fermentation process, the generated lactic acid is neutralized *in situ* with calcium oxide or ammonia. When calcium oxide is used for the neutralization, calcium lactate is precipitated out.

Table 1.1 Formation of D- and L-lactic acids with different lactic acid-producing bacteria.

Strain	Source	Yield of L-lactic acid[a]
Lactobacillus casei sp. *Rhamnosus* LC0001	IFO3425	97.5
Lactobacillus bulgarics LB0004	IAM1120	98.8
Lactobacillus delbrueckii LD0008	AHU1056	96.6
Lactobacillus delbrueckii LD0012	IAM1197	98.9
Lactobacillus delbrueckii LD0025	IFO3534	1.5
Lactobacillus delbrueckii LD0028	IFO3202	0.6
Lactobacillus helvetics LH0030	–	49.5
Lactococcus thermophillus LT	ATCC19987	97.0
Lactococcus lactis LL0005	ATCC8000	99.0
Lactococcus lactis LL0016	AHU1101	98.6
Lactococcus lactis LL0018	IFO3443	97.1
Sporolactobacillus inulinus SI0073	ATCC15538	1.1
Sporolactobacillus inulinus SI0074	ATCC15538	1.1

[a]$L/(D+L)\times100$.

Table 1.2 Saccharides assimilated by the representative lactic acid bacteria.

Strain	Glucose	Fructose	Maltose	Sucrose	Cellobiose
Lactobacillus casei sp. *lactis*	+	+	+	+	±
Lactobacillus bulgarics	+	+	+	–	–
Lactobacillus acidophilus	+	+	+	+	+
Lactobacillus delbrueckii LD0025	+	+	+	+	+
Lactobacillus helvetics LH0030	+	±	+	–	+
Sporolactobacillu sinulinus SI0073	+	+	–	–	–
Sporolactobacillus inulinus SI0074	+	+	–	–	–

+: assimilative, –: non-assimilative, ±: assimilative under specific conditions.

$$H_3C\text{-}CH\text{-}CO_2^-NH_4^+ \ + \ CH_3(CH_2)_3OH \longrightarrow H_3C\text{-}CH\text{-}CO_2\text{-}(CH_2)_3CH_3 \ + \ NH_3 \ + \ H_2O$$
$$\quad\ \ OH \qquad\qquad\qquad\qquad\qquad\qquad\qquad\qquad\quad OH$$

Scheme 1.2 Formation of butyl lactate from ammonium lactate.

This salt is isolated by filtration in the final step, washed with water and acidified with sulfuric acid to liberate free lactic acid with formation of calcium sulfate as solids. When ammonia is used for the neutralization, the ammonium lactate is formed and directly converted into butyl lactate by esterification with n-butanol, as shown in Scheme 1.2.[32] Here, the ammonia is recovered and recycled. The following distillation and hydrolysis of butyl lactate gives an aqueous lactic acid with high efficiency. The lactic acid obtained by this method has higher purity than that obtained by the calcium salt method. The technologies for the above lactic acid fermentation and purification have well been established, and the production of both D- and L-lactic acids is conducted industrially in a plant scale of 100000 ton year^{-1}.

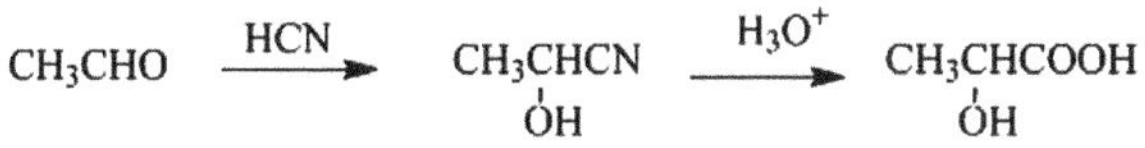

$$CH_3CHO \xrightarrow{\text{HCN}} \underset{\underset{OH}{|}}{CH_3CHCN} \xrightarrow{H_3O^+} \underset{\underset{OH}{|}}{CH_3CHCOOH}$$

Scheme 1.3　Chemical synthesis of lactic acid *via* lactonitrile.

1.2.4　Chemical Synthesis of Lactic Acids

Racemic DL-lactic acid can be synthesized by fermentation using appropriate bacteria (*Lactobacillus helvetics* in Table 1.1), but it is more easily synthesized by following the chemical process shown in Scheme 1.3.[33] Here, the DL-lactic acid is produced by hydrolysis of lactonitrile that is generally formed by the addition reaction of acetaldehyde and hydrogen cyanide. Industrially, the lactonitrile is obtained as a by-product of acrylonitrile production (Sohio process).[34] The lactic acid thus prepared is purified by distillation of its ester as described above.

1.3　Synthesis of Lactide Monomers

1.3.1　Stereoisomers of Lactides

Scheme 1.4 shows three lactides consisting of different stereoisomeric lactic acid units. L- and D-lactides consist of two L- and D-lactic acids, respectively, while meso-lactide consists of both D- and L-lactic acids. Racemic lactide (rac-lactide) is an equimolar mixture of D- and L-lactides. The melting points (T_m) of these lactides are compared in Table 1.3. Note that the T_m is higher in rac-lactide and is lower in meso-lactide.

1.3.2　Synthesis and Purification of Lactides

Each of the aforementioned lactides is usually synthesized by depolymerization of the corresponding oligo(lactic acid) (OLLA) obtained by

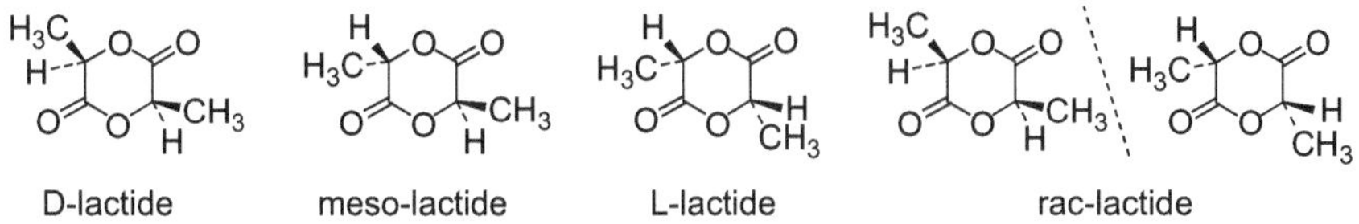

Scheme 1.4　Structures of various lactides.

Table 1.3　Thermal properties of lactides.

	Melting point (°C)
L-lactide	95–98
D-lactide	95–98
meso-lactide	53–54
rac-lactide	122–126

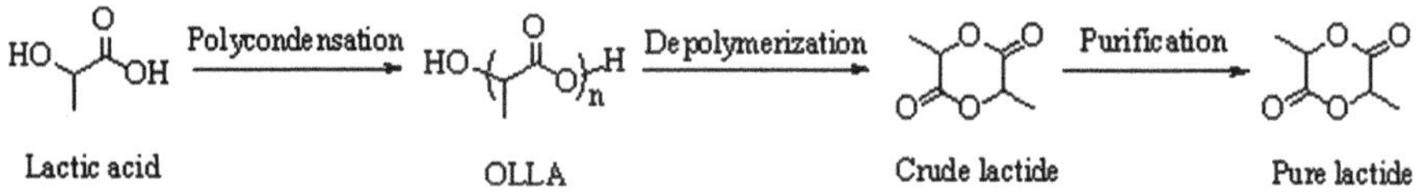

Scheme 1.5   Synthetic route to lactide from lactic acid *via* oilgolactide.

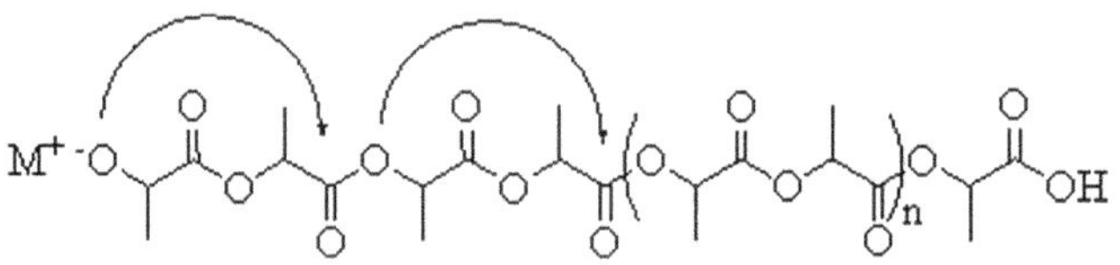

Scheme 1.6   Expected formation mechanism of lactide (back-biting mechanism).

polycondensation of relevant lactic acid, as shown in Scheme 1.5.[32] Because of the ring-chain equilibrium between lactide and OLLA, unzipping depolymerization generates lactide through the back-biting mechanism involving the -OH terminals of OLLA as the active site as shown in Scheme 1.6.[35] This reaction is well catalyzed by metal compounds involving Sn, Zn, Al and Sb ions, *etc.* The crude lactide can be purified by melt crystallization or ordinary recrystallization from solution.

1.4   Polymerization of Lactide Monomers

1.4.1   Structural Diversities of the Polylactides

As shown in Scheme 1.7, there are two major synthetic routes to PLA polymers: direct polycondensation of lactic acid and ring-opening polymerization (ROP) of lactide. Industrial production of PLA mostly depends on the latter route. The polymerization of optically pure L- and D-lactides gives isotactic homopolymers of PLLA and PDLA, respectively. Both PLLA and PDLA are crystalline, showing a T_m around 180 °C. Their crystallinity and T_m usually decrease with decreasing optical purity (OP) of the lactate units.[36] Optically inactive poly(DL-lactide) (PDLLA), prepared from rac- and meso-lactides, is an amorphous polymer, having an atactic sequence of D and L units. However, crystalline polymers can be obtained when the sequence of both D and L units are stereo-regularly controlled.[37] The most interesting issue comes from the fact that mixing of isotactic PLLA and PDLA in 1 : 1 ratio affords stereo-complex crystals (sc-PLA) whose T_m is 50 °C higher than that of PLLA or PDLA.[22,38–45] This sc-PLA is formed by co-crystallization of the helical macromolecular chains having opposite senses. Stereo-block copolymers (sb-PLA) consisting of isotactic PLLA and PDLA sequences are also synthesized by stereo-regular polymerization techniques involving block copolymerization.[45] These structural diversities of PLA polymers provide a broad range of physicochemical properties for PLA materials when processed.

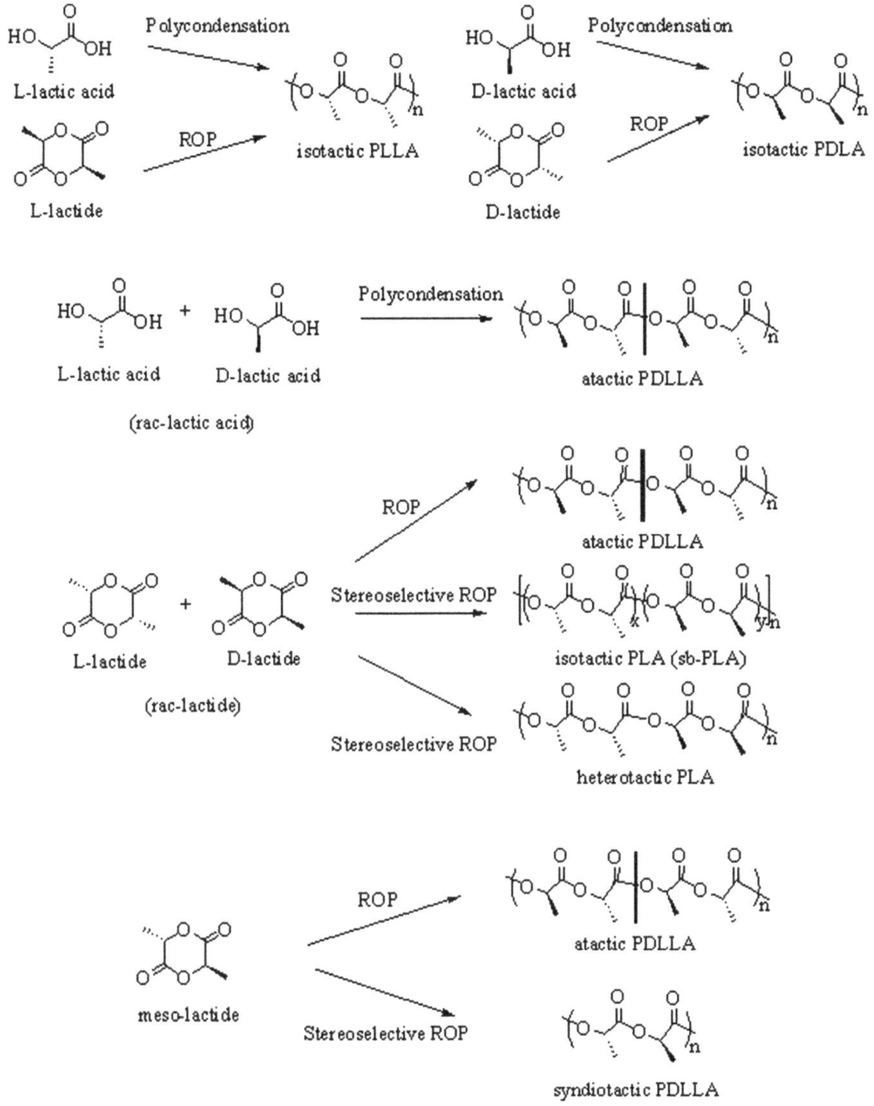

Scheme 1.7 A variety of microstructures of lactides and PLAs.

1.4.2 Thermodynamics for the Polymerization of D- and L-Lactides

The heat capacities and enthalpies of combustion were measured to analyze the thermodynamics of polymerization of D- and L-lactides into their polymers.[46,47] The enthalpies and entropies of the lactide polymerization determined from these data are as follows: $\Delta H_p = -27.0$ kJ mol^{-1} and $\Delta S_p = -13.0$ J mol^{-1} K^{-1} at 400 K, indicating an exothermic reaction. The kinetics of ROP of lactide have also been studied with various catalysts, showing that the polymerization rate is in first-order of each of the monomer and catalyst concentrations.

Witzke proposed a reversible kinetic model for the melt polymerization of L-lactide in the presence of tin(II) octoate as the catalyst and determined the following parameters:[48] $E_a = 70.9 \pm 1.5$ kJ mol^{-1}, $\Delta H_p = -23.3 \pm 1.5$ kJ mol^{-1} and $\Delta S_p = -22.0 \pm 3.2$ J mol^{-1} K^{-1}, and ceiling temperature (T_c) = 786 ± 87 °C. Model equations for monomer concentration and conversion as a function of time were derived as follows:

$$M_t = M_{eq} + \left(M_0 - M_{eq}\right) exp(-K_p It) \qquad (1)$$

$$X_t = \left(1 - \frac{M_{eq}}{M_0}\right)\left(1 - e^{(-K_p It)}\right) \qquad (2)$$

$$K_p = 86.0 exp\{(-E_a/R)(1/T - 0.00223)\} \qquad (3)$$

where

$M_{eq} = (\Delta H_p/RT - \Delta S/R)$
M_t = monomer concentration at time
M_{eq} = equilibrium monomer concentration
M_0 = initial monomer concentration
K_p = propagation rate constant in (1/cat.mole %-hr)
I = catalyst concentration in (mole %)
t = time in hours
X_t = monomer to polymer conversion at time t
E_a = activation energy
R = gas constant
T = polymerization temperature in Kelvin
ΔH = enthalpy of polymerization
ΔS = entropy of polymerization

Three reaction mechanisms have been proposed thus far for ROP of lactide: anionic, cationic and coordination mechanisms.[15,16] In the anionic polymerization, undesirable reactions such as racemization, back-biting reaction and other side reactions are often caused by the highly active anionic reactants that hinder the chain propagation. In the cationic polymerization, undesirable side reactions and racemization likely occur because of the nucleophilic attacks on the activated monomers and the propagating species. The decreases in molecular weight and optical purity lower the crystallinity and mechanical properties of the obtained polymers. On the contrary, coordination polymerization with metal catalysts (mostly alkoxides) can give a large molecular weight with the high optical purity maintained. Therefore, a variety of catalysts have been studied. The following sections deal with these polymerization aspects in detail.

1.4.3 Metal Catalysts

Metal complexes of Al, Mg, Zn, Ca, Sn, Fe, Y, Sm, Lu, Ti and Zr have been widely used as the catalysts in the ROP of various lactone monomers involving lactides. The standard catalyst system utilized for lactide polymerization is tin(II) octoate (stannous bis(2-ethylhexanoate): $Sn(Oct)_2$),[49–52] to which lauryl alcohol (1-dodecanol) is usually added as a real initiator. This catalyst system has many advantages over the other systems in that it is highly soluble in organic solvents and molten lactide in bulk state and very stable on storage. It also shows excellent catalytic activity to give high molecular weight of PLLA.[16] The most important characteristic is that this catalyst is biologically safe and approved by the FDA (the US Food and Drug Administration) for use in medical and food applications, although the approval is dependent on empirical safety data.[53] With these characters, $Sn(Oct)_2$ has been used as the catalyst in the industrial production of PLAs.

The mechanism of this tin-catalyzed polymerization of lactide has been disputed for a long time, *i.e.* discussing whether the polymerization is induced by "insertion-coordination mechanism" or "monomer activation mechanism".[54,55] Duda and Penzek proposed a comprehensive polymerization scheme based on the insertion-coordination mechanism.[51] In the ordinary lactide polymerization catalyzed by $Sn(Oct)_2$, a hydroxyl compound (alcohol) is added as the real initiator. The alcohol initiator first reacts with $Sn(Oct)_2$ to generate a tin alkoxide bond by ligand exchange. In the next stage, one of the exocyclic carbonyl oxygen atoms of the lactide temporarily coordinates with the tin atom of the catalyst having the alkoxide form. This coordination enhances the nucleophilicity of the alkoxide part of the initiator as well as the electrophilicity of the lactide carbonyl group. In the next step the acyl-oxygen bond (between the carbonyl group and the endocyclic oxygen) of the lactide is broken, making the lactide chain opened to insert into the tin-oxygen bond (alkoxide) of the catalyst. The following propagation is induced by identical mechanism and continues as additional lactide molecules are inserted into the tin-oxygen bond (Scheme 1.8).[56–59]

This mechanism was strongly supported by the MALDI-TOF mass spectrum showing molecular peaks that correspond to the oligomeric PLLA chains connecting with the tin residue, which are propagating species formed with the $Sn(Oct)_2$/lauryl alcohol system. Since the polymerization is pseudo-living, the molecular weight can be relatively well controlled. However, in the last stage of propagation where the monomer concentration becomes significantly lower, the reverse depolymeriztion by back-biting mechanism as well as intermolecular trans-esterification that is referred to chain transfer or polymer interchange reaction becomes evident to broaden the molecular weight distribution.[60] Despite the presence of this mechanism, the degrees of racemization and chain scrambling are much lower than those with anionic or cationic catalysis.

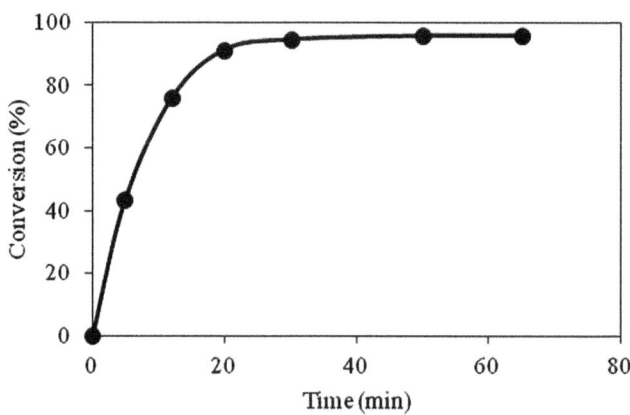

Scheme 1.8 ROP of lactide with tin octoate by coordination-insertion mechanism.

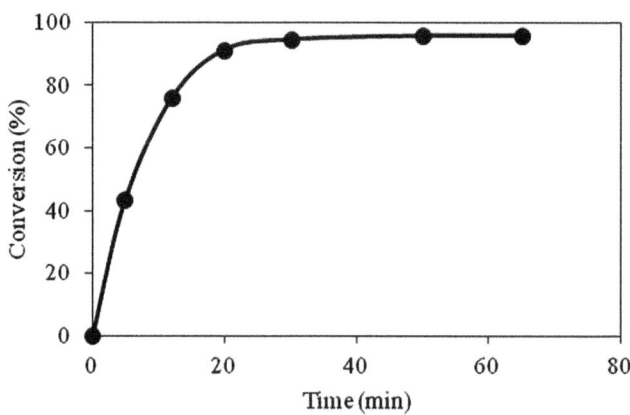

Figure 1.1 A typical time-conversion curve in the tin-catalyzed ROP of L-lactide.

Figure 1.1 shows a typical time-conversion curve in the melt-polymerization of L-lactide at 190 °C in the presence of $Sn(Oct)_2$ (1.78×10^{-5} mol%) and lauryl alcohol (0.205 mol%). It is shown that at the beginning of the polymerization process, the monomer consumption is fast to reach 94.7% within 30 min. The plateau observed thereafter indicates that the polymerization has reached the ring-chain equilibrium state where the amount of remaining monomer corresponds to the equilibrium monomer concentration. Here, the theoretical M_n estimated from the monomer/initiator ratio was 70 kDa, being identical to the M_n value of the final product. The weight average molecular weights (M_w) of the PLLA polymers thus obtained reach 500–700 kDa at maximum, if the monomer/initiator ratio is properly adjusted.

The catalyst and monomer remaining in the PLA polymers are problematic, because they are likely to cause the degradation of PLA polymers.[61] In particular, the catalyst promotes chain breakage of PLA polymers during the melt processing. Since the catalyst is not easily removed, it is generally deactivated by adding a deactivator such as sodium phosphonate in the final step of the ROP. As noted by Figure 1.1, the remaining monomer present in the final polymer reaches *ca.* 5 wt% when melt polymerization of lactide is conducted with $Sn(Oct)_2$/lauryl alcohol system at 180–200 °C. This lactide monomer is quickly hydrolyzed and stimulates the hydrolytic degradation of PLA polymers. It also plasticizes the PLA materials to lower the T_g and T_m values.[62] Therefore, the remaining monomer is removed by vacuum evaporation to a concentration lower than 0.1 wt%. The remaining monomer concentration can be also decreased by performing the ROP in solid-state. When the polymer melt is crystallized during polymerization, the monomer and catalyst are concentrated in amorphous domains to make the ring-chain equilibrium decline to the polymer side. Figure 1.2 shows typical results of the ROP of L-lactide conducted at 120, 140 and

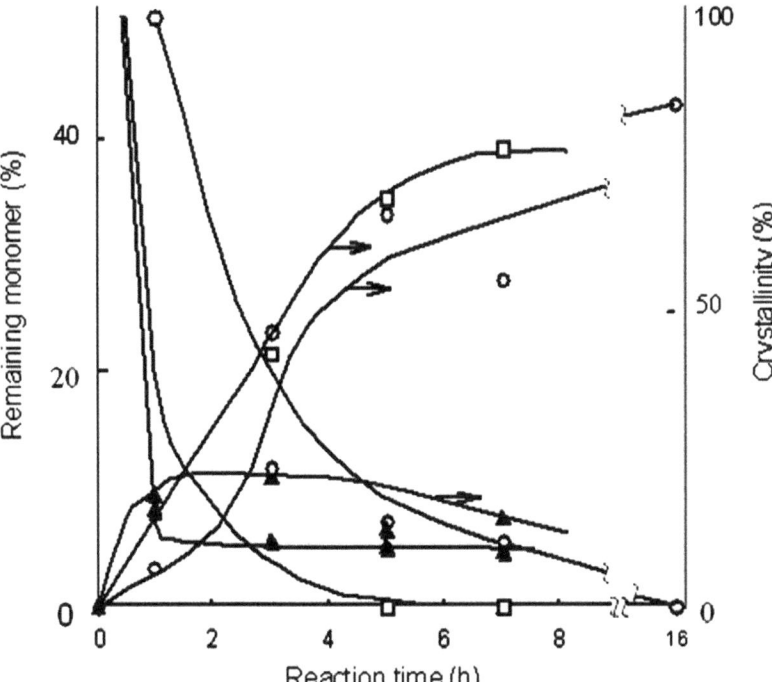

Figure 1.2 Changes in monomer consumption and polymer crystallinity during lactide polymerization at ($\bigcirc$) 120 °C (crystallization), ($\square$) 140 °C (crystallization) and ($\blacktriangle$) 160 °C (melt state).

160 °C.[57] The concentration of the remaining monomer decreases to zero at 120 and 140 °C where the polymerization systems turn to solid state, whereas at 160 °C the monomer consumption reaches a plateau with the melt-state retained. The same technique is utilized to increase the molecular weight of PLA in the solid-state polycondensation of lactic acid (*vide infra*).

Aluminium alkoxides have also been utilized as efficient catalysts for the lactide polymerization. The most popular one is aluminium isopropoxide $(Al(O^iPr)_3)$.[63] Duda verified that $Al(O^iPr)_3$ exists in two types of aggregates: a trimer (A_3) and a tetramer (A_4) as shown in Scheme 1.9.[64] The A_3, consisting of penta-coordinate aluminium ion, can initiate the lactide polymerization whilst A_4, having hexa-coordinate aluminium ion, is ineffective for the ROP. Since A_4 is equilibrated with A_3, the ROP of lactide is slowly initiated in the presence of A_4. All of the alkoxyl ligands in the aluminium alkoxides are involved in the initiation because interligand exchange of the propagating metal alkoxides is fast.

Other metal alkoxides are also effective for the ROP of lactides.[65] Depending on the Lewis acidity of the metal ion (or the availability of open coordination sites), the metal alkoxides can activate the monomer by binding to the carbonyl.[49,66–70] In case the trans-esterification of the propagating metal alkoxide is slow, the metal alkoxide functions as an initiator, whereas it works as a catalyst when the metal alkoxide reacts with alcohols to regenerate a new metal alkoxide. It is known that the active alkoxide end-groups quickly interchange with the other polymer chains and affect the molecular weight distribution of the final polymer. Furthermore, the metal alkoxides, including those formed from $Sn(Oct)_2$ and alcoholic initiators, remain attached to the propagating species, making it difficult to be removed from the final polymer. They ought to affect the thermal properties of the final polymers, in particular causing deterioration of their melt processability at high temperature even though they have been deactivated. With increasing applications of PLA polymers in packaging, biomedical and microelectronic fields, development of more biocompatible metal catalysts[65,71] or metal-free organic catalysts (below) is needed.

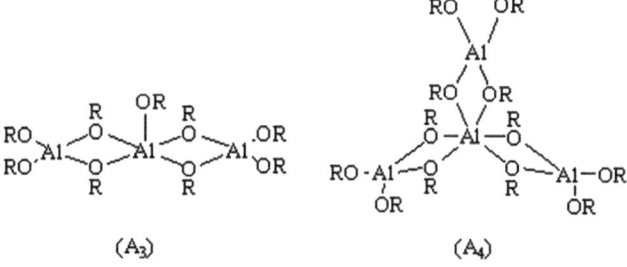

Scheme 1.9 Aggregation structures of a trimer (A_3) and a tetramer (A_4) of aluminium triisopropoxide $(R = CH(CH_3)_2)$.

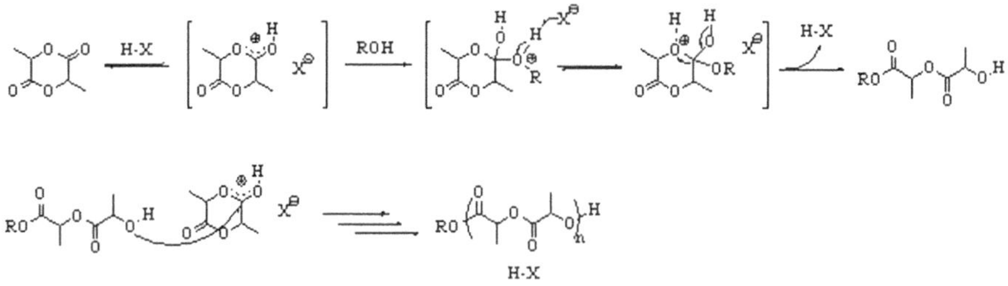

Scheme 1.10 Mechanism of the cationic ROP of lactide (monomer activation mechanism).

1.4.4 Cationic Catalysts

The ROP of lactide can be induced by the catalysis of strong organic acids.[72] In the ordinary cationic ROP, strong organic acids, such as super acids, are utilized as the catalysts as shown in Scheme 1.10. In the initiation step, the monomer activated by the protonation is ring-opened by the carbonyl attack of an alcohol initiator to form the lactyl alcohol with recovery of acidic proton. Successive attack of this lactyl alcohol on the protonated monomers propagates the polymer chain. Namely, the terminal hydroxyl group acts as the propagating species reacting with the protonated monomer, which corresponds to the monomer activation mechanism. Since the acid catalyst is free from the propagating polymer, it is readily removed, and fewer than one catalyst per monomer chain is needed.[73] This feature is an advantage of the cationic catalysts over the metal alkoxides, which ought to remain attached to the propagating chains as described above. In the acid-catalyzed ROP, however, the propagation is likely contaminated by chain termination or transfer reactions that may be related to the reactivity of the protonated monomer.[74]

The acid-catalyzed polymerization of lactide (LA) with tri-fluoro-methanesulfonic acid (HOTf) can be highly controlled in the presence of an alcohol initiator.[58] The low molecular weights, slow rates and high catalyst loadings associated with organic acids may be compensated by the operational simplicity of this process. It is also confirmed that the polymerization of L-lactide is highly stereospecific.[58,74] On the other hand, cationic ROP of lactide with strong methylating agents, such as methyl tri-fluoro-methanesulfonate (MeOTf)[72] has not yet been optimized in terms of polymerization mechanism.

1.4.5 Organic Catalysts

As mentioned above, much effort has been put in to developing more biocompatible metal-free organic catalysts for the use of PLA materials in biomedical and microelectronic applications. Scheme 1.11 summarizes the organic catalysts recently developed for the ROP of various lactone monomers.

The first ROP of lactide with organic bases was reported with 4-(dimethylamino) (DMAP) and 4-pyrrolidino-pyridines (PPY) as the catalysts.[75–80]

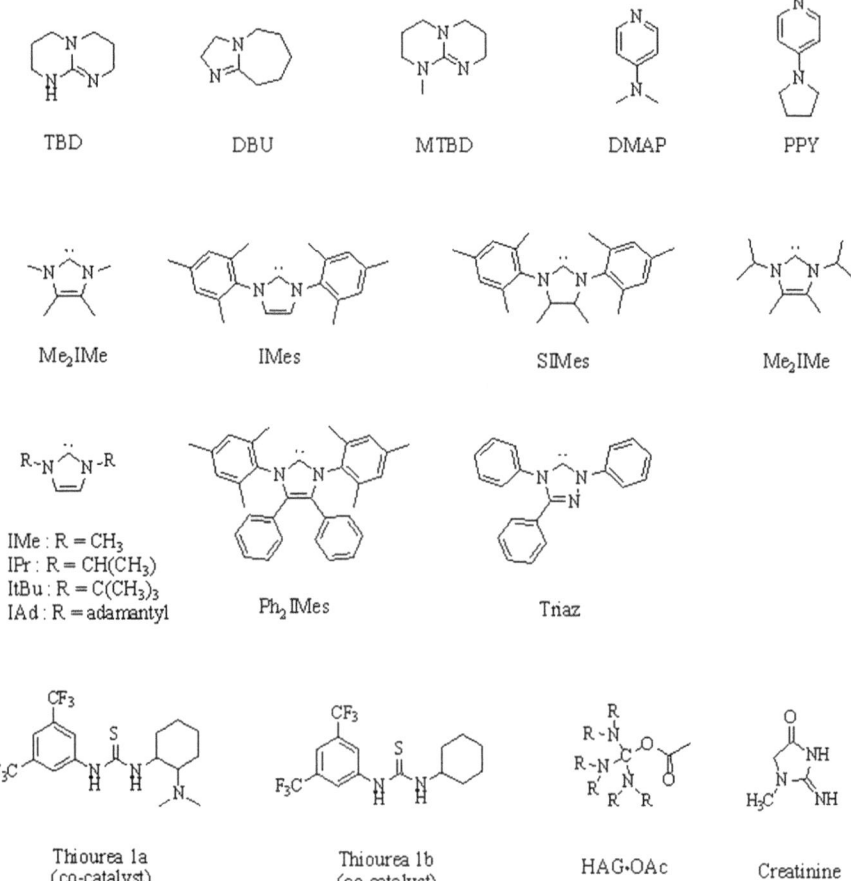

Scheme 1.11   Representative organic catalysts and initiators.

DMAP produces PLAs having a degree of polymerization (DP) of 100 with narrow dispersity (PDI < 1.13) in solution, while PPY was effective only for the ROP of lactide in the melt state. DMAP is believed to activate an alcoholic initiator which may successively react with lactide monomers.[81]

N-heterocyclic carbenes (NHCs), having higher basic character than DMAP, are also effective for the ROP of lactide.[82–84] For example, an aryl-substituted carbene IMes is very active for lactide polymerization but less active for the polymerization of ε-caprolactone (CL). The alternative more basic carbenes Me$_2$IMe and Me$_2$IPr are more effective than IMes, polymerizing both lactide and CL.[85] The polymerization of lactide is extremely rapid with IMes in which a feature of living polymerization is retained. In the absence of alcohol initiators this carbene forms a zwitterionic active species to generate a cyclic polylactide.[86]

Guanidine and amidine derivatives, such as *N*-methyl-1,5,7-tri-azabicy-clododecene (MTBD) and di-azabicycloundecene (DBU), show similar catalytic activities for the polymerization of lactide,[87,88] producing polymers with a DP reaching 500 and narrow dispersity (PDI < 1.1). The polymerization finishes within an hour although trans-esterification involving the chain scrambling occurs at a considerable rate. An alcohol-activated mechanism is proposed. These catalysts are not effective for the ROP of butyrolactone, valerolactone or CL at up to 20 mol% catalyst loading.

Hong Li *et al.* demonstrated that simple 1,1,2,2,3,3-hexa-alkylguanidinium acetate (HAG · OAc) and creatinine lactate exhibit excellent catalytic activity in the ROP of lactides and produce PLA polymers with a DP of 140 and narrow dispersity (PDI < 1.1).[89,90] Living polymerization driven by a carboxylate-activation mechanism is supported by the kinetic studies on the HAG · OAc-initiated ROP. The proposed mechanisms for the ROP initiated with HAG · OAc and creatinine are shown in Schemes 1.12 and 1.13, respectively. The use of bio-originated creatinine may guarantee the bio-safety required in biomedical application.

A combination of a thiourea (TU) and a tertiary amine (TA) gives an active catalyst for the ROP of lactide.[91] The catalytic activity is modulated by changing the thiourea. In a typical run, PLA having a narrow PDI (<1.08) is produced with TU with living polymerization features because M_n is critically correlated to the monomer-to-initiator ratio. The thioureas are thought to activate the monomer carbonyls through the hydrogen bonding. Similar combined catalysts TU/MTBD and TU/DBU are also highly effective for the ROP of various lactone monomers including lactide.[92]

A bicyclic guanidine, 1,5,7-tri-azabicyclododecene (TBD), shows much higher activity than DBU and MTBD in the ROP of cyclic monomers.[93] The ROP of lactide finishes within 1 min. with 0.1 mol% of TBD[88] whereas *ca.* 30 min. is required in the presence of 0.5 mol% MTBD. TBD shows a living polymerization nature, but induces trans-esterification at high conversion, resulting in the broadening of dispersity.

Alkyl and alcohol adducts of saturated N-heterocyclic carbenes (NHCs) are also used in the ROP of lactide. In these systems, NHC catalysts are generated *in situ* at elevated temperatures (65–144 °C) to polymerize lactide in the presence of an alcohol initiator.[94–96] For example, alcohol adducts of SIMes act as single-component catalyst/initiators for the ROP of lactide. As shown in Scheme 1.14, they reversibly liberate the alcohol initiator with the carbene

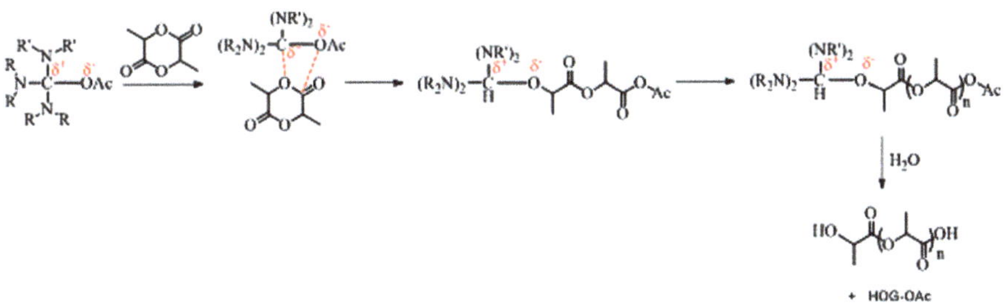

Scheme 1.12 Proposed mechanism for the ROP of lactides initiated with HAG · OAc.

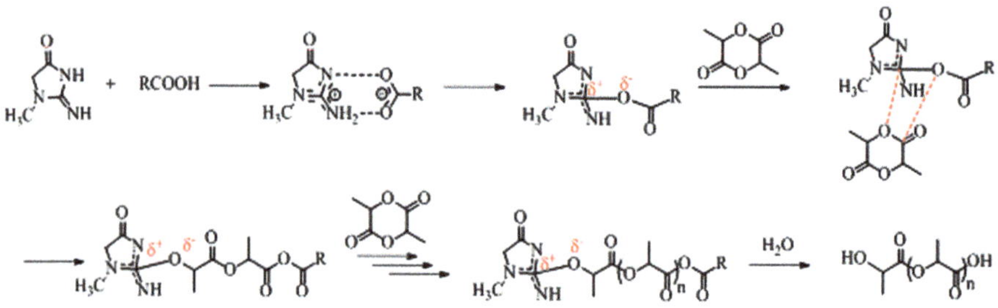

Scheme 1.13 Proposed mechanism for the ROP of lactides initiated with creatinine.

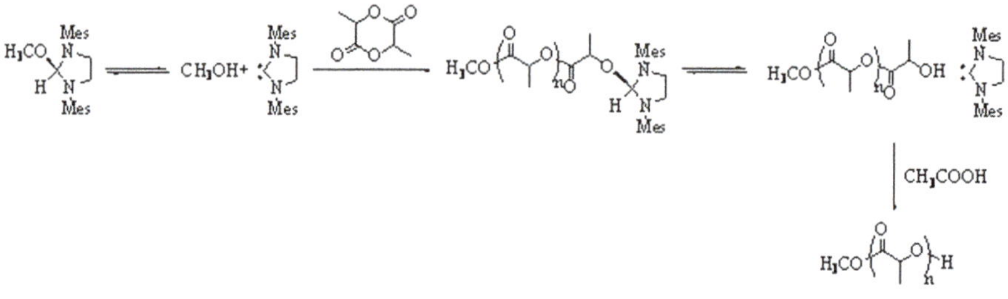

Scheme 1.14 Polymerization of lactide by reversible activation and deactivation of SIMes.

catalyst at room temperature to induce ROP of lactide. PLA having narrow polydispersity is obtained within minutes in high yields.

On the other hand, alcohol adducts of triazol-5-ylidene are stable at room temperature and reversibly dissociate into the alcohol and carbine only at 90 °C.[97–100] The triazolylidene catalysts are inactive at room temperature even in the presence of alcohols, but at 90 °C they polymerize lactide into PLA of narrow dispersity (Scheme 1.15). The reversible formation of the active and dormant carbene species is the key factor to control the polymerization.

1.4.6 Stereo-controlled Polymerization

Stereochemistry is one of the most important factors that determine the physical and chemical properties of polymeric materials. Spassky *et al.* developed the first catalyst system for the stereoselective polymerization of DL-lactide into isotactic PLA.[101] They used Al complexes containing achiral salen ligands to synthesize a multi-block PLLA and PDLA stereo-copolymer (Scheme 1.16). Spassky also reported a one-step synthesis of gradient PLA from DL-lactide by using homochiral catalysts.[102] Since then, many other single-site catalysts supported by various kinds of multi-valent ligands were made and used for well-controlled or stereoselective polymerizations of lactide to demonstrate the synthetic possibilities from DL-lactide to stereo-regular PLA materials.[101–114] In general, the single-site catalysts used thus far function with two different mechanisms: (1) chain-end control

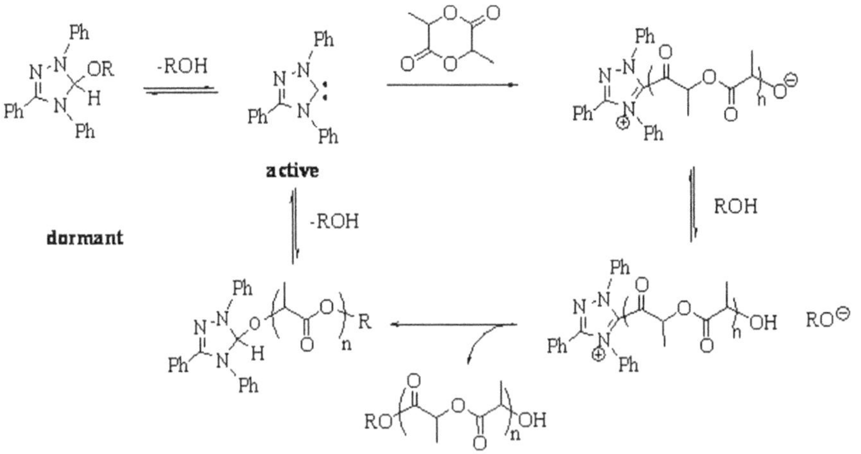

Scheme 1.15 Polymerization of lactide by reversible activation of triazolylidene carbenes.

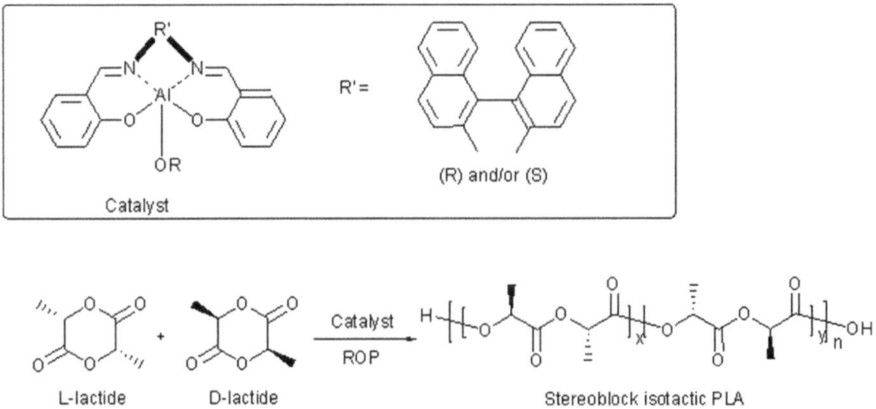

Scheme 1.16 Synthesis of multi-sb-PLA by stereoselective ROP of racemic DL-lactides.

mechanism[103,110] and (2) site control mechanism.[113] In the chain-end control mechanism, the configuration of the next inserted monomer is selected by the stereo-genic centre in the last repeating unit in the propagating chain. In the site control mechanism, the configuration of the inserted monomer is determined by the configuration of the ligands surrounding the catalyst site. If the DL-lactide polymerization follows either the chain-end control mechanism or the site control mechanism, only isotactic or heterotactic (having iso-syndio repeating units) PLA can be obtained.

Table 1.4 summarizes the representative aluminium catalysts reported thus far. Although many other metal catalysts have also been reported, they are not included here because of their lower catalytic activities.[65]

Table 1.4 Synthesis of multi-sb-PLA by ROP of rac-lactide with various aluminium catalysts.

Structure of Initiator/Catalyst					Solvent	Temp. (°C)	Time (h)	M_n [GPC]	M_w/M_n [GPC]	T_m (°C)	P_{meso}	Block length	Ref.
Type	R1	R2	R3	X									
1	H	H	OCH3	A	Toluene	70	281	12,700		187			1
	H	H	OCH3	(CH2)2	CH2Cl2	70	95	17,300	1.2	164		7.6	2
	H	H	OiPr	(CH2)2	(Solution)				1.05	191			3
	H	H	OiPr	(CH2)2	Toluene	70	40	22,600	1.09	179		11	4
	tBu	tBu	OiPr	B	Toluene	70	288	7,700	1.06	183.5	0.93		5
	tBu	tBu	OiPr		(Bulk)	130	48	24,900	1.37		0.88		5
	H	ph	Et	(CH2)3	(Solution)	70	1.3	20,000	1.1	170	0.81		6
	tBu	tBu	Et	(CH2)3	(Solution)	70	14	22,400	1.06	192	0.91		8
	tBu	tBu	OBn	CH2C(CH3)2CH2	(Bulk)	130	0.5	14,300	1.05	169	0.91		8
	tBu	tBu	Et	CH2C(CH3)2CH2	Toluene	70	17.25	14,900	1.05	193	0.90		9
	tBu	tBu	OiPr	CH2C(CH3)2CH2	Toluene	70	8.95	22,300	1.04	196	0.90	20	10
	H	H	OiPr	A	THF	80	25	12,300	1.11	210			11
2	CH3	H	CH3	(CH2)2	Toluene	70	21	21,180	1.08		0.79		7
3	ph	H		CH2C(CH3)2CH2	Toluene	70	5.6	36,800	1.04		0.78		12

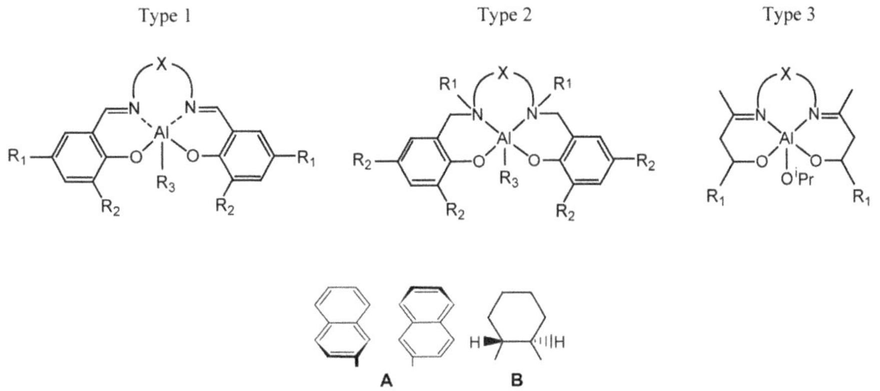

Type 1   Type 2   Type 3

1.4.7 Stereo-block Copolymerization

Mixing of enantiomeric PLLA and PDLA generates stereo-complex (sc) crystals that exhibit high T_m about 230–240 °C, being 50 °C higher than those of the single PLLA or PDLA polymers.[22,38–45] The resultant stereo-complex-type poly(lactic acid)s (sc-PLA) ought to be more thermally stable, possessing wider application. However, the melt-blending of PLLA and PDLA having high molecular weight is likely accompanied by their homo-chiral (hc) crystallization together with the sc crystallization, resulting in a mixed crystalline state and a deteriorated performance of the processed materials. Therefore, various trials have been made for improving the miscibility of PLLA and PDLA.[115,116] Among them, making block copolymers of PLLA and PDLA, *i.e.* stereo-block-type poly(lactic acid)s (sb-PLA), has been found to be highly effective for generating the stereo-complex, because the neighbouring D- and L-stereo-sequences can readily interact with each other to fall into molecular mixing state.[27–29]

In 1990, Yui *et al.* first prepared a diblock copolymer of PLLA-*b*-PDLA by step-wise ROP of L- and D-lactides (Scheme 1.17) to demonstrate its easy sc formation with little hc crystallization.[115] However, the highest molecular weight (M_n) obtained was as low as 20.1 kDa, with which the sc

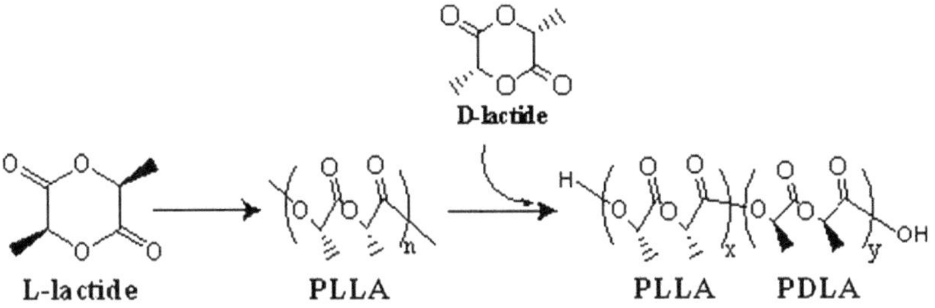

Scheme 1.17 Two-step ROP to synthesize diblock sb-PLAs.

Table 1.5   Diblock sb-PLA obtained by the controlled two-step ROP.

Sample No.	Feed ratio	M_w(GPC) Da	ΔH_{ms} (J g^{-1})	X_s (%)	T_m (°C)
PLA 20/80	L 20/D 80	141,000	44.2	100	210.0
PLA 80/20	D 20/L 80	143,000	36.2	100	208.5
PLA 35/65	L 35/D 65	166,000	40.4	100	212.4
PLA 65/35	D 35/L 65	159,000	38.5	100	208.5

ΔH_{ms}: Heat of melting of stereo-complex crystal. X_s: Percentage of stereo-complex crystal against homo-chiral crystal in sb-PLA.

crystallization is known to predominantly occur even by mixing PLLA and PDLA homopolymers.

Later, the synthesis of sb-PLA having defined block length and sequence was performed by the stepwise ROP method,[27,28] consisting of three steps: 1) polymerization of either L- or D-lactide to obtain PLLA or PDLA having molecular weight of 7–100 kDa (preferably lower than 50 kDa), 2) purification of the obtained polymer to remove the residual lactide and 3) polymerization of the enantiomeric lactide in the presence of the purified polymer. The purification step is important to avoid racemization of the second block. Table 1.5 shows some results together with the characteristics of the sb-PLA polymers produced. The preferable PLLA/PDLA ratio is in the range from 85/15 to 15/80 where the resultant diblock sb-PLAs show almost exclusive sc crystallization without the hc crystallization occurring. The shorter and longer blocks of these copolymers must be synthesized in steps 1 and 3, respectively, for the pre-polymer first prepared in step 1 can be well solvated by the larger amount of lactide monomer with opposite chirality and promote the chain elongation from its terminal in step 3. When the longer block is first made, the chain extension in the second ROP becomes imperfect to retard the block copolymerization. When the PLLA/PDLA ratio is higher than 90/10 and lower than 10/90, the sc crystallinity of the resultant sb-PLA becomes lower than 80%. The sb-PLAs prepared by this method are useful not only because preferential sc crystallization is possible even at non-equivalent PLLA/PDLA ratios, but also because the more expensive D-lactide monomer is required in a smaller amount for obtaining the sc materials.

1.4.8   Copolymerization

Various lactate copolymers have been developed thus far to tune the properties of PLLA and cope with different applications.[117] Two synthetic approaches are available for the copolymerization: (1) ring-opening copolymerization of L-lactide with other monomers such as ε-caprolactone, glycolide and depsipeptides; and (2) use of new cyclic monomers consisting of lactate and other monomer units. In the former approach, random copolymers are usually prepared and the unit compositions of the comonomers are limited to 10 mol% to retain the polymer crystallinity. For example, copolymerization of L-lactide and ε-caprolactone (CL) gives a copolymer

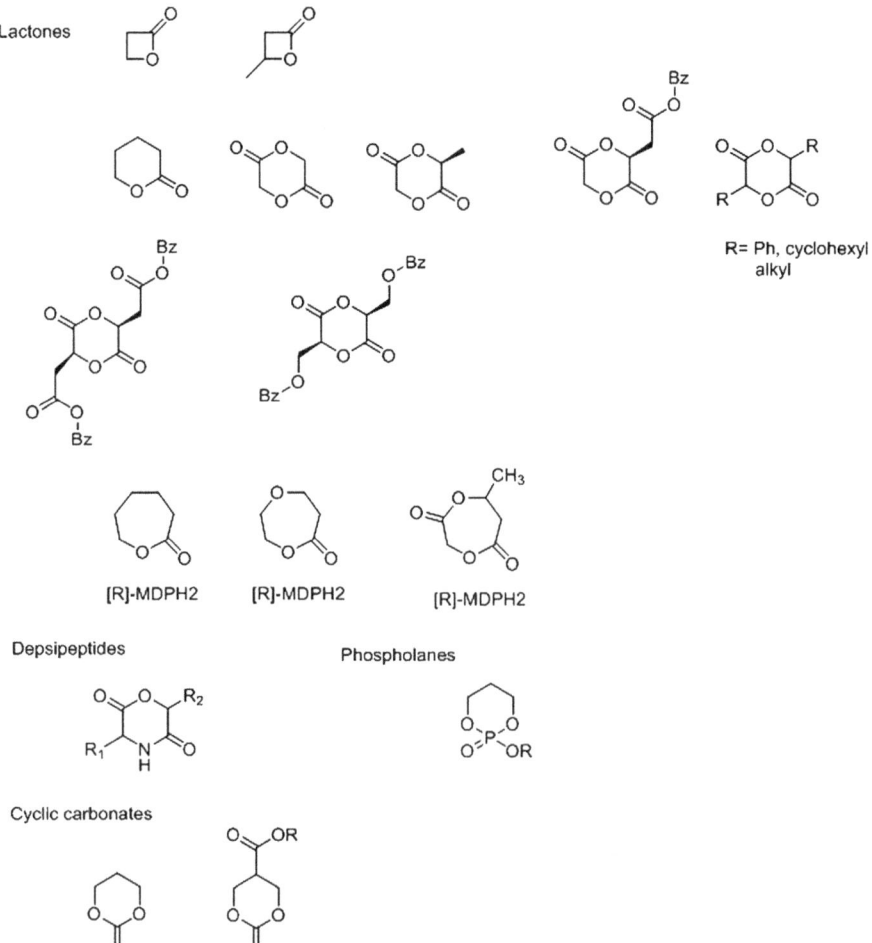

Scheme 1.18 Monomers copolymerized with L-lactide for synthesizing functional PLA polymers.

showing T_m around 150 °C at a unit composition of 90/10.[118] Various co-polymers of glycolide have been synthesized by this technique for making bio-absorbable sutures, such as polyglactin (glycolide/L-lactide)[119] and polyglyconate (glycolide/trimethylene carbonate).[120] Scheme 1.18 summarizes the monomers copolymerized with L-lactide for synthesizing functional PLA polymers.

In the second approach, new cyclic diester monomers, 3S-(benzyloxymethyl)-6S-methyl-1,4-dioxane-2,5-dione (BMD)[121] consisting of both L-lactate and α-O-benzyl-glycerate units and 3S-phenyl-6S-methyl-1,4-dioxane-2,5-dione (PMD)[122] consisting of mandelate and lactate units, have been synthesized by base-catalyzed cyclization of the coupling products of the corresponding hydroxyl-acids with 2-bromo-propionyl chloride. These monomers can be homo-polymerized and copolymerized with other cyclic monomers to obtain polymers having improved properties. The glyceric acid-containing polymers prepared from BMD can be utilized as functional biomedical materials (Scheme 1.19).[123]

Many block copolymers consisting of PLLA or PDLA have also been prepared.[124] Variation of the chain length of the PLLA or PDLA and modification of the copolymerized chains allow us to tune the mechanical properties, degradation rates and crystallinity. For example, the ordinary ROP of L- or D-lactide initiated with poly(ethylene glycol) (PEG) and poly(ethylene glycol) monomethyl ether (MePEG) can give amphiphilic PLA-PEG block copolymers with different types.[124] Scheme 1.20 compares a set of the related copolymers of ABA, BAB and AB types.[125] The BAB triblock copolymers can be obtained by the coupling of the AB diblock copolymers with hexamethylene diisocyanate.[126] These amphiphilic copolymers readily form core-shell type micelles in water by placing the hydrophobic PLA and hydrophilic PEG segments in the core and shell, respectively, as shown in Scheme 1.20. The average hydrodynamic diameters of the micelles measured by dynamic light scattering (DLS) are in the range of 20–30 nm for 1 wt%

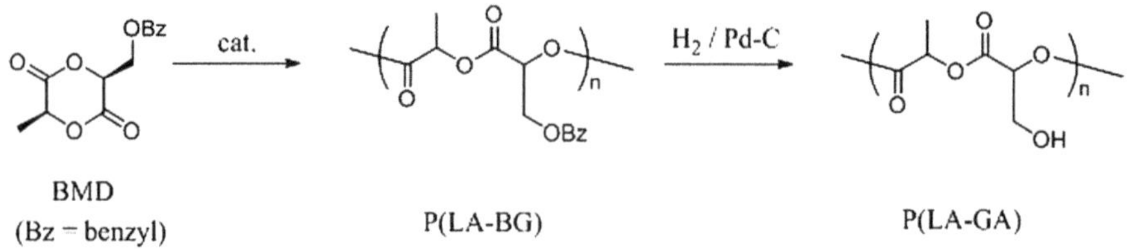

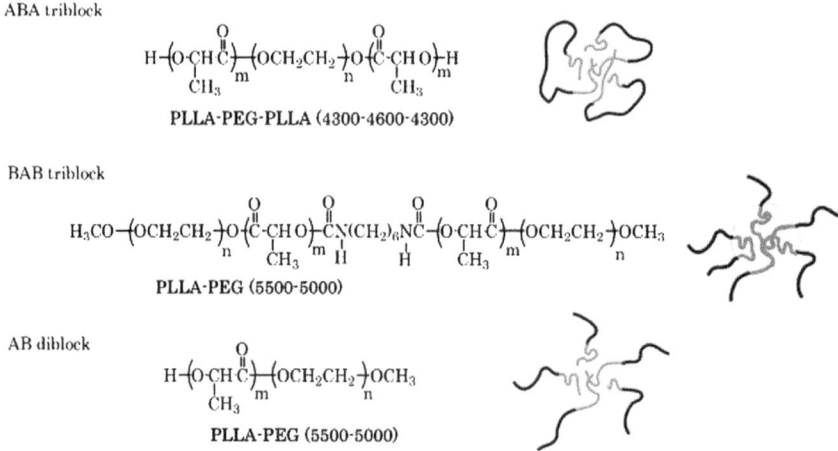

Scheme 1.19   Synthesis of glyceric acid containing PLA by ROP of BMD and subsequent hydrogenolysis for deprotection.

ABA triblock

PLLA-PEG-PLLA (4300-4600-4300)

BAB triblock

PLLA-PEG (5500-5000)

AB diblock

PLLA-PEG (5500-5000)

Scheme 1.20   Polymer structures of PLLA-PEG block copolymers and schematics of their micelles formed in aqueous media.

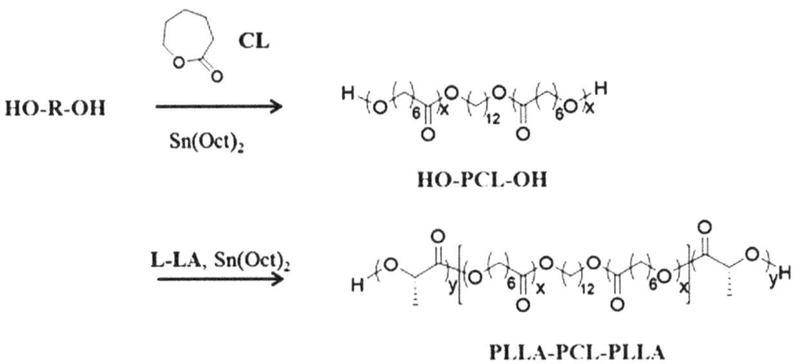

Scheme 1.21 Synthesis of a triblock copolymer PLLA-PCL-PLLA (as an example utilizing 1,6-hexanediol as the initiator for the ROP of CL).

solutions of the respective copolymers. These micelles can be utilized in drug delivery and cell-growth as matrices.[124–126]

When aliphatic polyester pre-polymers are used as the macro-initiators of the lactide polymerization, various block copolymers consisting of soft (polyester) and hard (PLA) blocks can be obtained.[127] Scheme 1.21 shows a typical synthetic route to the block copolymers (PLA-PCL) of poly(ε-caprolactone) (PCL) and polylactide (PLA) in which Sn(Oct)$_2$ is used as the catalyst for the ROP of CL and lactides.[128] The di- and triblock copolymers can be obtained by using the mono-ol- and diol-type PCL pre-polymers as the macro-initiators in the lactide polymerization. Here, the PCL pre-polymers can be prepared by using a mono-ol (*e.g.* 1-dodecanol and benzyl alcohol) and a diol (*e.g.* 1,6-hexanediol) as the initiators of the CL polymerization, respectively. The resultant di- and triblock copolymers show two endothermic peaks at 40–55 °C and 140–170 °C on their DSC curves because of the separate crystallization of PLA and PCL having poor compatibility.[129,130] Furthermore, many studies have been done on the stereo-complexation between various sets of block copolymers comprising PLLA and PDLA, particularly for the AB- and ABA-type block copolymers where A is PLLA or PDLA and B is its enantiomer.[130] Mixing of the above enatiomeric copolymers PLLA-PCL and PDLA-PCL affords sc crystals by which T_m becomes higher than 200 °C.

1.5 Polycondensation of Lactic Acids

No one had believed that direct polycondensation of L-lactic acid can give a high molecular weight of PLLA, until 1995 when Mitsui Chemical Co. first succeeded in synthesizing such a PLLA by using a special solution polycondensation technique.[17] With this epoch-making success, many researchers including the present authors made efforts to establish the direct polycondensation of L-lactic acid. As shown in Scheme 1.22, two equilibria exist in the dehydrative polycondensation of L-lactic acid: one is hydration/

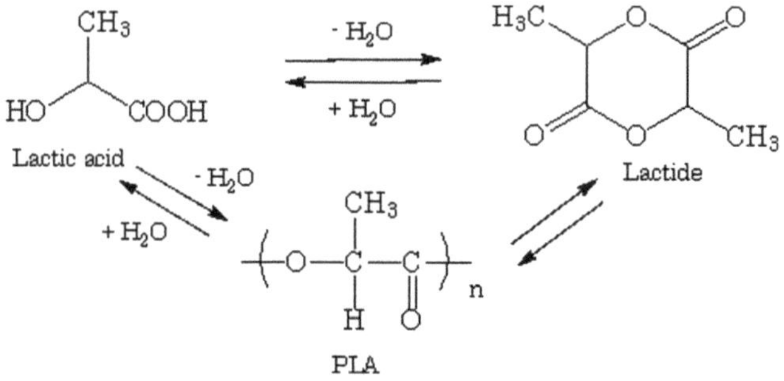

Scheme 1.22 Equilibrium reactions of PLLA and its monomers, L-lactic acid and L-lactide.

dehydration equilibrium of carboxyl and hydroxyl terminals, and the other is ring/chain equilibrium between L-lactide and PLLA.

The former equilibrium constant K, defined by the ordinary equilibrium equation (eqn (5)), can be determined by measuring the amounts of hydroxyl and carboxyl groups formed by the acid hydrolysis of PLLA as well as the remaining water (eqn (4)).[131] Assuming that the relation [COOH] = [OH] is always maintained, the average degree of polymerization (DP) can be correlated with K when x is the molar ratio of water relative to ester, *i.e.* $x = [H_2O]/[-COO-]$.

$$-COOH + HO- \leftrightarrow -COO- + H_2O \qquad (4)$$

$$K = [COOH][OH]/[-COO-][H_2O] \qquad (5)$$

$$DP\text{-}1 = [-COO-]/[COOH] = (1/xK)^{1/2} \qquad (6)$$

The K values were roughly determined to be 1.5 at room temperature and 0.5 at 50 °C, being comparable with those for the ordinary ester formation reactions. Substituting the K value of eqn (6) by these values, the change in M_n of PLLA can be deduced as a function of x. Figure 1.3 shows a typical plot for the M_n change. Since this change is based on the equilibrium constant at room temperature, the real M_n value would be slightly different at higher equilibrium temperatures. It is, however, evident that the water content should be less than 1 ppm to make the M_n value of PLLA exceed 50 kDa that corresponds to 100 kDa in weight-average molecular weight (M_w) assuming $M_w/M_n = 2.0$ in the ordinary polycondensation.

The other ring-chain equilibrium constant is provided by the equilibrium monomer (L-lactide) concentration $[M]_e$ that is affected by the back-biting reaction of the hydroxyl terminal. As mentioned in the preceding section, the $[M]_e$ are lower than 1.0 wt% below 120 °C and higher than 5 wt% above 180 °C in the ROP of L-lactide. High evacuation, needed to remove condensed water, is likely to result in the removal of L-lactide from the system

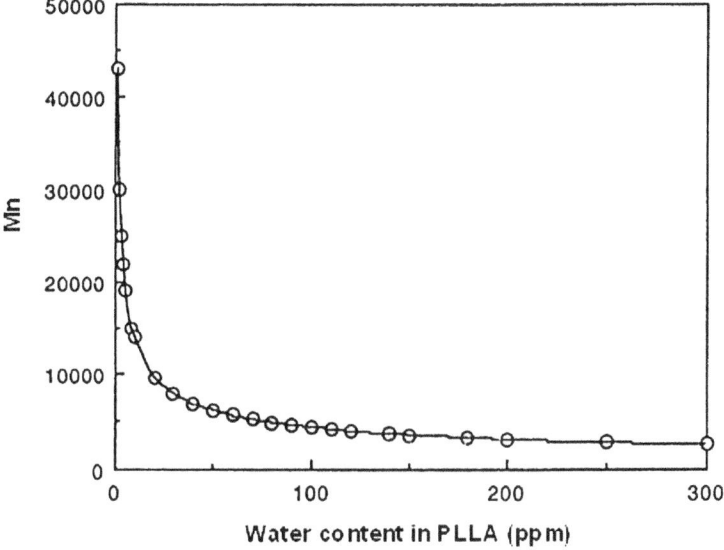

Figure 1.3 Relation of the molecular weight (M_n) and water content of PLLA in the polycondensation.

and reduce the polymer yield. Therefore, reaction conditions must be controlled to obtain high molecular weight of PLLA by direct polycondensation.

The above thermodynamic analysis of the polycondensation reveals that PLLAs having high molecular weights may be produced when the condensed water is efficiently removed to a level of 1 ppm from the polymerization system without evaporation of the L-lactide monomer present in equilibrium. The ordinary reaction conditions that may allow the effective removal of the water may involve: (1) a temperature range of 180–200 °C; (2) a low pressure below 5 torr; and (3) a long reaction time in the presence of an appropriate catalyst and, in some cases, azeotropic solvent for removing water efficiently.[57]

1.5.1 Solution Polycondensation

In 1995, as mentioned above, Ajioka *et al.* first succeeded in synthesizing PLLA having M_w higher than 100 kDa by using diphenyl ether as an azeotropic solvent.[17] In this polycondensation system the high boiling solvent was refluxed at reduced pressure to azeotropically distil the condensed water and to make it adsorb on molecular sieves. This method is especially effective for the co-polycondensation of L-lactic acid with other monomers (hydroxyl-acids and diol/dicarboxylic acid combinations).

1.5.2 Melt/Solid Polycondensation

Following the success of the solution polycondensation, bulk polycondensation of L-lactic acid was studied to eliminate the use of solvent.

The simplest melt-polycondensation can be obtained by continuous de-hydration of oligo(L-lactic acid) with various ionic metal catalysts.[18,132] This process, however, is likely accompanied by unfavourable discoloration of the product as well as by the depolymerization to L-lactide. Kimura *et al.* found out that use of a bi-component catalyst system comprising tin dichloride hydrate and *p*-toluenesulfonic acid (TSA) is effective not only for increasing the molecular weight but also for preventing the discoloration of PLLA in the melt polycondensation of L-lactic acid.[133] As a result, a high molecular weight of PLLA ($M_w \geq 100\,000$ Da) can be obtained within a short reaction time. Bi-component catalyst systems comprising Sn(II)-Ge(IV) and Sn(II)-Si(IV) are also demonstrated to be effective as the catalysts for producing high-molecular-weight PLLA.[134]

The alternative melt/solid polycondensation of L-lactic acid was also established by utilizing the bi-component catalyst comprising Sn(II) and TSA.[133,135] First, L-lactic acid is thermally dehydrated to prepare an oligo-meric PLLA having a degree of polymerization of *ca.* 8. It is then mixed with the bi-component catalyst and subjected to melt-polycondensation at 180 °C for 5 h to obtain a melt-polycondensate having an average molecular weight of 13 000 Da. This melt-polycondensate is finally heat-treated at 105 °C for crystallization and subjected to solid-state polycondensation (SSP) for chain extension at a temperature of 140 or 150 °C. The molecular weight of the SSP products increases above 500 kDa in a relatively short reaction time. In the SSP, the polymer tails and catalysts are allowed to concentrate in the amorphous domain with the polymer crystallization, inducing the chain extension efficiently. Figure 1.4 shows the typical changes in M_w as a func-tion of crystallinity of PLLA at two different temperatures. It is clearly shown that the time-dependent increase in crystallinity is well correlated with the increase in molecular weight. Since the SSP can be conducted on a large

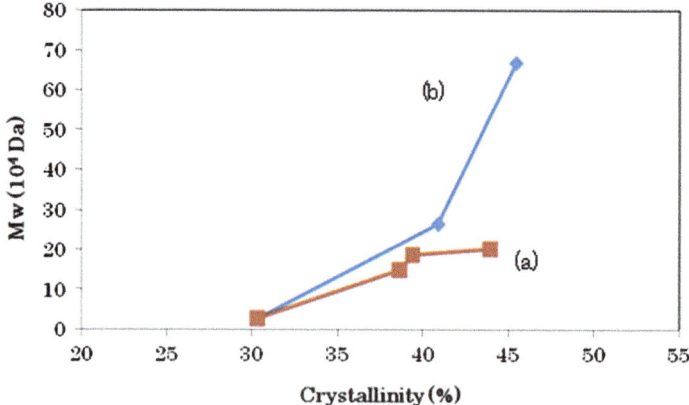

Figure 1.4   Changes in M_w of PLLA as a function of polymer crystallinity in the solid-state polycondensation at (a) 140 °C and (b) 150 °C.

scale in a continuous way, it will afford a facile industrial route to the synthesis of PLLA.

The direct polycondensation technique can be applied to the syntheses of various copolymers that should have different properties and degradability. For example, copolymers of L-lactic acid and phenyl-substituted α-hydroxy acids, such as L- and D-mandelic acids (L-, D-MA) and phenyl-lactic acid (Phe-LA) have been prepared.[136] Polyglactin is also obtainable by the direct co-polycondensation of glycolic acid and L-lactic acid.[137,138]

1.5.3   Stereo-block Polycondensation

The above direct polycondensation can be a practical route to sb-PLA, replacing the above ROP route.[139] Namely, both PLLA and PDLA having medium molecular weight (20–40 kDa) are first prepared by the above melt-polycondensation method. They are mixed in melt state to easily form stereo-complex crystals. SSP of this mixture can yield sb-PLA having high molecular weight in which the stereo-block sequences are generated by the cross propagation of the PLLA and PDLA chains.

Figure 1.5 shows the increase in M_w of sb-PLA as a function of the reaction time through the whole process where the SSP is conducted at 170 °C using a 1:1 melt-blend of PLLA and PDLA having initial molecular weights of 40 kDa.[140] It is shown that the M_w of the resultant sb-PLA reaches 100 kDa, although the molecular weight increase is somewhat limited. Furthermore, the polymer recovery becomes significantly lower with increasing the reaction temperature (43% at 200 °C) because of the increased

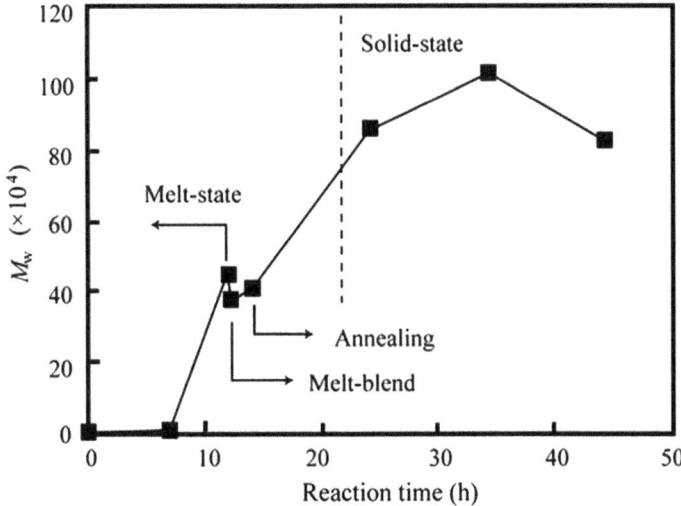

Figure 1.5   Changes in M_w of sb-PLA as a function of the reaction time through the whole process involving the solid-state polycondensation conducted at 170 °C (Refs. 21,26, with permission).

depolymerization to lactides. In this equivalent stereo-mixture of PLLA and PDLA, the diad sequences formed by their hetero-chain couplings are racemic, and the crystallization of the extended chains is retarded. The polymer tails excluded out of the sc domain are likely to lose their mobility due to its strong bond to crystal lattices, having a retarded reactivity in the amorphous domain. Consequently, a high reaction temperature is needed for the polycondensation of the chain ends to trigger trans-esterification and depolymerization by which the randomized sequences are formed together with sublimation of lactides.

The enhancement of the crystallization-induced polycondensation is easier when homopolymer domains that can induce crystallization of the extended chains are increased in ratio. For this purpose, the blending ratio of PLLA and PDLA pre-polymers is changed from $1:1$.[25,141] When the SSP is conducted at 140–160 °C with different PLLA-to-PDLA compositions, the molecular weight increases to higher than 100 kDa with a polymer recovery as high as 70–85%. Figure 1.6 shows the typical results for M_w changes of the final sb-PLAs as a function of PDLA ratio in the feed. It is clearly shown that the M_w becomes higher with the PLLA/PDLA ratio deviating from $1:1$ or L- or D-rich compositions.

The obtained sb-PLAs having PLLA-rich and PDLA-rich compositions exhibit good stereo-complexity in spite of involving long homopolymer sequences,[26] and the properties of the sb-PLAs can be widely tuned by changing the PLLA/PDLA composition and the length of the homopolymer sequences. This structural variety of sb-PLA will guarantee a wide range of properties and applications of PLA materials.

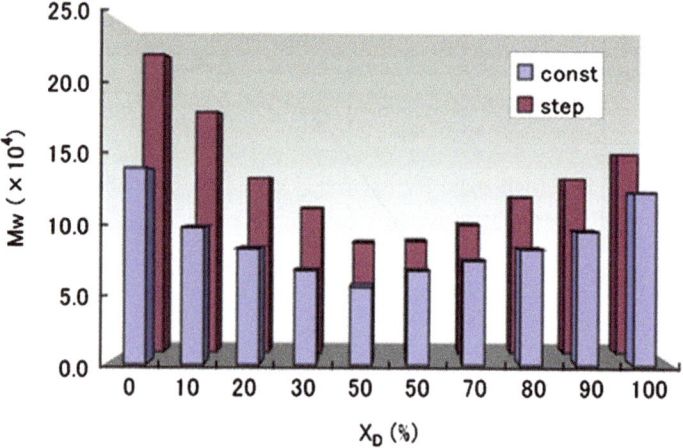

Figure 1.6   Effect of the PDLA composition (X_D) on the M_w of the final sb-PLAs. The PLLA-rich $(0 \leq X_D \leq 50)$ and PDLA-rich $(50 \leq X_D \leq 100)$ systems were separately examined with different melt-blends. The SSP was conducted at constant temperature (150 °C) for 30 h and with stepwise temperature increase (at 140 °C, 150 °C and 160 °C each for 10 h).

References

1. H. Benninga, *A History of Lactic Acid Making*, Springer, New York, 1990.
2. W. H. Carothers, G. L. Dorough and F. J. van Natta, *J. Am. Chem. Soc.*, 1932, **54**, 761.
3. T. Ouch, T. Saito, T. Kontani and Y. Ohya, *Macromolecular Biosciences*, 2004, **4**, 458.
4. Y. Hu, X. Jiang, Y. Ding, L. Zhang, C. Yang, J. Zhang, J. Chen and Y. Yang, *Biomaterials*, 2003, **24**, 2395.
5. T. Chandy, G. S. Das, R. F. Wilson and G. H. R. Rao, *J. Appl. Polymer Sci.*, 2002, **86**, 1285.
6. T. G. Park, M. J. Alonso and R. Langer, *J. Appl. Polymer Sci.*, 1994, **52**, 1797.
7. K. E. Gonsalves, S. Jin and M. I. Baraton, *Biomaterials*, 1998, **19**, 1501.
8. H. Arimura, Y. Ohya, T. Ouchi and H. Yamada, *Macromolecular Biosciences*, 2003, **3**, 18.
9. X. Deng, Y. Liu, M. Yuan, X. Li, L. Liu and W. X. Jia, *J. Appl. Polymer Sci.*, 2002, **86**, 2557.
10. M. Pluta, A. Galeski, M. Alexandre, M. A. Paul and P. Dubois, *J. Appl. Polymer Sci.*, 2002, **86**, 1497.
11. S. Li, *Macromolecular Biosciences*, 2003, **3**, 657.
12. R. Zhang and P. X. Ma, *Macromolecular Biosciences*, 2004, **4**, 100.
13. K. Kesenci, L. Fambri, C. Migliare and E. Piskin, *J. Biomater. Sci. Polymer Ed.*, 2000, **11**, 617.
14. J. A. Roether, A. R. Boccaccini, L. L. Hench, V. Maquet, S. Gautier and R. Jérome, *Biomaterials*, 2002, **23**, 3871.
15. D. Garlotta, *J. Polymer Environ.*, 2001, **9**, 63.
16. A. P. Gupta and V. Kumar, *Eur. Polym. J.*, 2007, **43**, 4053.
17. M. Ajioka, K. Enomoto, K. Suzuki and A. Yamaguchi, *Bull. Chem. Soc. Jpn*, 1995, **68**, 2125.
18. S. I. Moon, C. W. Lee, M. Miyamoto and Y. Kimura, *J. Polymer Sci. Polymer Chem.*, 2000, **38**, 1673.
19. T. Maharana, B. Mohanty and Y. S. Negi, *Progr. Polymer. Sci.*, 2009, **34**, 99.
20. S. H. Hyon, K. Jamshidi and Y. Ikada, *Biomaterials*, 1997, **18**, 1503.
21. A. Kowalski, J. Libiszowski, A. Duda and S. Penczek, *Macromolecules*, 2000, **33**, 1964.
22. Y. Ikada, K. Jamshidi, H. Tsuji and S. H. Hyon, *Macromolecules*, 1987, **20**, 906.
23. B. P. De Boer, M. J. Teixeira de Mattos and O. M. Neijssel, *Appl. Microbiol. Biotechnol.*, 1990, **34**, 149.
24. K. Fukushima, K. Sogo, S. Miura and Y. Kimura, *Macromolecular Biosciences*, 2004, **4**, 1021.
25. K. Fukushima and Y. Kimura, *J. Polymer Sci. Polymer Chem.*, 2008, **46**, 3714.

26. M. Hirata and Y. Kimura, *Polymer*, 2008, **49**, 2656.
27. M. Hirata, K. Kobayashi and Y. Kimura, *J. Polymer Sci. Polymer. Chem.*, 2010, **48**, 794.
28. K. Masutani, C. W. Lee and Y. Kimura, *Macromol. Chem. Phys.*, 2012, **213**, 695.
29. K. Masutani, C. W. Lee and Y. Kimura, *Polymer*, 2012, **53**, 6053.
30. A. Vaidya, R. Pandey, S. Mudliar, M. Kumar, T. Chakrabarti and S. Devotta, *Crit. Rev. Environ. Sci. Tech.*, 2005, **35**, 429.
31. H. W. Doelle, *Bacterial Metabolism*, Academic Press, New York, 1969, pp. 332–340.
32. E. M. Filaclvone and E. J. Costello, *Ind. Eng. Chem.*, 1952, **44**, 2189.
33. S. K. Bhattacharyya, S. K. Palit and A. R. Das, *Ind. Eng. Chem. Prod. Res. Dev.*, 1970, **9**, 92.
34. The ACS Office of Public Outreach, *The SOHIO Acrylonitrile*, BP Chemicals Inc., Warrensville Heights, Ohio, American Chemical Society, 1996.
35. T. Mori, H. Nishida, Y. Shirai and T. Endo, *Polymer Degrad. Stabil.*, 2003, **81**, 515.
36. H. Urayama, S. I. Moon and Y. Kimura, *Macromol. Mater. Eng.*, 2003, **288**, 137.
37. M. J. Stanford and A. P. Dove, *Chem. Soc. Rev.*, 2010, **39**, 486.
38. H. Tsuji, S. H. Hyon and Y. Ikada, *Macromolecules*, 1991, **24**, 5652.
39. H. Tsuji, S. H. Hyon and Y. Ikada, *Macromolecules*, 1991, **24**, 5657.
40. H. Tsuji, F. Horii and S. H. Hyon, *Macromolecules*, 1991, **24**, 2719.
41. T. Okihara, M. Tsuji, A. Kawaguchi, K. Katayama, H. Tsuji, S. H. Hyon and Y. Ikada, *J. Macromol. Sci. Phys.*, 1991, **30**, 119.
42. H. Tsuji, F. Horii, M. Nakagawa, Y. Ikada, H. Odani and R. Kitamura, *Macromolecules*, 1992, **25**, 4114.
43. D. Brizzolara, H. J. Cantow, K. Diederichs, E. Keller and A. J. Domb, *Macromolecules*, 1996, **29**, 191.
44. L. Cartier, T. Okihara and B. Lotz, *Macromolecules*, 1997, **30**, 6313.
45. M. Kakuta, M. Hirata and Y. Kimura, *J. Macromol. Sci. Polymer Rev.*, 2009, **49**, 107.
46. I. C. McNeill and H. A. Leiper, *Polymer Degrad. Stabil.*, 1985, **11**, 309.
47. I. G. Barskaya, Y. B. Lyudvig, R. R. Shifrina and A. L. Izyumnikov, *Polymer Sci. USSR*, 1983, **25**, 1491.
48. D. R. Witzke, R. Narayan and J. J. Kolstad, *Macromolecules*, 1997, **30**, 7075.
49. P. Degee, P. Dubois, R. Jerome, S. Jacobsen and H. Fritz, *Macromol. Symp.*, 1999, **144**, 289.
50. A. Kowalski, A. Duda and S. Penczek, *Macromolecules*, 2000, **33**, 689.
51. A. Kowalski, A. Duda and S. Penczek, *Macromolecules*, 2000, **33**, 7539.
52. S. Penczek, A. Duda, A. Kowalski, J. Libiszowski, K. Majerska and T. Biela, *Macromol. Symp.*, 2000, **157**, 61.
53. US Food and Drug Administration, Code of Federal Regulations Title 21.

54. E. F. Connor, G. W. Nyce, M. Myers, A. Mock and J. L. Hedrick, *J. Am. Chem. Soc.*, 2002, **124**, 914.
55. D. Bourissou, B. Martin-Vaca, A. Dumitrescu, M. Graullier and F. Lacombe, *Macromolecules*, 2005, **38**, 9993.
56. H. R. Kricheldorf and A. Serra, *Polymer Bull.*, 1985, **14**, 497.
57. K. Shinno, M. Miyamoto, Y. Kimura, Y. Hirai and H. Yoshitome, *Macromolecules*, 1997, **30**, 6438.
58. M. Bero, P. Dobrzynski and J. Kasperczyk, *J. Polymer Sci. Polymer Chem.*, 1999, **37**, 4038.
59. M. Ryner, K. Stridberg and A. C. Albertsson, *Macromolecules*, 2001, **34**, 3877.
60. S. Penczek, R. Szymanski, A. Duda and J. Baran, *Macromol. Symp.*, 2003, **201**, 261.
61. Y. Fana, H. Nishida, T. Moria, S. Hoshiharaa, Y. Fana, Y. Shirai and T. Endo, *Polymer Degrad. Stabil.*, 2004, **84**, 143.
62. S. Jacobsen and H. G. Fritz, *Polymer Eng. Sci.*, 1999, **39**, 1303.
63. A. Kowalski, A. Duda, S. Penczek and J. Baran, *Macromolecules*, 1998, **31**, 2114.
64. N. Ropson, P. Dubois, R. Jerome and P. Teyssie, *Macromolecules*, 1995, **28**, 7589.
65. R. H. Platel, L. M. Hodgson and C. K. Williams, *Polymer Rev.*, 2008, **48**, 11.
66. V. C. Gibson and E. L. Marshall, *Compr. Coord. Chem.*, 2004, **9**, 1.
67. H. R. Kricheldorf, M. Berl and N. Scharnagl, *Macromolecules*, 1988, **21**, 286.
68. B. J. O'Keefe, M. A. Hillmyer and W. B. Tolman, *J. Chem. Soc. Dalton Trans.*, 2001, **1**, 2215.
69. O. Dechy-Cabaret, B. Martin-Vaca and D. Bourissou, *Chem. Rev.*, 2004, **104**, 6147.
70. A. C. Albertsson and I. K. Varma, *Biomacromolecules*, 2003, **4**, 1466.
71. H. R. Kricheldorf, *Chem. Rev.*, 2009, **109**, 5579.
72. H. R. Kricheldorf and R. Dunsing, *Macromol. Chem. Phys.*, 1986, **187**, 1611.
73. T. Saegusa, *Angew. Chem. Int. Ed.*, 1977, **16**, 826.
74. D. Bourissou, S. Moebs-Sanchez and B. Martin-Vaca, *C. R. Chimie*, 2007, **10**, 775.
75. H. R. Kricheldorf, *Angew. Chem. Int. Ed.*, 2006, **45**, 5752.
76. F. Nederberg, E. F. Connor, T. Glausser and J. L. Hedrick, *Chem. Comm.*, 2001, **3**, 2066.
77. F. Nederberg, E. F. Connor, M. Moller, T. Glauser and J. L. Hedrick, *Angew. Chem. Int. Ed.*, 2001, **40**, 2712.
78. H. R. Kricheldorf, N. Lomadze and G. Schwarz, *Macromolecules*, 2008, **41**, 7812.
79. H. R. Kricheldorf, N. Lomadze and G. Schwarz, *Macromolecules*, 2007, **40**, 4859.
80. H. R. Kricheldorf, C. Von Lossow and G. Schwarz, *J. Polymer Sci. Polymer Chem.*, 2006, **44**, 4680.

81. C. Bonduelle, B. Martin-Vaca, F. P. Cossio and D. Bourissou, *Chem. Eur. J.*, 2008, **14**, 5304.

82. E. F. Connor, G. W. Nyce, M. Myers, A. Mock and J. L. Hedrick, *J. Am. Chem. Soc.*, 2002, **124**, 914.

83. G. W. Nyce, T. Glauser, E. F. Connor, A. Mock, R. M. Waymouth and J. L. Hedrick, *J. Am. Chem. Soc.*, 2003, **125**, 3046.

84. A. P. Dove, R. C. Pratt, B. G. G. Lohmeijer, D. A. Culkin, E. C. Hagberg, G. W. Nyce, R. M. Waymouth and J. L. Hedrick, *Polymer*, 2006, **47**, 4018.

85. N. E. Kamber, W. Jeong, S. Gonzalez, J. L. Hedrick and R. M. Waymouth, *Macromolecules*, 2009, **42**, 1634.

86. D. A. Culkin, W. Jeong, S. Csihony, E. D. Gomez, N. P. Balsara, J. L. Hedrick and R. M. Waymouth, *Angew. Chem. Int. Ed.*, 2007, **46**, 2627.

87. R. C. Pratt, B. G. Lohmeijer, D. A. Long, R. M. Waymouth and J. L. Hedrick, *J. Am. Chem. Soc.*, 2006, **128**, 4556.

88. B. G. G. Lohmeijer, R. C. Pratt, F. Leibfarth, J. W. Logan, D. A. Long, A. P. Dove, F. Nederberg, J. Choi, C. Wade, R. M. Waymouth and J. L. Hedrick, *Macromolecules*, 2006, **39**, 8574.

89. H. Li, S. Zhang, J. Jiao, Z. Jiao, L. Kong, J. Xu, J. Li, J. Zuo and X. Zhao, *Biomacromolecules*, 2009, **10**, 1311.

90. Z. Pang, H. Li, P. He, Y. Wang, H. Ren, H. Wang and X. X. Zhu, *J. Polymer Sci.*, 2012, **50**, 4004.

91. A. P. Dove, R. C. Pratt, B. G. G. Lohmeijer, R. M. Waymouth and J. L. Hedrick, *J. Am. Chem. Soc.*, 2005, **127**, 13798.

92. L. Zhang, R. C. Pratt, F. Nederberg, H. W. Horn, J. E. Rice, R. M. Waymouth, C. G. Wade and J. L. Hedrick, *Macromolecules*, 2010, **43**, 1660.

93. M. K. Kiesewetter, M. D. Scholten, N. Kirn, R. L. Weber, J. L. Hedrick and R. M. Waymouth, *J. Org. Chem.*, 2009, **74**, 9490.

94. G. W. Nyce, S. Csihony, R. M. Waymouth and J. L. Hedrick, *Chem. Eur. J.*, 2004, **10**, 4073.

95. S. Csihony, D. A. Culkin, A. C. Sentman, A. P. Dove, R. M. Waymouth and J. L. Hedrick, *J. Am. Chem. Soc.*, 2005, **127**, 9079.

96. T. M. Trnka, J. P. Morgan, M. S. Sanford, T. E. Wilhelm, M. Scholl, T. L. Choi, S. Ding, M. W. Day and R. Grubbs, *J. Am. Chem. Soc.*, 2003, **125**, 2546.

97. D. Enders, K. Breuer, G. Raabe, J. Runsink, J. H. Teles, J. P. Melder, K. Ebel and S. Brode, *Angew. Chem. Int. Ed.*, 1995, **34**, 1021.

98. O. Coulembier, A. P. Dove, R. C. Pratt, A. C. Sentman, D. A. Culkin, L. Mespouille, P. Dubois, R. M. Waymouth and J. L. Hedrick, *Angew. Chem. Int. Ed.*, 2005, **44**, 4964.

99. H. C. Kolb, M. G. Finn and K. B. Sharpless, *Angew. Chem. Int. Ed.*, 2001, **40**, 2004.

100. O. Coulembier, B. G. G. Lohmeijer, A. P. Dove, R. C. Pratt, L. Mespouille, D. A. Culkin, S. J. Benight, P. Dubois, R. M. Waymouth and J. L. Hedrick, *Macromolecules*, 2006, **39**, 5617.

101. M. Wisniewski, A. LeBorgne and N. Spassky, *Macromol. Chem. Phys.*, 1997, **198**, 1227.
102. N. Spassky, M. Wisniewski, C. Pluta and A. LeBorgne, *Macromol. Chem. Phys.*, 1996, **197**, 2627.
103. A. Dove, V. Gibson, E. Marshall, H. Rzepa, A. White and D. Williams, *J. Am. Chem. Soc.*, 2006, **128**, 9834.
104. N. Nimitsiriwat, V. Gibson, E. Marshall and M. Elsegood, *Inorg. Chem.*, 2008, **47**, 5417.
105. P. Cameron, D. Jhurry, V. Gibson, A. White, D. Williams and S. Williams, *Macromol. Rapid Comm.*, 1999, **20**, 616.
106. M. Cheng, A. Attygalle, E. Lobkovsky and G. Coates, *J. Am. Chem. Soc.*, 1999, **121**, 11583.
107. A. Bhaw-Luximon, D. Jhurry and N. Spassky, *Polymer Bull.*, 2000, **44**, 31.
108. T. Ovitt and G. Coates, *J. Polymer Sci. Polymer Chem.*, 2000, **38**, 4686.
109. C. Radano, G. Baker and M. Smith, *J. Am. Chem. Soc.*, 2000, **122**, 1552.
110. N. Nomura, R. Ishii, M. Akakura and K. Aoi, *J. Am. Chem. Soc.*, 2002, **124**, 5938.
111. T. Ovitt and G. Coates, *J. Am. Chem. Soc.*, 2002, **124**, 1316.
112. K. Majerska and A. Duda, *J. Am. Chem. Soc.*, 2004, **126**, 1026.
113. M. Chisholm, N. Patmore and Z. Zhou, *Chem. Comm.*, 2005, **1**, 127.
114. X. Pang, H. Du, X. Chen, X. Wang and X. Jing, *Chem. Eur. J.*, 2008, **14**, 3126.
115. N. Yui, P. J. Dijkstra and J. Feijen, *Makromol. Chem.*, 1990, **191**, 481.
116. K. Masutani, S. Kawabata, T. Aoki and Y. Kimura, *Polymer Int.*, 2010, **59**, 1526.
117. Y. Kimura, *Polymer J.*, 2009, **41**, 797.
118. S. Pensec, M. Leory, H. Akkouche and N. Spassky, *Polymer Bull.*, 2000, **45**, 373.
119. D. K. Gilding and A. M. Reed, *Polymer*, 1979, **20**, 1459.
120. C. Jie, K. J. Zhu and Y. Shilir, *Polymer Int.*, 1996, **41**, 369.
121. M. Leemhuis, C. F. van Nostrum, J. A. W. Kruijtzer, Z. Y. Zhong, M. R. Ten Breteler, P. J. Dijkstra, J. Feijen and W. E. Hennink, *Macromolecules*, 2006, **39**, 3500.
122. F. Jing and M. A. Hillmyer, *J. Am. Chem. Soc.*, 2008, **130**, 13826.
123. M. Leemhuis, C. F. Nostrum, J. A. W. Kruijtzer, Z. Y. Zhong, M. R. Breteler, P. J. Dijkstra, J. Feijen and W. E. Hennink, *Macromolecules*, 2006, **39**, 3500.
124. K. L. Wooley, *J. Polymer Sci. Polymer Chem.*, 2000, **38**, 1397.
125. T. Fujiwara and Y. Kimura, *Macromolecular Biosciences*, 2002, **2**, 11.
126. T. Mukose, T. Fujiwara, J. Nakano, I. Taniguchi, M. Miyamoto, Y. Kimura, I. Teraoka and C. W. Lee, *Macromolecular Biosciences*, 2004, **4**, 361.
127. A. C. Albertsson, I. K. Varma, B. Lochab, A. F. Wistrand and K. Kumar, *Poly(lactic acid)*, John Wiley & Sons, Inc., New Jersey, 2010, pp. 43–76.

128. C. Jacobs, P. Dubois, R. Jerome and P. Teyssie, *Macromolecules*, 1991, **24**, 3027.
129. C. X. Song and X. D. Feng, *Macromolecules*, 1984, **17**, 2764.
130. W. M. Stevels, M. J. K. Ankone, P. J . Dijkstra and J. Feijen, *Macromol. Symp.*, 1996, **102**, 107.
131. Y. Kimura, *Green PLA J.*, 2004, **13**, 114.
132. T. Fukushima, Y. Sumihiro, K. Koyanagi, N. Hashimoto, Y. Kimura and T. Sakai, *Int. Polymer Process.*, 2001, **4**, 380.
133. S. I. Moon, C. W. Lee, I. Taniguchi, M. Miyamoto and Y. Kimura, *Polymer*, 2001, **42**, 5059.
134. S. I. Moon and Y. Kimura, *Polymer Int.*, 2003, **52**, 299.
135. S. I. Moon, C. W. Lee, I. Taniguchi, M. Miyamoto, Y. Kimura and C. W. Lee, *High Perform. Polymer.*, 2001, **13**, S189.
136. S. I. Moon, H. Urayama and Y. Kimura, *Macromolecular Biosciences*, 2003, **3**, 301.
137. S. I. Moon, K. Deguchi, M. Miyamoto and Y. Kimura, *Polymer Int.*, 2004, **53**, 254.
138. K. Takahashi, I. Taniguchi, M. Miyamoto and Y. Kimura, *Polymer*, 2000, **41**, 8725.
139. K. Fukushima, Y. Furuhashi, K. Sogo, S. Miura and Y. Kimura, *Macromolecular Bioscience*, 2005, **5**, 21.
140. K. Fukushima and Y. Kimura, *Macromol. Symp.*, 2005, **224**, 133.
141. K. Fukushima, M. Hirata and Y. Kimura, *Macromolecules*, 2007, **40**, 3049.

CHAPTER 2

Polylactide Stereo-complex: From Principles to Applications

SUMING LI*[a] AND YANFEI HU[b]

[a] Institut Européen des Membranes, UMR CNRS 5635, Université Montpellier II, Place Eugene Bataillon, 34095, Montpellier Cedex 5, France; [b] Institut des Biomolécules Max Mousseron, UMR CNRS 5247, Université Montpellier I, 15 Avenue Charles Flahault, BP 14491, 34093, Montpellier Cedex 5, France
*Email: sli@univ-montp2.fr

2.1 Introduction

Biodegradable polymers have shown great potential in the biomedical field.[1,2] Among them, polylactide (PLA) has been extensively investigated for biomedical and pharmaceutical applications, such as controlled drug release systems, medical implants and scaffolds in tissue engineering due to its outstanding degradability and biocompatibility.[3–7] In fact, the final degradation product of PLA, lactic acid, is a metabolite and can be easily eliminated from the human body *via* the Krebs cycle. Moreover, lactic acid can be obtained from renewable resources, which is of great importance for the development of various biomedical applications.

High molar mass PLA is generally obtained by ring-opening polymerization of lactide, the cyclic dimer of lactic acid. The latter is a chiral molecule and exists in two enantiomeric forms, *i.e.* L-lactic acid and D-lactic acid. Accordingly, there exist three diastereoisomers of lactide, *i.e.* L-lactide composed of two L-lactyl units, D-lactide composed of two D-lactyl units and meso-lactide composed of one L-lactyl and one D-lactyl unit in the cycle. The 50/50 mixture

RSC Polymer Chemistry Series No. 12
Poly(lactic acid) Science and Technology: Processing, Properties, Additives and Applications
Edited by Alfonso Jiménez, Mercedes Peltzer and Roxana Ruseckaite

of D-lactide and L-lactide, namely DL-lactide or racemic lactide, is also used in the synthesis of PLA. Thus, PLA with various stereostructures can be obtained by ring-opening polymerization of different lactide feeds, including poly(L-lactide) (PLLA), poly(D-lactide) (PDLA), poly(DL-lactide) (PDLLA) and various stereo-copolymers poly(L-lactide-*co*-D-lactide). The chirality of lactides is a key parameter which allows tailoring the degradation rate as well as the physical and mechanical properties of PLA-based polymers by varying the D/L ratio.

Polymer stereo-complexation results from stereoselective interactions between two complementing stereo-regular polymers which interlock to form a new material with altered physical properties in comparison with the parent polymers.[8] The interactions are mainly stereoselective van der Waals forces. Stereo-complexation between PLLA and PDLA was first reported by Ikada *et al.* in 1987, and is now a well-known phenomenon for optically active PLA stereo-copolymers.[8-11] In the past decades, the properties and potential applications of PLA stereo-complex have been extensively investigated.[4,12,13]

Stereo-complex enhances the thermal resistance, mechanical properties and the hydrolytic stability of PLA-based materials. These improvements result from the strong interaction between L-lactyl and D-lactyl sequences. Stereo-complexation provides a new route to conceive biomedical applications, such as stereo-complexation-induced nanotubes, microparticles, micelles and hydrogels. Copolymers of PLLA and PDLA with poly(ethylene glycol) (PEG) are susceptible to form micelles by self-assembly in an aqueous solution, and hydrogels can be obtained by mixing PLLA/PEG and PDLA/PEG aqueous solutions. Both micelles and hydrogels present great interest as carriers for controlled drug delivery.

PLA homo-stereo-complexation can take place both in enantiomeric mixture and in stereo-blocks of PLA-based polymers. The key parameter affecting stereo-complexation is the mixing ratio of PLLA to PDLA sequences, the equimolar mixture yielding the optimal degree of stereo-complexation. PLA stereo-complex is usually prepared from co-precipitation of PLLA and PDLA in solution or through cooling from a melted mixture of both polymers.[14-18] Hetero-stereo-complexation between PLA and other optically active materials was also reported.[19,20]

In this contribution we intend to provide a comprehensive review on PLA stereo-complex, including preparation, structure-properties and biomedical applications. The effect of stereo-complex on the thermal dynamics, hydrolytic degradation, micellization and hydrogelation will be discussed.

2.2   Synthesis and Structure-Properties of PLA Stereo-complex

2.2.1   Synthesis of PLA Homopolymers and Stereo-copolymers

PLA synthesis is realized by ring-opening polymerization of lactide, as shown in Scheme 2.1. Many initiation systems have been investigated in the past decades. Among them, two compounds are industrially used, namely tin (II)

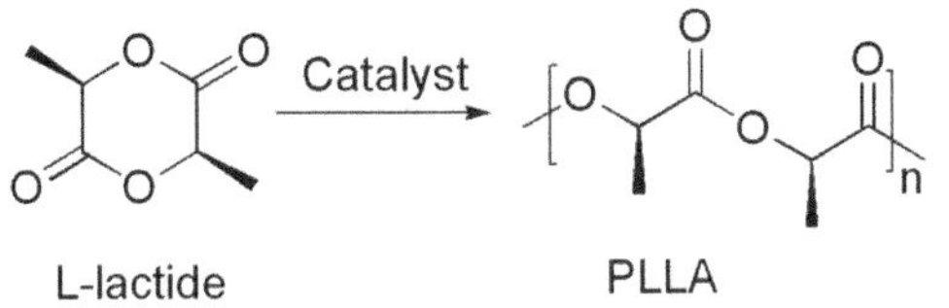

Scheme 2.1   Synthesis of poly(L-lactide) (PLLA) by ring-opening polymerization of L-lactide.

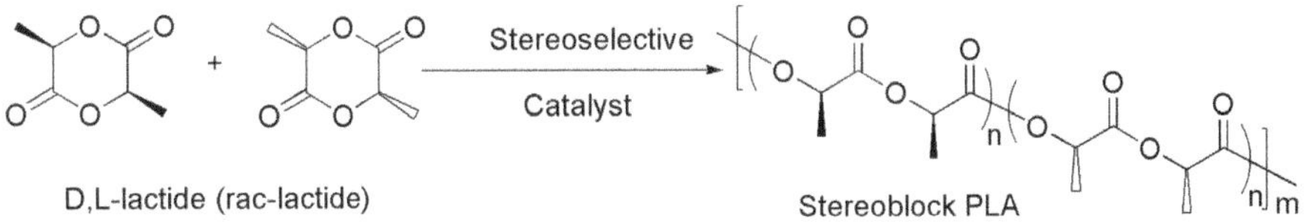

Scheme 2.2   Synthesis of PLA multiblock stereo-copolymers by ring-opening polymerization of DL-lactide using stereoselective catalyst.

2-ethyl hexanoate or stannous octoate (SnOct$_2$), and zinc lactate (ZnLac$_2$). Approved by the US Food and Drug Administration for surgical and pharmacological applications, stannous octoate is the most widely utilized initiator due to its high efficiency allowing high molar mass polymers to be achieved under relatively mild conditions. Zinc lactate is used as an alternative to stannous octoate although it leads to slower polymerization. In fact, zinc is an oligoelement with daily allowance for the metabolism of mammalian bodies and Zn ions present bacteriostatic properties. Small alcohols, such as ethylene glycol, 2-methoxyethanol, 1,2-butanediol and lauryl alcohol are commonly used as initiator of the ring-opening polymerization.

The ring-opening polymerization of lactides proceeds *via* pair addition of repeat units. The stereochemical chain structure of PLA can be controlled by the feed composition (ratio of L-lactide, D-lactide and DL-lactide diastereoisomers) and/or by the stereochemical preference of the initiator/catalyst system. In fact, stereoselective catalysts such as aluminium alkoxides (SBO$_2$Al-OR) or lithium *tert*-butoxide (K$^+$(CH$_3$)$_3$CO$^-$) bearing Schiff's base ligands are susceptible to initiate the ring-opening polymerization of DL-lactide, leading to multiblock stereo-copolymers due to a chain-end control or an enantiomorphic site control of the monomer addition as shown in Scheme 2.2.[21–28] The resulting stereo-copolymers are able to form a stereo-complex due to van der Waals interactions between homo-chiral PLA sequences of opposite configuration.

2.2.2   PLA Stereo-complex by Co-crystallization

PLA stereo-complexes are obtained by co-crystallization of PLLA and PDLA in bulk or in organic solvents such as chloroform or acetonitrile. It has been shown that the minimal degree of polymerization (DP) of PLLA or PDLA oligomers for crystallization is 11, with corresponding melting temperature (T_m) at 59 °C and melting enthalpy (ΔH_m) of 39 J g^{-1}. In contrast, in blends of PLLA and PDLA oligomers, crystallization or stereo-complexation is already observed with a DP of 7, with corresponding T_m at 54 °C and ΔH_m of 65 J g^{-1}.[29] On the other hand, stereo-complex of PLLA and PDLA oligomers with a DP of 11 exhibits a T_m at 119 °C with ΔH_m of 69 J g^{-1}.[29] Higher molar mass yields a stereo-complex with higher T_m and ΔH_m.[8–11]

Different procedures are used for co-crystallization or stereo-complexation in bulk from melt: (1) crystallization at a fixed temperature directly from the melt; (2) crystallization at a fixed temperature quenching from the melt; (3) crystallization during cooling down from the melt; and (4) crystallization during heating of a quenched melt.[11,30]

Together with stereo-complexation, homo-crystallites only composed of either PLLA or PDLA can also be formed. The formation of stereo-complex crystallites and/or homo-crystallites depends on parameters, such as the blending ratio,[9,10,31–35] the molar mass and optical purity,[10,31–37] the co-crystallization conditions, such as the temperature and time,[10,32–35,37] the nature of solvents,[31,33,38] *etc.* In general, equimolar blending or low molar mass tends to favour exclusive formation of stereo-complex without formation of homo-crystallites. The blending mode and nature of solvents also strongly affect stereo-complexation.

Moreover, stereo-complexation can occur in block copolymer mixtures provided that the copolymers contain L-LA and D-LA sequences of sufficient lengths. PLA multiblock stereo-copolymers can easily form a stereo-complex as mentioned above.[21–28] Examples also include stereo-complexation between block copolymers of L-lactide or D-lactide with ε-caprolactone,[39] between L-lactide-rich PLA and D-lactide-rich PLA,[36,37] between copolymers of L-lactide or D-lactide with glycolide[40] and between block copolymers of L-lactide or D-lactide with PEG.[41,42]

Interestingly, PLA stereo-complex can also be obtained during *in vitro* degradation of poly(DL-lactide), which is an intrinsically amorphous polymer.[13,43–45] In fact, poly(DL-lactide) contains isotactic L-LA and D-LA sequences along polymer chains due to the pair-addition mechanism of the ring-opening polymerization of DL-lactide. Once released by degradation, these sequences tend to co-crystallize, yielding an oligomeric stereo-complex with melting temperature at *ca.* 115–125 °C.[13,43–45]

2.2.3 Properties of PLA Stereo-complex

The crystal structure of PLLA/PDLA stereo-complex is trigonal, consisting of sub-cells with two enantiomorphous, antiparallel chains.[46,47] PLLA and PDLA stems are arranged alternately taking 3_1 helical conformation in the stereo-complex.[46,48] Van der Waals forces between the hydrogen atoms of CH_3 and the oxygen of O=C in PLA chains with opposite configurations have been suggested to induce chain packing for stereo-complexation. Recently, according to FTIR analysis by Zhang *et al.*, the CH_3•••O=C hydrogen bonding contributes to the interaction between chains in PLLA/PDLA stereo-complex, and thus constitutes the driving force for the nucleation of stereo-complexation.[49] Brizzolara *et al.* proposed a possible growing mechanism of triangular lamellar crystals.[48] At the beginning, a triangular nucleus is formed by PLLA chains in the shape of 3_1-helix, then a PDLA layer grows onto the PLLA layer. PLLA and PDLA grow at the same time, yielding a racemic crystal formed by packing β-form 3_1-helices of opposite absolute configurations alternately. Helices of identical absolute configuration are packed differently from pairs of enantiomer α-helices.[48,50,51] The stereo-complexation can be characterized using a number of analytical techniques, in particular differential scanning calorimetry (DSC), polarization optical microscopy (POM), infrared (IR) and Raman spectroscopy, ^{1}H and ^{13}C NMR spectroscopy and wide angle X-ray diffraction (WAXD).

Figure 2.1 presents the typical thermal behaviours of PLA stereo-complex obtained by mixing PLLA and PDLA solutions with weight average molar mass $M_w = 36\,000$ and $19\,000$ Da, respectively.[47] In the first run, only a T_m is detected at 215.9 °C with ΔH_m of 64.2 J g^{-1}, in agreement of exclusive formation of stereo-complex crystallites. In the second heating process after quenching of the melt, T_g is detected at 53.1 °C, followed by cold crystallization at 95.9 °C and melting at 186.8 °C with ΔH_m of 26.0 J g^{-1}. In the third heating run, T_g appears at 50.9 °C, followed by a very weak cold crystallization around 130 °C and a very weak melting at 173.0 °C with ΔH_m of *ca.* 1.0 J g^{-1}. The T_m values obtained in the second and third runs (186.8 and 173.0 °C) are higher than those of PLLA and PDLA (165.8 °C and 160.3 °C,

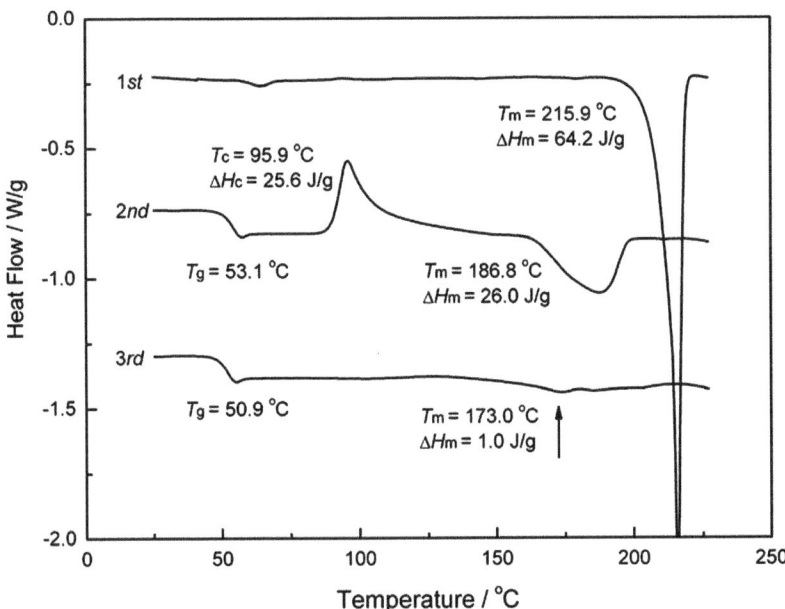

Figure 2.1 DSC thermograms of the PLLA/PDLA blend at a heating rate of 10 °C min^{-1}. The first run corresponds to the blended sample from solution blending and drying, the second and third ones to samples after quenching in liquid nitrogen from the melt. Before quenching, the sample was kept at 230 °C for 1 min. to eliminate all nuclei.

respectively). This finding indicates that an imperfect stereo-complex is formed instead of homo-crystallites. In fact, the corresponding WAXD spectrum (Figure 2.2) exhibits three main diffraction peaks with θ values of 5.9, 10.3 and 11.9°, respectively, which are characteristic of PLA stereo-complex crystallites in a trigonal unit cell of dimensions: $a = b = 1.498$ nm, $c = 0.870$ nm, $\alpha = \beta = 90°$ and $\gamma = 120°$.[46]

The much lower T_m and ΔH_m values obtained in the second and third heating runs are attributed to highly depressed stereo-complex crystallization.[48] In fact, the crystallization behaviour of PLLA/PDLA blends depends on the initial melt state. Stereo-complexation is strongly depressed when higher temperature and longer melting period are applied. It is assumed that PLLA and PDLA chain couples would preserve their interactions (melt memory) when the stereo-complex crystal melts smoothly, thus resulting in a heterogeneous melt which can easily re-crystallize. The melt could gradually become homogeneous at higher temperature or longer melting time. The strong interactions between PLLA and PDLA chain segments are randomly distributed in a homogeneous melt, thus preventing subsequent stereo-complex crystallization. However, the homogeneous melt can recover its ability to crystallize *via* dissolution in a solvent.

The depressed stereo-complex crystallization at different temperatures can be illustrated by non-isothermal crystallization experiments.[47] Figure 2.3A

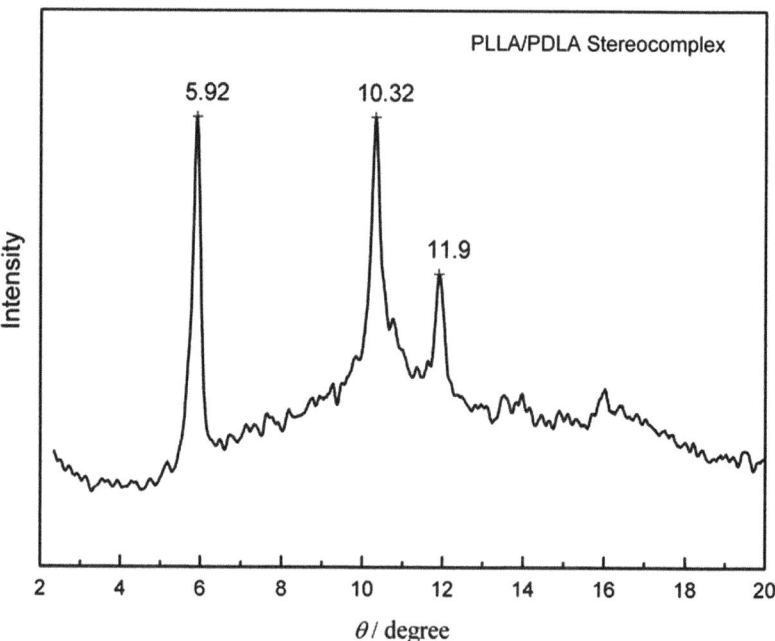

Figure 2.2 WAXD spectrum of the PLLA/PDLA blend after crystallization in the second cycle of Figure 2.1 (the sample was taken out at 140 °C for WAXD measurements).

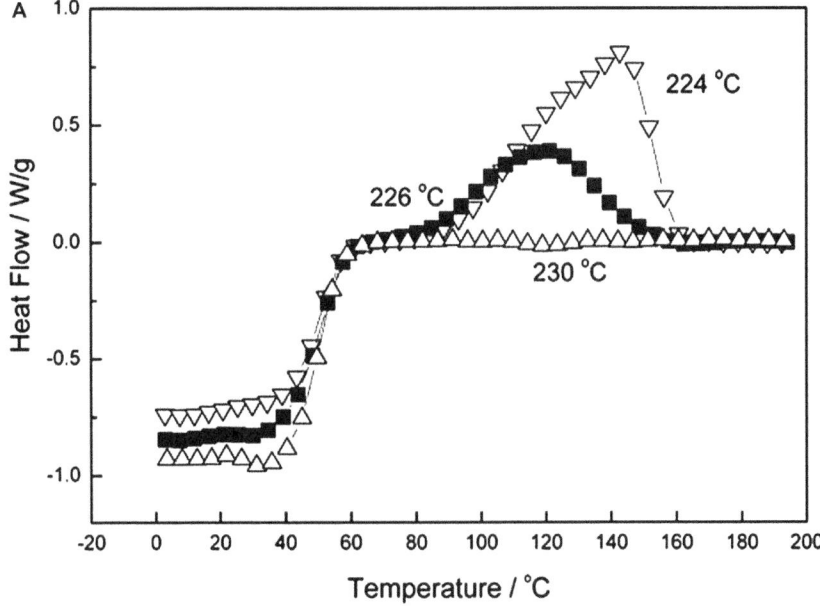

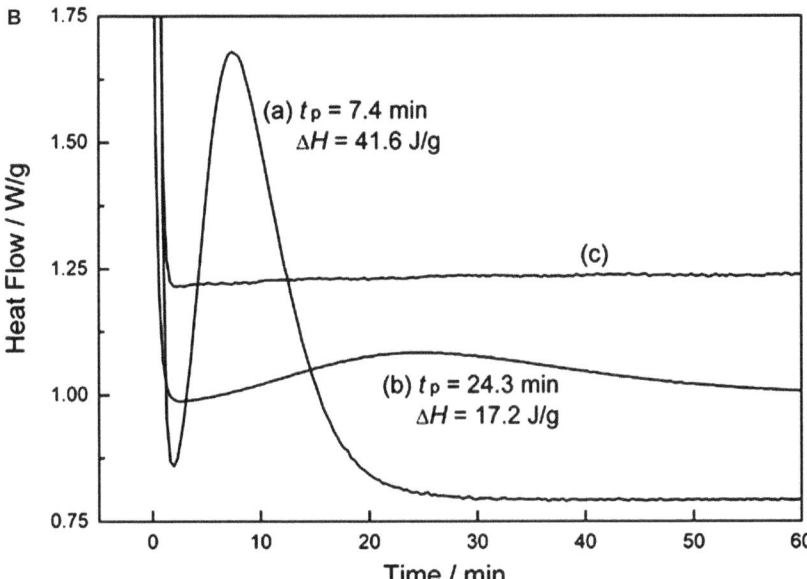

Figure 2.3 (A) DSC thermograms of the PLLA/PDLA blend at a cooling rate of 10 °C min^{-1} from the melt after 1 min. at 230 °C, 226 °C and 224 °C. (B) Isothermal crystallization at 145 °C after melting at 230 °C for different periods (a – 1 min.; b – 2 min.; c – 3 min.).

exhibits the cooling heat flows of the above-mentioned PLLA/PDLA blend after 1 min. melting at 224 °C, 226 °C and 230 °C, respectively. With melting at 224 °C, the melt crystallization temperature (T_{mc}) appears at 144.0 °C with crystallization enthalpy (ΔH_c) of 23.6 J g^{-1}. Melting at 226 °C leads to lower T_{mc} at 120.0 °C with lower ΔH_c of 10.7 J g^{-1}. Finally with melting at 230 °C, the melt crystallization is hardly detectable. The T_g remains at *ca.* 51 °C in all cases, while the thermal capacity (ΔC_p) increases from 0.39 J g^{-1} K^{-1} at 224 °C to 0.48 J g^{-1} K^{-1} at 226 °C, and to 0.63 J g^{-1} K^{-1} at 230 °C. A higher ΔC_p value implies lower crystallinity and a higher fraction of amorphous region remaining in the blend. Therefore, melt crystallization in the cooling process is strongly depressed when the blend is melted at higher temperatures.

The depressed stereo-complex crystallization can also be illustrated by isothermal crystallization for various time periods.[47] Figure 2.3B shows the DSC curves of the PLLA/PDLA blend during crystallization at 145 °C after melting at 230 °C for 1, 2 and 3 minutes, respectively. An intense crystallization peak is detected with a peak time (t_p) of 7.4 min. and ΔH_c of 41.6 J g^{-1} for 1 minute melting. After 2 min. melting, however, the crystallization ability strongly decreases with a prolonged t_p of 24.3 min. and lower ΔH_c of 17.2 J g^{-1}. Finally after 3 min. melting at 230 °C, isothermal crystallization is no longer detectable.

2.2.4   Degradation of PLA Stereo-complex

Enzymatic degradation provides a means of choice to detect morphologies in different PLLA/PDLA blends. Figure 2.4 illustrates the morphology of enzymatically degraded samples corresponding to those in Figure 2.3B.[47] As reported previously, proteinase K selectively degrades L-LA segments of PLA-based polymers, but not D-LA ones.[52] On the other hand, only amorphous regions of semicrystalline PLA can be degraded by proteinase K.[53] Figure 2.4A clearly reveals the crystal morphology corresponding to sample (a) in Figure 2.3B. Lamellae in the spherulites seem to organize in a unique manner different from the PLLA crystallization.[54] The boundary between spherulites is well degraded and removed. Figure 2.4B presents a smaller number and size of spherulites, in agreement with strongly depressed isothermal crystallization shown in Figure 2.3B. In contrast, Figure 2.4C exhibits a network-like morphology after degradation, in agreement with homogeneous degradation of the amorphous blend as previously observed for amorphous PLA stereo-copolymers.[54-57]

Tsuji *et al.* investigated the effects of molar mass, L-LA content and average L-LA and D-LA sequence length on the enzymatic degradation of various PLLA/PDLA blends in the presence of proteinase K.[58] The authors observed that enzymatic hydrolysis of PLLA proceeds *via* both endo and exo chain scission. Non-blended PLA films are enzymatically hydrolyzable when L-LA content and L-LA sequence length are higher than 0.3 and 3, respectively, and D-LA sequence length is lower than 10.

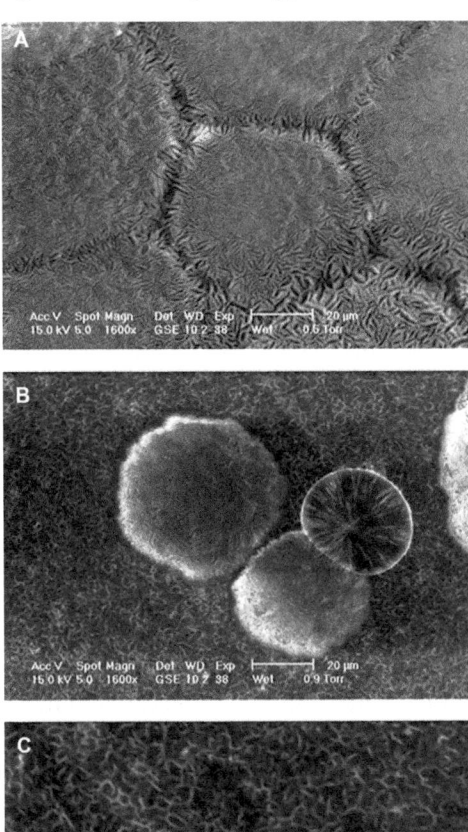

Figure 2.4 ESEM micrographs of the enzymatically degraded PLLA/PDLA blend which was melted at 230 °C for 1 min. (A); 2 min. (B) and 3 min. (C), followed by isothermal crystallization at 145 °C for 60 min.

The degradation of aliphatic polyesters by pure hydrolysis is a complex process involving several phenomena, namely water absorption, ester bond cleavage, loss of properties and diffusion and solubilization of soluble oligomers. It is known that degradation is catalyzed by carboxyl end-groups produced by chain cleavage,[59–61] and that amorphous regions are preferentially degraded.[62–64] Important advances have been accomplished by Li *et al.* in the understanding of the hydrolytic degradation characteristics, in particular faster internal degradation and degradation-induced morphological and compositional changes as previously described.[65–71]

In the case of PLA stereo-complex, the hydrolytic degradation has also been investigated. Tsuji *et al.* studied the *in vitro* degradation of stereo-complexed blends and non-blended PLA films.[72] The rates of molar mass decrease and loss of mechanical properties is lower for stereo-complexed 1/1 blend films than for non-blended films. Similarly, the stereo-complex requires a longer induction period before the onset of decrease in mechanical properties and mass loss, suggesting that the stereo-complex is more hydrolysis-resistant than the non-blended films due to the strong interaction between PLLA and PDLA sequences in the amorphous region and/or their three-dimensional structure in the blend film. The change in the molar mass distribution and surface morphology of the blend film after hydrolysis shows that the hydrolytic degradation proceeds homogeneously *via* bulk erosion mechanism. The authors also considered the degradation of well-homo-crystallized blend and non-blended films.[73] Well-homo-crystallized blend is obtained using heat treatment conditions, which exclusively induce homo-crystallization in equimolar blends of PLLA and PDLA having high molar masses from the melt.[10] In the first 12 months, little difference is detected between the PLLA/PDLA blend and non-blended films. In the period of 12–24 months, degradation is significantly retarded for the PLLA/PDLA blend film due to the enhanced interaction between shortened PLLA and PDLA chains after chain cleavage. Tsuji *et al.* also investigated the accelerated degradation of PLA stereo-complex at elevated temperatures from 70 to 97 °C up to the late stage.[74] Similar to the degradation of pure PLLA or PDLA samples, the degradation of stereo-complexed PLLA/PDLA blends slows down at the late stage when most of the amorphous chains are removed. The estimated activation energy for the degradation of stereo-complex crystallites (97.3 kJ mol^{-1}) is significantly higher than that reported for α-form PLLA crystallites (75.2 kJ mol^{-1}), thus indicating the higher hydrolysis resistance of stereo-complex as compared to PLLA crystallites. The hydrolytic degradation of stereo-multiblock and diblock PLA with block lengths ranging from 3.8 to 61.9 was investigated in comparison with quenched neat PLLA, PDLA and PLLA/PDLA blends.[75] The chain cleavage was also studied by ^{13}C NMR, and spectra showed that chain cleavage occurs on the atactic segments connecting relatively long isotactic L-LA and D-LA segments. The rate of hydrolytic degradation of stereo-block PLA decreases with an increase in the average stereo-block length. Stereo-complex crystallization occurs during degradation of stereo-block PLA with average stereo-block lengths above 7 lactyl units. The crystallization rate of stereo-block PLA during degradation increases with increasing the block length, but it is slower than that of the PLLA/PDLA blend.

Andersson *et al.* studied the degradation of 9 PLA-based materials: PLLA and 8 blends of PLLA with different PDLA oligomers.[76] Stereo-complexation of PLLA with star-shaped PDLA oligomers with different architectures and end-groups clearly affects the degradation rate and degradation product patterns. Stereo-complexes containing star shaped PDLA oligomers with 4 alcoholic end-groups undergo rather slow hydrolytic degradation, and those with linear PDLA oligomers exhibit similar mass loss but release more acidic degradation products. The increase in the fraction of linear or star-shaped PDLA oligomers results in slower mass loss due to higher stereo-complexation degree. The opposite results are obtained in the case of PDLA oligomers with carboxylic chain-ends. These materials demonstrate larger mass loss and molar mass loss, and also larger release of degradation products due to lower stereo-complexation degree.

Lee *et al.* used the Langmuir film balance technique to determine the hydrolytic kinetics of stereo-complex monolayers formed from PLLA/PDLA mixtures spread at the air-water interface.[77] The hydrolysis of the mixture monolayers under basic conditions is slower than that of individual PLLA or PDLA monolayers, depending on the composition or the degree of complexation. In the presence of proteinase K, the hydrolysis rate of mixture monolayers with >50 mol% PLLA is much slower than that of the single-component PLLA monolayer. The monolayers formed from mixtures with ≤50 mol% L-PLA do not show any degradation. It is concluded that the slower hydrolysis of mixture monolayers is mainly due to the strong interaction between PLLA and PDLA chains, which prevents the penetration of water or enzyme into the bulk. In an *in vivo* study on the biocompatibility of PLLA and stereo-complexed nanofibres by subcutaneous implantation in rats, Ishii *et al.* also observed that stereo-complexed nanofibres exhibit slower degradation than PLLA.[78]

2.3 Biomedical Applications of PLA Stereo-complex

PLLA/PDLA stereo-complex allows enhancing the thermal-resistance and mechanical properties as well as the hydrolytic stability of PLA-based materials. In the past decades, a great deal of work has been performed to exploit potential biomedical applications of PLA stereo-complex as shown in Table 2.1.[78–86]

Stereo-complex films were reported by Masutani *et al.*[79] By combining bifunctional PLLA and PDLA pre-polymers, multi-stereo-block copolymers having different block lengths and sequences are obtained. The resultant copolymers are readily fabricated into transparent films by hot-pressing. The films present excellent thermal stability and thermo-mechanical properties because of the easy formation of stereo-complex crystals. This synthetic method based on the dual terminal couplings allows obtaining stereo-block copolymers of PLLA and PDLA showing excellent thermo-mechanical properties and melt processability.

Watanabe *et al.* reported stereo-complex material by enantiomeric PLA graft-type phospholipid copolymers as tissue engineering scaffolds.[80] The copolymers as cell-compatible materials are prepared using a phospholipid copolymer composed of 2-methacryloyloxyethyl phosphorylcholine, *n*-butyl methacrylate, and enantiomeric PLLA and PDLA macromonomers. The porous scaffold presents advantages for tissue engineering, including cell compatibility using phospholipid copolymer, adequate cell adhesion by PLA

Table 2.1 Biomedical applications of PLA-based stereo-complex materials.

Stereo-complex type	Structure	Size	Application	Reference
PLLA/PDLA	Nanofibre	400–1400 nm	Tissue engineering	Tsuji *et al.*[83]
PLLA/PDLA	Porous scaffold	200 μm	Tissue engineering	Watanabe *et al.*[80]
PLLA/PDLA	Nanofibre	300 nm	Tissue engineering	Ishii *et al.*[78]
PDLA/insulin	Microparticle	300–3000 nm	Insulin release	Slager *et al.*[84]
PDLA/leuprolid PDLA/vapreotide	Microparticle	300–3000 nm	Leuprolide and peptide release	Slager *et al.*[19,20,85,86]
PDLA/L-octreotide	Microparticle	1.5–4 μm	Octreotide release	Bishara *et al.*[85]
PLLA-PEG/PDLA-PEG	Micelle	40–120 nm	Rifampin release	Chen *et al.*[81]
PLLA-PEG-PLLA/ PDLA-PEG-PDLA	Micelle	100–200 nm	Paclitaxel release	Kang *et al.*[87] Yang *et al.*[16]
PLLA-PEG-PLLA/ PDLA-PEG-PDLA	Hydrogel	–	Thymopentin release	Zhang *et al.*[14]
PLLA-dextran/ PDLA-dextran	Hydrogel	–	IgG and lysozyme release	de Jong *et al.*[109]
PLLA-chitosan/ PDLA-chitosan	Hydrogel	–	Thymopentin release	Hu *et al.*[119]

and complete disappearance of scaffold by dissociation of stereo-complex. Recently, stereo-complex-induced nanofibres, microparticles, micelles and hydrogels attracted much attention as carriers for controlled drug delivery.[16,81–87]

2.3.1 PLA Stereo-complex Nanofibre Scaffolds

Nanofibre scaffolds present fine pores and grooves with a width of a few micrometres. Such fine structural features facilitate the adhesion and proliferation of cells. Nanofibre scaffolds should have sufficient strength to support regenerating tissues and to be degraded after completion of the tissue regeneration process. PLA fibres possess sufficient mechanical properties for tissue engineering applications. Nevertheless, PLA is not suitable for long-time applications under physiological conditions. Therefore, PLLA/ PDLA stereo-complex nanofibres are developed to improve the hydrolytic stability of scaffolds in long-time implantations.

Tsuji *et al.* prepared stereo-complexed nanofibres by electrospinning of PLLA/PDLA solutions (4 g dL^{-1}) in chloroform.[83] Stereo-complex nanofibres are obtained with negligible homo-crystallites despite the high molar masses of PLLA and PDLA. It is assumed that the orientation caused by electrically induced high shearing force during electrospinning enhances the formation and growth of stereo-complex crystallites.

Ishii *et al.* evaluated the *in vivo* tissue response and degradation behaviour of PLLA and stereo-complexed PLA nanofibres.[78] Both nanofibres were prepared by electrospinning of PLLA and/or PDLA solutions (1 wt%) in 1,1,1,3,3,3-hexafluoro-2-propanol (HFIP), followed by annealing at 100 °C for 8h. Degradation of PLLA/PDLA nanofibres is slower than that of PLLA nanofibres, and thus their shape is conserved after prolonged implantation in rats due to the presence of more stable stereo-complex crystals. The stereo-complex nanofibres caused mild inflammatory reaction as compared to PLLA nanofibres. Thus the potential use of PLLA and stereo-complexed PLA nanofibres can be considered as a biomaterial for short-term and long-term tissue regeneration, respectively.

2.3.2 PLA Stereo-complex Microparticles

Stereo-complexation occurs not only in PLLA/PDLA blends, but also in PLA/chirality systems, such as PDLA/peptides. Homo-stereo-complex of PLLA and PDLA consists of 3_1-helix crystallites. Similarly, hetero-stereo-complex is composed of intertwined helices formed by PDLA blocks and peptides, which presents great interest for the sustained release of peptides.

Stereo-complex microparticles are generally prepared by free-spay, precipitation or solvent evaporation.[84] Domb *et al.* investigated PLLA/PDLA homo-stereo-complex and PDLA/L-octreotide as carrier of L-octreotide.[85] Hetero-stereo-complexes are obtained by spray freezing of a solution containing PDLA and octreotide, as evidenced by DSC measurements. The spray freezing method leads to compact and smooth microparticles compared to those obtained by precipitation. The particle size of stereo-complexes is in the range of 1.5 to 4 µm. Increasing peptide concentration in the stereo-complex particles enhances the release rate of the peptide and the polymer degradation rate.[85]

Domb *et al.* also reported on PDLA/insulin, PDLA/L-leuprolide and PDLA/vapreotide hetero-stereo-complex microparticles for controlled drug release.[19,20,84,86] PDLA/insulin complexes are spontaneously formed as porous microparticles of 1–3 µm when insulin and PDLA are mixed together in acetonitrile solution.[19,84,86] Insulin is constantly released for a few weeks when the complexes are placed in a buffer solution of pH 7.4 at 37 °C. PLLA and PDLLA cannot form a precipitate with insulin, which indicates the stereospecificity for the complex formation. The authors also reported stereo-complexes of PDLA with L-configured leuprolide – a luteinizing hormone releasing hormone (LHRH) nonapeptide analogue – and vapreotide – a cyclic octapeptide somatostatin analogue.[19,20,84,86] The high porosity of the particles of 1–5 µm diameter provides a relatively long release period over several months. The release kinetics from stereo-complex microparticles is close to first order without burst release. It is suggested that the release from these microparticles is not diffusion dependent, but relies on the detachment of the complex between PDLA blocks and peptides.[86] The influencing factors include the molar mass of PDLA blocks, content of peptide and other components (PEG or opposite configuration PLA).[19] Higher molar mass of PLA blocks and low peptide content lead to faster release, which is enhanced when using PLA-PEG block copolymers or mixing low molar mass PEG into the microparticles. On the other hand, the release from PDLA/peptide is faster than that from PDLA/PLLA/peptide due to faster degradation as evidenced by faster lactic acid release and mass loss from PDLA/peptide.[19] Therefore, these hetero-stereo-complexes present great interest for the development of novel controlled release systems for optically active peptides and proteins.[19,20,84,86]

2.3.3 PLA Stereo-complex Micelles

Copolymerization of PLA with hydrophilic polymers, such as PEG, is a commonly used method to obtain amphiphilic PLA-based copolymers, which are widely investigated as drug delivery systems (DDS).[88–91] PLA-PEG diblock copolymers are synthesized by ring-opening polymerization of L-lactide or D-lactide in the presence of a monomethoxy-PEG, as shown in Scheme 2.3. Similarly, PLA-PEG-PLA triblock copolymers are obtained by ring-opening polymerization of L-lactide or D-lactide using dihydroxyl PEG as initiator. For the sake of clarity, triblock copolymers are named as $L_xEO_yL_x$ or $D_xEO_yD_x$, and diblock copolymers as L_xEO_y or D_xEO_y. In these acronyms, L, D and EO represent PLLA, PDLA and PEG blocks, respectively, x and y representing the number-average degree of polymerization of corresponding blocks. The water solubility of PLA/PEG copolymers depends on the EO/LA ratio and the molar mass. The higher the EO/LA ratio, the better the water solubility. On the other hand, low molar mass also favours the solubility of PLA/PEG copolymers in water.

The most common methods to prepare micelles of water insoluble copolymers are nanoprecipitation, solvent evaporation/film hydration and dialysis.[92,93] However, all the organic solvents involved in these methods, such as dichloromethane, acetone, dimethylformamide (DMF) and acetonitrile are more or less toxic, and residual solvents might cause side effects to the human body.

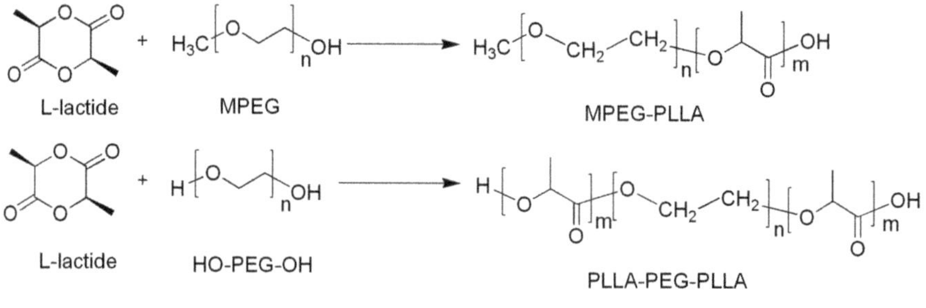

Scheme 2.3   Synthesis of PLLA-PEG diblock and PLLA-PEG-PLLA triblock copolymers by ring-opening polymerization of L-lactide in the presence of a monomethoxy PEG or dihydroxyl PEG.

Polymeric micelles with core-shell architecture can be obtained by directly dissolving water soluble copolymers in aqueous medium: the hydrophobic segments aggregate to form a core able to encapsulate hydrophobic drugs with improved solubility; and the hydrophilic shell consisting of a brush-like protective corona stabilizing the micelles in aqueous solution.[15,94–96] In fact, micelles constitute a dynamic system with permanent exchanges between them and free molecules in solution, continuously breaking and re-forming.[97] Polymeric micelles as novel drug carriers present numerous advantages, such as reduced side effects of drugs, targeted delivery and prolonged blood circulation.[96,98] Furthermore, polymeric micelles possess a nanoscale size range with narrow distribution, which allows better accumulation at the target site to be achieved through enhanced permeation retention effect (EPR effect).[99]

Stereo-complexation not only occurs in melt and in organic solution, but also in aqueous solution. When a water soluble PLLA/PEG block copolymer is mixed with a similar PDLA/PEG in dilute solution, the strong interactions between PLLA and PDLA blocks lead to the formation of stereo-complex micelles.

Kang *et al.* reported on single micelles and stereo-complex micelles prepared by nanoprecipitation of PLA-PEG diblock copolymers composed of a monomethoxy PEG with $M_n = 5400$ Da and PLLA or PDLA blocks with DP from 29 to 104.[87] The micelles possess a partially crystallized core and mean hydrodynamic diameters ranging from 31 to 56 nm, depending on the PLA block length. Stereo-complex micelles exhibit better kinetic stability and re-dispersion properties as compared to single micelles, thus showing the advantages of stereo-complex formation in the design of stabilized water-soluble micelles.

Yang *et al.* systemically investigated the preparation, aggregation properties and drug release behaviour of PLA/PEG stereo-complex micelles.[15,16,100] Predetermined amounts of PLLA/PEG or PDLA/PEG copolymer are separately dissolved in distilled water or phosphate buffered saline (PBS). The two solutions with equal molar concentrations are then mixed to yield a micellar solution by self-assembly due to interactions between L-LA and D-LA blocks.[15]

Polymeric micelles can be formed only when the block copolymer concentration is higher than the critical micellar concentration (CMC) which characterizes the micelle stability.[101] The CMC of PLA/PEG copolymers depends on the EO/LA ratio, molar mass and chain structure, as shown in Table 2.2. Lower EO/LA ratios lead to lower CMC since copolymers are more hydrophobic, and thus can self-assemble more easily to form micelles. Higher molar mass of PEG blocks leads to higher CMC value for similar hydrophilicity or EO/LA ratio. It is also of interest to compare the micellization properties of diblock and triblock copolymers. In fact, triblock copolymers are more inclined to form micelles than diblock copolymers due to the different structures. In dilute aqueous solutions, diblock copolymers tend to form micelles with a stick-like chain structure, while triblock chains

Table 2.2 CMC values of PLA/PEG diblock, triblock copolymers and mixed solutions at 20 °C in water, at 20 °C in 0.1 M NaCl aqueous solution, and at 37 °C in water.

			CMC (g l^{-1})		
Acronym	Copolymer	EO/LA	Water (20 °C)	0.1 M NaCl (20 °C)	Water (37 °C)
1L	$L_{12}EO_{104}L_{12}$	4.2	0.050	0.046	0.045
1D	$D_{13}EO_{104}D_{13}$	4.1	0.050	0.052	0.045
1-mixed^a			0.040	0.048	0.037
2L	$L_{21}EO_{454}L_{21}$	11.0	0.106	0.100	0.108
2D	$D_{22}EO_{454}D_{22}$	10.5	0.110	0.100	0.110
2-mixed			0.090	0.090	0.090
3L	$L_{12}EO_{45}$	3.7	0.073	0.060	0.054
3D	$D_{11}EO_{45}$	4.2	0.080	0.080	0.066
3-mixed			0.068	0.060	0.050
4L	$L_{19}EO_{45}$	2.3	0.050	0.050	0.040
4D	$D_{17}EO_{45}$	2.6	0.050	0.052	0.040
4-mixed			0.045	0.048	0.032
5L	$L_{18}EO_{113}$	6.3	0.100	0.100	0.085
5D	$D_{18}EO_{113}$	6.3	0.097	0.091	0.082
5-mixed			0.086	0.073	0.070
6L	$L_{25}EO_{113}$	4.5	0.090	0.090	0.076
6D	$D_{26}EO_{113}$	4.3	0.090	0.090	0.070
6-mixed			0.080	0.080	0.065

^{a}The mixed micellar solutions of corresponding PLLA/PEG and PDLA/PEG copolymers.

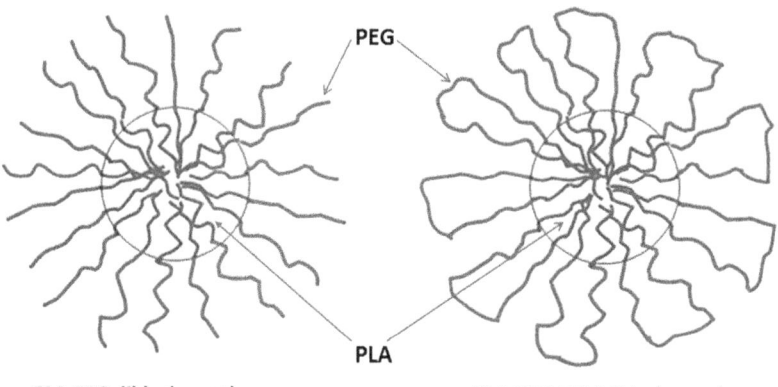

PLA-PEG diblock copolymer   PLA-PEG-PLA triblock copolymer

Figure 2.5 Schematic presentation of micelles obtained from PLA-PEG diblock (left) and PLA-PEG-PLA triblock copolymers (right).

have to fold so as to insert the two hydrophobic PLA blocks inside micelles, as shown in Figure 2.5.[102,103] Hydrophilic PEG blocks of diblock copolymers can more easily interact with water molecules by hydrogen bonding than those of the triblocks, leading to higher CMC values. PLLA-PEG and PDLA-PEG mixed solutions present lower CMC than separate PLLA-PEG or PDLA-PEG solutions since mixed micellar solutions are more stable than separate polymers due to stronger interactions between PLLA and PDLA blocks. In contrast, there is no significant difference between the CMC values of copolymers in pure water or in NaCl solutions. The insensitivity of the CMC to salt addition is probably due to the non-ionic nature of the polymers.[93] This finding indicates that polymeric micelles present good stability after salt addition, which is of major importance for intravenous injection. Finally, the CMC values of copolymers at 37 °C are slightly lower than those at 20 °C (Table 2.2). This result is assigned to the increased hydrophobicity or loss of polarity of PEG blocks at elevated temperatures, thus leading to dehydration of PEG chains and decrease in the CMC.[104] On the other hand, polymeric micelles are in dynamic equilibrium with free polymer chains by continuous exchanges. Chain mobility is improved at higher temperatures and, thus, the probability for hydrophobic PLA segments to meet each other and further assemble to form the inner core of micelles is enhanced.

The size distribution of micelles is determined by dynamic light scattering (DLS). Figure 2.6 shows the size distributions of $L_{21}EO_{451}L_{21}$ single micelles and $L_{21}EO_{451}L_{21}/D_{22}EO_{451}D_{22}$ stereo-complex micelles at a concentration of 0.2 g L^{-1}. Two peaks are observed for single micelles: a small peak ranging from 20–40 nm and a large peak ranging from 100–400 nm with a maximum at 210 nm. In the case of stereo-complex micelles, a small peak with size ranging from 20–33 nm and a large one ranging from 100–200 nm with a maximum at 154 nm are observed. The smaller size of stereo-complex micelles is assigned to the more compact structure due to a strong interaction between PLLA and PDLA blocks.[100]

The morphology of $L_{21}EO_{451}L_{21}/D_{22}EO_{451}D_{22}$ stereo-complex micelles was examined by transmission electron microscopy (TEM). As shown in Figure 2.7, micelles are clearly distinguished as dark spots. The size can be estimated from the micrographs with the scale bar. The size of stereo-complex micelles ranges from 70–100 nm, which is smaller than that determined by DLS (Figure 2.6). This difference could be attributed to the experimental conditions. In fact, DLS determines the hydrodynamic diameter of micelles in aqueous solution, whereas TEM shows the dehydrated solid state of micelles.

Yang *et al.* investigated the *in vitro* and *in vivo* release properties of PLA/PEG micelles loaded with paclitaxel, an efficient anticancer drug.[16] The encapsulation efficiency (EE) and loading content (LC) of $L_{12}EO_{104}L_{12}/D_{13}EO_{104}D_{13}$ stereo-complex micelles are 52.2% and 5.1%, respectively, which are higher than those of $L_{12}EO_{104}L_{12}$ single micelles (EE = 32.1%, LC = 3.2%). The stereo-complexation strongly affects the drug release behaviours of micelles. As shown in Figure 2.8, similar parabolic release profiles are observed for single and stereo-complex micelles. However, single micelles exhibit a faster release rate of paclitaxel as compared to stereo-complex micelles. In fact, stereo-complex micelles have more compact structure due to strong interactions between PLLA and PDLA blocks, which disfavours the drug diffusion leading to slow drug release. *In vivo* studies show that paclitaxel-loaded stereo-complex micelles exhibit better

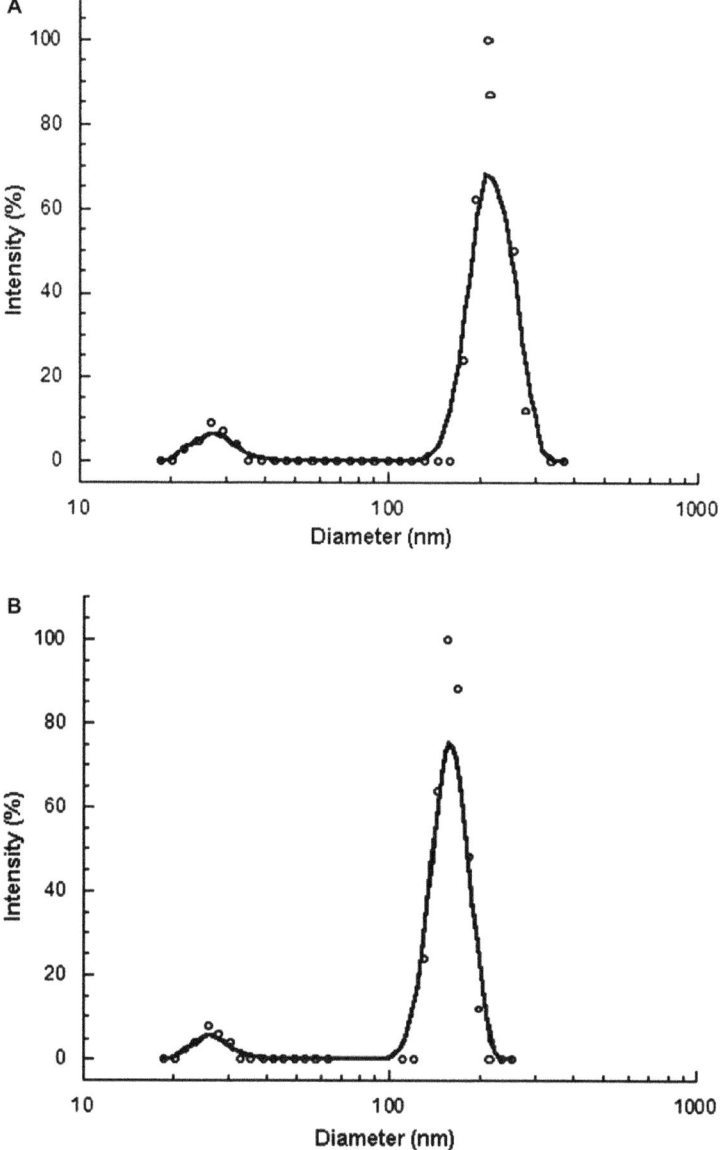

Figure 2.6   DLS graphs of (A) $L_{21}EO_{451}L_{21}$ micelles and (B) $L_{21}EO_{451}L_{21}/D_{22}EO_{451}D_{22}$
stereo-complex micelles.

antitumour efficiency as compared to the clinical formulation Taxol®
consisting of 50 : 50 Cremophor EL and dehydrated alcohol.

Chen *et al.* investigated comparatively single micelles and stereo-complex
micelles prepared by nanoprecipitation of PLA-PEG diblock copolymers
composed of a monomethoxy PEG with $M_n = 5000$ Da and PLLA or PDLA
blocks with DP of 14, 35 and 70.[81] The CMC and micelle sizes of single

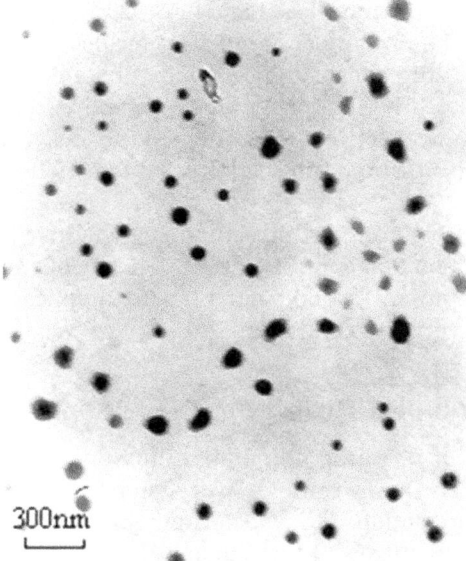

Figure 2.7 TEM images of $L_{21}EO_{451}L_{21}/D_{22}EO_{451}D_{22}$ stereo-complex micelles.

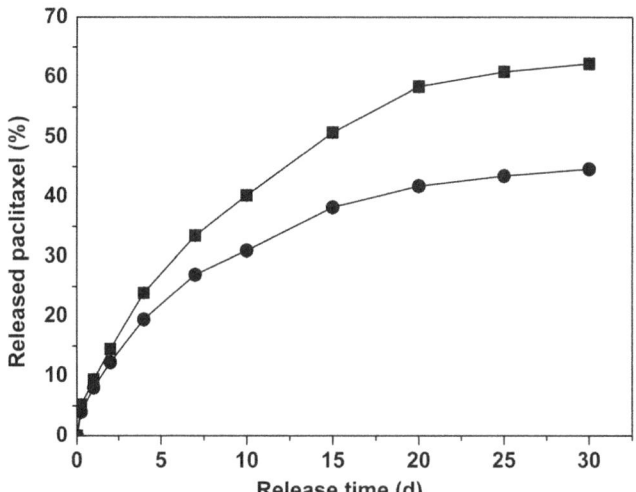

Figure 2.8 Paclitaxel release from $L_{12}EO_{104}L_{12}$ micelles (■) and $L_{12}EO_{104}L_{12}/D_{13}EO_{104}L_{13}$ stereo-complex micelles (●) in phosphate buffered saline at 37 °C.

micelles are higher than those of stereo-complex micelles, which corroborates the results reported by Yang *et al.*[15,16] The authors also studied the *in vitro* release of rifampin, a bactericidal antibiotic drug, from PLA/PEG micelles.[81] The encapsulation efficiency (EE) and loading content (LC) of stereo-complex micelles are 63% and 10.6%, respectively, which are higher than those of PLLA-PEG micelles (EE = 45%, LC = 7.5%). This is in agreement with the results of Yang *et al.* on the encapsulation of paclitaxel, indicating enhanced drug encapsulation ability of stereo-complex micelles.[81] The release of initial 50% rifampin from single and stereo-complex micelles takes 2.5 and 7.5 h, respectively. The much slower release profile of stereo-complex micelles is attributed to the strong interaction between PLLA and PDLA blocks.

Therefore, the more compact structure, higher drug loading efficiency, higher loading content and controlled release behaviour suggest that stereo-complex micelles present a good potential for applications as carrier of drugs, especially hydrophobic drugs.

2.3.4   Stereo-complex Hydrogels

Hydrogels consist of a hydrophilic matrix crosslinked chemically through covalent bonds or physically through hydrogen bonds, crystallized domains or hydrophobic interactions.[105–107] The high water content and water insolubility endow hydrogels with outstanding biocompatibility, hydrophilicity, low adsorption of proteins and similar mechanical properties to soft tissues, making them promising candidates for biomedical applications as drug carrier or tissue engineering scaffold.

De Jong *et al.* first reported stereo-complexation-induced physical hydrogels prepared from dextran-*g*-PLLA and dextran-*g*-PDLA graft copolymers as protein carriers.[108] PLLA and PDLA oligomers are attached to the dextran backbone *via* their terminal hydroxyl groups. The resulting graft copolymers are susceptible to form a hydrogel by mixing dextran-*g*-PLLA and dextran-*g*-PDLA aqueous solutions at room temperature. Rheological measurements show that the hydrogels are thermo-reversible, and that the storage modulus depends on the degree of polymerization of PLA chains and on the degree of substitution of PLA on the dextran backbone. Protein-loaded hydrogels are prepared by dissolving the protein in the dextran-*g*-PLA solutions prior to mixing. The entrapped model proteins (IgG and lysozyme) are released in a period of 6 days. The release kinetics depends on the hydrogel characteristics, including the PLA block length and initial water content. Last but not least, the proteins are quantitatively released from the hydrogel with preservation of the enzymatic activity of lysozyme, in agreement with the protein-friendly preparation method of stereo-complex hydrogels.[109]

Stereo-complex hydrogels have also been obtained from PLLA/PEG and PDLA/PEG block copolymers. Grijpma *et al.* reported the formation of hydrogels by stereo-complexation of water soluble PLA/PEG copolymers.[110,111] Water soluble 8-arm PLLA-PEG and PDLA-PEG star block copolymers are synthesized by ring-opening polymerization of L-lactide or D-lactide using a single-site ethyl zinc complex and 8-arm PEG as catalyst and initiator, respectively.[111] The resulting PEG-(PLA)$_8$ copolymers are water soluble when the number of LA units per PLA block is below 14 or 17 for PEG with M_n of 21 800 Da or 43 500 Da, respectively. Stereo-complex hydrogels are prepared by mixing aqueous solutions with equimolar amounts of PEG-(PLLA)$_8$ and PEG-(PDLA)$_8$ in a concentration range of 5–25 w/v% for PEG. Rheological measurements show that hydrogels with storage modulus up to 14 kPa in phosphate buffered saline at 37 °C and gelation time up to 1 h can be obtained by varying the PLA block length and polymer concentration.

Fujiwara *et al.* reported on the preparation of PLA-PEG-PLA stereo-complex hydrogels.[112] The block copolymers are composed of PEG block ($M_n = 4600$ Da) and PLLA blocks with $M_n = 1300$ Da or PDLA blocks with $M_n = 1100$ Da, and are not water soluble. Aqueous suspensions of the co-polymers are obtained by dissolving the copolymers in tetrahydrofuran (THF), followed by dispersion in water under ultrasonic waves. The mixed dispersion is heated up to 75 °C to induce the spontaneous gelation. This system is characterized by a temperature-dependent sol-gel transition around 37 °C. Later on, the same authors reported similar hydrogels from mixed suspension of enantiomeric BAB triblock copolymers, PEG-PLLA-PEG and PEG-PDLA-PEG. Diblock copolymers MePEG-PLLA and MePEG-PDLA are first prepared by ring-opening polymerization of L-lactide or D-lactide in the presence of monomethoxy MePEG ($M_n = 2000$ Da). The resultant diblock copolymers are then allowed to react with hexamethylene diisocyanate (HMDI), yielding a BAB-type PEG-PLA-PEG triblock copolymer. A similar procedure is used to prepare hydrogels by heat treatment up to 75 °C. Reversible gel to sol transition is observed, depending on the polymer concentration and temperature. The storage modulus (G′) and loss modulus (G″) of these gels are dramatically improved in comparison with those of ABA-type triblock copolymers. It is assumed that PEG chains are involved in the helix formation of PLLA and PDLA to form helices with opposite helical senses and that the particles are coagulated by the interdigitation of the helical PEG chains to lead gelation.[113] Recently, Abebe *et al.* reported on stereo-complex hydrogels with improved physical and mechanical properties from PLA-PEG-PLA micelles.[114] A unique micelle structure comprising two different sizes of hydrophilic corona blocks is achieved from PLA-PEG-PLA triblock copolymers. Hydrogels prepared from the racemic mixture of these hybrid micelles can be tailored to have sol to gel transition at a desired temperature by simply adjusting the copolymer ratio and concentration. However, in all these systems, THF is used as solvent to dissolve the block copolymers.[112–114]

Li *et al.* reported on stereo-complex hydrogels from totally aqueous PLLA/PEG and PDLA/PEG solutions.[82,115–118] The copolymers are synthesized by ring-opening polymerization of L-lactide or D-lactide in the presence of mono- or dihydroxyl PEG, using zinc lactate as catalyst. Hydrogels are obtained from aqueous solutions containing equimolar amounts of PLLA/PEG and PDLA/PEG block copolymers due to the stereo-complexation between PLLA and PDLA blocks.

Figure 2.9 shows the modulus changes of 14% $L_{19}EO_{227}L_{19}/D_{20}EO_{227}D_{20}$ mixture as a function of time at 25 °C and at a frequency of 1 Hz. The mixture is initially an aqueous solution since the storage modulus (G′) is lower than the

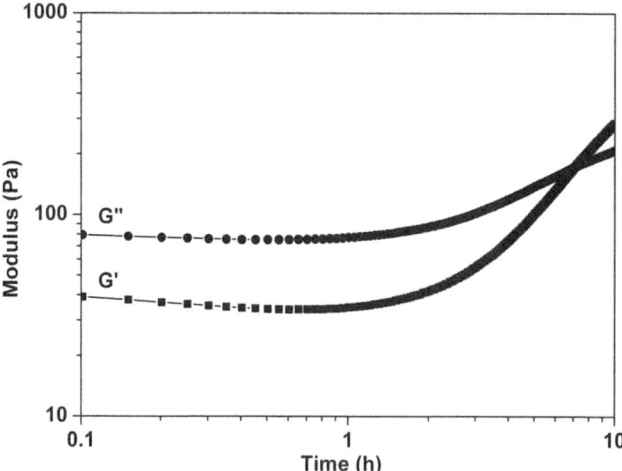

Figure 2.9   Modulus changes of 14% $L_{19}EO_{227}L_{19}/D_{20}EO_{227}D_{20}$ as a function of time at 25 °C.

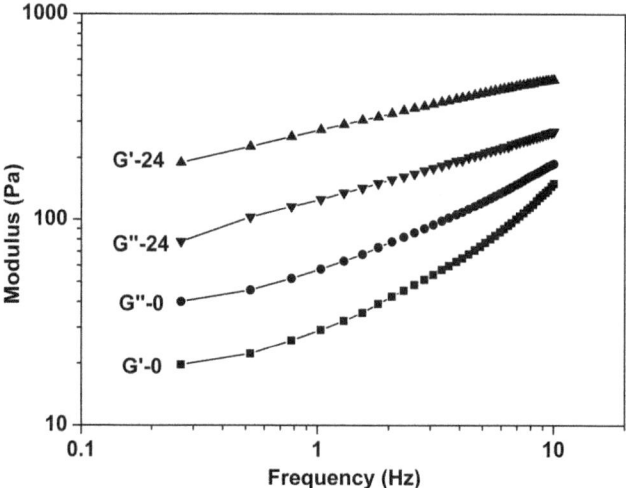

Figure 2.10   Modulus changes of 14% $L_{19}EO_{227}L_{19}/D_{20}EO_{227}D_{20}$ as a function of frequency after 0 and 24 h at 25 °C.

loss modulus (G″). Both G′ and G″ remain almost constant during the first 60 min. Beyond, the moduli increase continuously, G′ increasing faster than G″. A crossover point is observed at *ca.* 7 h, indicating the sol to gel transition.[116]

Figure 2.10 illustrates the modulus changes of 14% $L_{19}EO_{227}L_{19}/D_{20}EO_{227}D_{20}$ mixture as a function of frequency after 0 and 24 h at 25 °C, using logarithmic axes. At t = 0, the storage modulus G′ is higher than the loss modulus G″, and both moduli increase almost linearly with frequency,

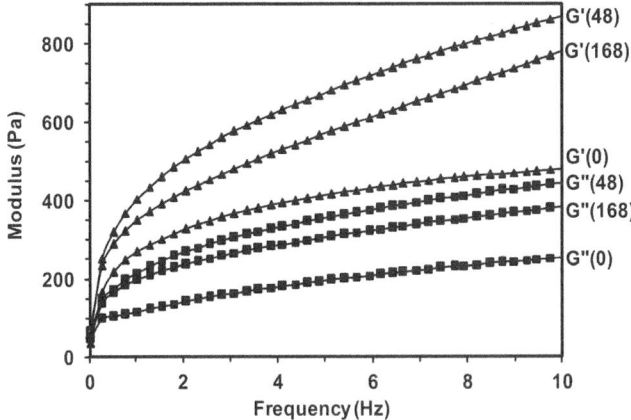

Figure 2.11 Modulus changes of 20% $L_{28}EO_{113}/D_{27}EO_{113}$ mixture as a function of frequency at 37 °C and at t = 0, 48 and 168 h.

which is characteristic of a viscoelastic liquid-like state. In contrast, at t = 24 h, G' became higher than G", indicating formation of a hydrogel.[116]

Degradation also occurs during the gelation process of stereo-complex hydrogels. Figure 2.11 shows the modulus changes of a $L_{28}EO_{113}/D_{27}EO_{113}$ sample as a function of frequency at 37 °C after 0, 48 and 168 h gelation. The sample is initially in the state of a gel, G' being largely higher than G". After 48 h, both moduli strongly increase, indicating that the gelation process continues at 37 °C and the gel becomes much more consistent. After 168 h, the moduli slightly decrease as compared to the values at 48 h, which can be assigned to the partial degradation of the copolymers. In fact, since the hydrogel is a dynamic system, gelation and degradation occur simultaneously.[116]

There are two types of ester bonds where hydrolysis occurs in the polymer chains: ester bonds within the PLA blocks and ester bonds linking PEG and PLA blocks. Degradation of PLA-PEG-PLA copolymers appears homogeneous without autocatalysis of carboxyl end-groups due to the highly swollen structure of hydrogels. The hydrolysis of two types of bonds proceeds with comparable rates. The length of PLA blocks decreases quickly and the degradation by-products become soluble in water and are removed from the hydrogels.

The influence of copolymer concentration on the drug release behaviour of stereo-complex hydrogels is also examined.[14,116] Figure 2.12 illustrates the release profiles of $L_{28}EO_{113}/D_{27}EO_{113}$ hydrogels containing *ca.* 40 mg of bovine serum albumin (BSA) after 24 h gelation at 50 °C, the copolymer concentration of the gel systems varying from 15% to 30%. Similar release curves are observed for different concentrations with almost constant release rate and little burst. Nevertheless, the release rate decreases with increasing copolymer concentration, which can be assigned to the fact that protein diffusion is enhanced with low polymer concentrations.

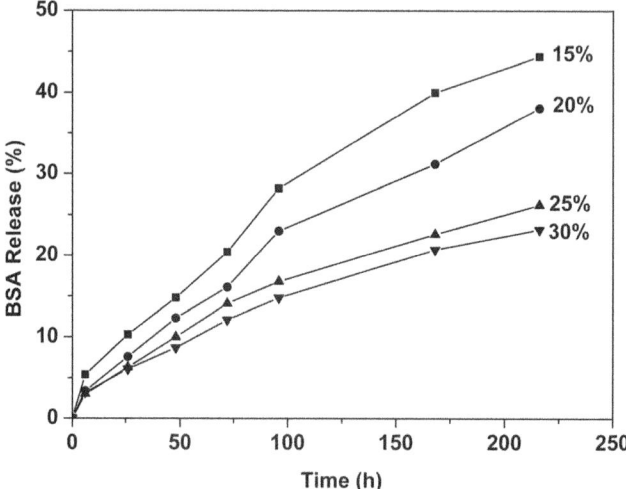

Figure 2.12   BSA release from $L_{28}EO_{113}/D_{27}EO_{113}$ hydrogels with different copolymer concentrations.

The release of thymopentin (TP5), an immune-stimulant from PLA-PEG-PLA stereo-complex hydrogels, was also reported.[14] Rheological measurements showed that the hydrogels present typical viscoelastic behaviours, although degradation could occur during the gelation process. Water soluble TP5 is successfully incorporated into the hydrogels by simply mixing before gelation. The release profiles are characterized by an initial burst followed by slower release. Higher copolymer concentration leads to slower release rate and less burst effect due to more compact structure which disfavours the drug diffusion. Similarly, higher molar mass of copolymers disfavours the release of TP5, and stereo-complex hydrogels present slower release rates than single copolymers. In contrast, drug load exhibits little influence on the release profiles due to the high water solubility of TP5. *In vivo* studies proved the immunization efficiency of the TP5 release systems based on PLA-PEG-PLA hydrogels.[14]

Li *et al.* also studied stereo-complex hydrogels based on chitosan-graft-polylactide (CS-*g*-PLA) copolymers prepared by grafting PLLA or PDLA precursors to the backbone of chitosan using N,N′-carbonyldiimidazole (CDI) as coupling agent.[119] Hydrogels were prepared by mixing water soluble CS-*g*-PLLA and CS-*g*-PDLA solutions. Gelation was evidenced by DSC and WAXD measurements. TP5 is taken as a model drug to evaluate the potential of CS-*g*-PLA hydrogels as drug carrier. An initial burst and a final release up to 82% of TP5 are observed from HPLC analysis.[119]

Drug release from hydrogels is generally driven by two mechanisms: diffusion and degradation.[116,120] All parameters which influence diffusion and degradation phenomena of stereo-complex hydrogels affect the drug release rates, including copolymer concentration, PLA block length, EO/LA ratio, molar mass, drug solubility, *etc.*

2.3.5   Other Stereo-complex Systems

Kondo *et al.* reported a novel fabrication method of polymer nanotubes by combination of an alternate layer-by-layer (LbL) assembly of PLLA and PDLA and the silica template method.[121] Silica nanoparticles with a diameter of 300 nm were alternately immersed in acetonitrile solutions of PLLA and PDLA at 50 °C. The immersion process was performed for 10 cycles to deposit 10 double layers of PLLA and PDLA. The resulting particles were then treated with 2.3% aqueous hydrofluoric acid to remove the silica core. The hollow capsules have a spherical shape with a diameter of 320 nm and a shell thickness of approximately 60 nm. Tubular assemblies with an average diameter of 300 nm and lengths of 2–5 µm are obtained by evaporating water at ambient temperature from a water dispersion of the hollow capsules on a polyethylene terephthalate (PET) substrate.

Liu *et al.* studied the synthesis and biocompatibility of pH-sensitive stereo-complex nanoparticles prepared from poly(ethylene glycol)-poly-(L-histidine)-poly(L-lactide) (PEG-PH-PLLA) and poly(ethylene glycol)-poly-(L-histidine)-poly(D-lactide) (PEG-PH-PDLA).[122] The two amphiphilic triblock copolymers are synthesized and mixed together to fabricate pH-sensitive stereo-complex nanoparticles by self-assembly. The stereo-complexation could not only stabilize the nanoparticles in size and morphology when the pH values are varied from 5.0 to 7.9, but also decrease the cytotoxicity of the nanoparticles originated from the PDLA segments. These findings show that PEG-PH-PLA stereo-complex nanoparticles are potential pH-sensitive drug carriers.

References

1. A. C. Albertsson and I. Varma, in *Degradable Aliphatic Polyesters*, Springer Berlin Heidelberg, 2002, vol. 157, p. 1.
2. L. S. Nair and C. T. Laurencin, *Progr. Polymer Sci.*, 2007, **32**, 762.
3. J. Pan, M. M. Zhao, Y. Liu, B. Wang, L. Mi and L. Yang, *J. Biomed. Mater. Res.*, 2009, **89A**, 160.
4. Y. K. Luu, K. Kim, B. S. Hsiao, B. Chu and M. Hadjiargyrou, *J. Contr. Release*, 2003, **89**, 341.
5. I. Barwal, A. Sood, M. Sharma, B. Singh and S. C. Yadav, *Colloids Surf. B*, 2013, **101**, 510.
6. Q. He, W. Wu, K. M. Xiu, Q. Zhang, F. J. Xu and J. S. Li, *Int. J. Pharmacol.*, 2013, **443**, 110.
7. K. Kim, M. Yu, X. Zong, J. Chiu, D. Fang, Y.-S. Seo, B. S. Hsiao, B. Chu and M. Hadjiargyrou, *Biomaterials*, 2003, **24**, 4977.
8. J. Slager and A. J. Domb, *Adv. Drug Deliv. Rev.*, 2003, **55**, 549.
9. Y. Ikada, K. Jamshidi, H. Tsuji and S. H. Hyon, *Macromolecules*, 1987, **20**, 904.
10. H. Tsuji and Y. Ikada, *Macromolecules*, 1993, **26**, 6918.
11. H. Tsuji, *Macromolecular Biosciences*, 2005, **5**, 569.

12. J. M. Anderson and M. S. Shive, *Adv. Drug Deliv. Rev.*, 2012, **64**, 72.
13. S. M. Li and S. McCarthy, *Biomaterials*, 1999, **20**, 35.
14. Y. Zhang, X. H. Wu, Y. R. Han, F. Mo, Y. R. Duan and S. M. Li, *Int. J. Pharmacol.*, 2010, **386**, 15.
15. L. Yang, Z. X. Zhao, J. Wei, A. El Ghzaoui and S. M. Li, *J. Colloid Interface. Sci.*, 2007, **314**, 470.
16. L. Yang, X. H. Wu, F. Liu, Y. R. Duan and S. M. Li, *Pharmaceut. Res.*, 2009, **26**, 2332.
17. P. J. Pan, J. J. Yang, G. R. Shan, Y. Z. Bao, Z. X. Weng, A. Cao, K. Yazawa and Y. Inoue, *Macromolecules*, 2012, **45**, 189.
18. M. Kakuta, M. Hirata and Y. Kimura, *Polymer Rev.*, 2009, **49**, 107.
19. J. Slager and A. J. Domb, *Eur. J. Pharmacol. Biopharmacol.*, 2004, **58**, 461.
20. J. Slager and A. J. Domb, *Biomacromolecules*, 2003, **4**, 1316.
21. N. Spassky, M. Wisniewski, C. Pluta and A. Le Borgne, *Macromol. Chem. Phys.*, 1996, **197**, 2627.
22. M. Wisniewski, A. L. Borgne and N. Spassky, *Macromol. Chem. Phys.*, 1997, **198**, 1227.
23. N. Nomura, R. Ishii, M. Akakura and K. Aoi, *J. Am. Chem. Soc.*, 2002, **124**, 5938.
24. Z. Zhong, P. J. Dijkstra and J. Feijen, *J. Am. Chem. Soc.*, 2003, **125**, 11291.
25. T. M. Ovitt and G. W. Coates, *J. Polymer Sci. Polymer Chem.*, 2000, **38**, 4686.
26. M. Cheng, A. B. Attygalle, E. B. Lobkovsky and G. W. Coates, *J. Am. Chem. Soc.*, 1999, **121**, 11583.
27. K. Majerska and A. Duda, *J. Am. Chem. Soc.*, 2004, **126**, 1026.
28. C. P. Radano, G. L. Baker and M. R. Smith, *J. Am. Chem. Soc.*, 2000, **122**, 1552.
29. S. J. de Jong, W. N. E. van Dijk-Wolthuis, J. J. Kettenes-van den Bosch, P. J. W. Schuyl and W. E. Hennink, *Macromolecules*, 1998, **31**, 6397.
30. H. Tsuji and Y. Ikada, *Polymer*, 1995, **36**, 2709.
31. H. Tsuji, S. H. Hyon and Y. Ikada, *Macromolecules*, 1991, **24**, 5657.
32. H. Tsuji, F. Horii, S. H. Hyon and Y. Ikada, *Macromolecules*, 1991, **24**, 2719.
33. H. Tsuji, S. H. Hyon and Y. Ikada, *Macromolecules*, 1991, **24**, 5651.
34. H. Tsuji, S. H. Hyon and Y. Ikada, *Macromolecules*, 1992, **25**, 2940.
35. S. Brochu, R. E. Prud'homme, I. Barakat and R. Jerome, *Macromolecules*, 1995, **28**, 5230.
36. H. Tsuji and Y. Ikada, *Macromolecules*, 1992, **25**, 5719.
37. H. Tsuji and Y. Ikada, *Macromol. Chem. Phys.*, 1996, **197**, 3483.
38. H. Tsuji and Y. Ikada, *Polymer*, 1999, **40**, 6699.
39. S. Pensec, M. Leroy, H. Akkouche and N. Spassky, *Polymer Bull.*, 2000, **45**, 373.
40. H. Tsuji and Y. Ikada, *J. Appl. Polymer Sci.*, 1994, **53**, 1061.
41. W. M. Stevels, M. J. K. Ankone, P. J. Dijkstra and J. Feijen, *Macromol. Chem. Phys.*, 1995, **196**, 3687.
42. D. W. Lim and T. G. Park, *J. Appl. Polymer Sci.*, 2000, **75**, 1615.

43. S. Li and M. Vert, *Macromolecules*, 1994, **27**, 3107.
44. S. Li and M. Vert, *Polymer Int.*, 1994, **33**, 37.
45. S. Li, S. Girod-Holland and M. Vert, *J. Contr. Release*, 1996, **40**, 41.
46. L. Cartier, T. Okihara and B. Lotz, *Macromolecules*, 1997, **30**, 6313.
47. Y. He, Y. Xu, J. Wei, Z. Fan and S. Li, *Polymer*, 2008, **49**, 5670.
48. D. Brizzolara, H.-J. Cantow, K. Diederichs, E. Keller and A. J. Domb, *Macromolecules*, 1996, **29**, 191.
49. J. Zhang, K. Tashiro, H. Tsuji and A. J. Domb, *Macromolecules*, 2007, **40**, 1049.
50. Z. J. Xiong, G. M. Liu, X. Q. Zhang, T. Wen, S. de Vos, C. Joziasse and D. J. Wang, *Polymer*, 2013, **54**, 964.
51. T. Wen, Z. J. Xiong, G. M. Liu, X. Q. Zhang, S. de Vos, R. Wang, C. A. P. Joziasse, F. S. Wang and D. J. Wang, *Polymer*, 2013, **54**, 1923.
52. M. S. Reeve, S. P. McCarthy, M. J. Downey and R. A. Gross, *Macromolecules*, 1994, **27**, 825.
53. S. Li and S. McCarthy, *Macromolecules*, 1999, **32**, 4454.
54. Y. He, T. Wu, J. Wei, Z. Fan and S. Li, *J. Polymer Sci. B Polymer Phys.*, 2008, **46**, 959.
55. S. Li, M. Tenon, H. Garreau, C. Braud and M. Vert, *Polymer Degrad. Stabil.*, 2000, **67**, 85.
56. L. Liu, S. Li, H. Garreau and M. Vert, *Biomacromolecules*, 2000, **1**, 350.
57. S. M. Li, A. Girard, H. Garreau and M. Vert, *Polymer Degrad. Stabil.*, 2001, **71**, 61.
58. H. Tsuji and S. Miyauchi, *Biomacromolecules*, 2001, **2**, 597.
59. R. A. Kenley, M. O. Lee, T. R. Mahoney and L. M. Sanders, *Macromolecules*, 1987, **20**, 2398.
60. C. G. Pitt, M. M. Gratzl, G. L. Kimmel, J. Surles and A. Schindler, *Biomaterials*, 1981, **2**, 215.
61. E. A. Schmitt, D. R. Flanagan and R. J. Linhardt, *Macromolecules*, 1994, **27**, 743.
62. E. W. Fischer, H. Sterzel and G. Wegner, *Kolloid Z. Z. Polym.*, 1973, **251**, 980.
63. S. Shalaby, A. Hoffman, B. Ratner, T. Horbett, B. Carter and G. Wilkes, in *Polymers as Biomaterials*, Springer US, 1984, p. 67.
64. R. J. Fredericks, A. J. Melveger and L. J. Dolegiewitz, *J. Polymer Sci. Polymer Phys. Ed.*, 1984, **22**, 57.
65. S. Li, H. Garreau and M. Vert, *J. Mater. Sci. Mater. Med.*, 1990, **1**, 123.
66. S. Li, H. Garreau and M. Vert, *J. Mater. Sci. Mater. Med.*, 1990, **1**, 131.
67. S. M. Li, H. Garreau and M. Vert, *J. Mater. Sci. Mater. Med.*, 1990, **1**, 198.
68. M. Vert, S. Li and H. Garreau, *J. Contr. Release*, 1991, **16**, 15.
69. M. Therin, P. Christel, S. Li, H. Garreau and M. Vert, *Biomaterials*, 1992, **13**, 594.
70. S. Li, *J. Biomed. Mater. Res.*, 1999, **48**, 342.
71. S. M. Li and M. Vert, in *Encyclopedia of Controlled Drug Delivery*, ed. E. Mathiowitz, John Wiley & Sons, 1999, p. 71.
72. H. Tsuji, *Polymer*, 2000, **41**, 3621.

73. H. Tsuji, *Biomaterials*, 2003, **24**, 537.
74. H. Tsuji and T. Tsuruno, *Polymer Degrad. Stabil.*, 2010, **95**, 477.
75. M. H. Rahaman and H. Tsuji, *Polymer Degrad. Stabil.*, 2013, **98**, 709.
76. S. R. Andersson, M. Hakkarainen, S. Inkinen, A. Sodergard and A. C. Albertsson, *Biomacromolecules*, 2012, **13**, 1212.
77. W.-K. Lee, T. Iwata and J. A. Gardella, *Langmuir*, 2005, **21**, 11180.
78. D. Ishii, T. H. Ying, A. Mahara, S. Murakami, T. Yamaoka, W.-K. Lee and T. Iwata, *Biomacromolecules*, 2008, **10**, 237.
79. K. Masutani, C. W. Lee and Y. Kimura, *Polymer*, 2012, **53**, 6053.
80. J. Watanabe, T. Eriguchi and K. Ishihara, *Biomacromolecules*, 2002, **3**, 1109.
81. L. Chen, Z. Xie, J. Hu, X. Chen and X. Jing, *J. Nanopart. Res.*, 2007, **9**, 777.
82. S. Li and M. Vert, *Macromolecules*, 2003, **36**, 8008.
83. H. Tsuji, M. Nakano, M. Hashimoto, K. Takashima, S. Katsura and A. Mizuno, *Biomacromolecules*, 2006, 7, 3316.
84. J. Slager and A. J. Domb, *Biomaterials*, 2002, **23**, 4389.
85. A. Bishara and A. J. Domb, *J. Contr. Release*, 2005, **107**, 474.
86. J. Slager, Y. Cohen, R. Khalfin, Y. Talmon and A. J. Domb, *Macromolecules*, 2003, **36**, 2999.
87. N. Kang, M. E. Perron, R. E. Prud'homme, Y. Zhang, G. Gaucher and J.-C. Leroux, *Nano Lett.*, 2005, **5**, 315.
88. Y. Yan, G. K. Such, A. P. R. Johnston, J. P. Best and F. Caruso, *ACS Nano*, 2012, **6**, 3663.
89. C. Oerlemans, W. Bult, M. Bos, G. Storm, J. F. Nijsen and W. Hennink, *Pharmaceut. Res.*, 2010, **27**, 2569.
90. S. Li, I. Rashkov, J. L. Espartero, N. Manolova and M. Vert, *Macromolecules*, 1996, **29**, 57.
91. I. Rashkov, N. Manolova, S. Li, J. L. Espartero and M. Vert, *Macromolecules*, 1996, **29**, 50.
92. K. Yasugi, Y. Nagasaki, M. Kato and K. Kataoka, *J. Contr. Release*, 1999, **62**, 89.
93. X. Zhang, J. K. Jackson and H. M. Burt, *Int. J. Pharmacol.*, 1996, **132**, 195.
94. T. Riley, C. R. Heald, S. Stolnik, M. C. Garnett, L. Illum, S. S. Davis, S. M. King, R. K. Heenan, S. C. Purkiss, R. J. Barlow, P. R. Gellert and C. Washington, *Langmuir*, 2003, **19**, 8428.
95. X. Shuai, T. Merdan, A. K. Schaper, F. Xi and T. Kissel, *Bioconjugate Chem.*, 2004, **15**, 441.
96. L. Liu, C. Li, X. Li, Z. Yuan, Y. An and B. He, *J. Appl. Polymer Sci.*, 2001, **80**, 1976.
97. S. A. Hagan, A. G. A. Coombes, M. C. Garnett, S. E. Dunn, M. C. Davis, L. Illum, S. S. Davis, S. E. Harding, S. Purkiss and P. R. Gellert, *Langmuir*, 1996, **12**, 2153.
98. E. Pierri and K. Avgoustakis, *J. Biomed. Mater. Res. A*, 2005, **75A**, 639.
99. K. Kataoka, A. Harada and Y. Nagasaki, *Adv. Drug Deliv. Rev.*, 2001, **47**, 113.

100. L. Yang, X. Qi, P. Liu, A. El Ghzaoui and S. Li, *Int. J. Pharmacol.*, 2010, **394**, 43.

101. M.-C. Jones and J.-C. Leroux, *Eur. J. Pharmacol. Biopharmacol.*, 1999, **48**, 101.

102. Z. Dai, L. Piao, X. Zhang, M. Deng, X. Chen and X. Jing, *Colloid Polymer Sci.*, 2004, **282**, 343.

103. S. K. Agrawal, N. Sanabria-DeLong, J. M. Coburn, G. N. Tew and S. R. Bhatia, *J. Contr. Release*, 2006, **112**, 64.

104. K. Letchford, J. Zastre, R. Liggins and H. Burt, *Colloids Surf. B*, 2004, **35**, 81.

105. A. S. Hoffman, *Adv. Drug Deliv. Rev.*, 2002, **54**, 3.

106. T. Vermonden, R. Censi and W. E. Hennink, *Chem. Rev.*, 2012, **112**, 2853.

107. I. Molina, S. Li, M. B. Martinez and M. Vert, *Biomaterials*, 2001, **22**, 363.

108. S. J. de Jong, S. C. De Smedt, M. W. C. Wahls, J. Demeester, J. J. Kettenes-van den Bosch and W. E. Hennink, *Macromolecules*, 2000, **33**, 3680.

109. S. J. de Jong, B. van Eerdenbrugh, C. F. van Nostrum, J. J. Kettenes-van den Bosch and W. E. Hennink, *J. Contr. Release*, 2001, **71**, 261.

110. D. W. Grijpma and J. Feijen, *J. Contr. Release*, 2001, **72**, 247.

111. C. Hiemstra, Z. Y. Zhong, L. B. Li, P. J. Dijkstra and J. Feijen, *Biomacromolecules*, 2006, 7, 2790.

112. T. Fujiwara, T. Mukose, T. Yamaoka, H. Yamane, S. Sakurai and Y. Kimura, *Macromolecular Biosciences*, 2001, **1**, 204.

113. T. Mukose, T. Fujiwara, J. Nakano, I. Taniguchi, M. Miyamoto, Y. Kimura, I. Teraoka and C. Woo Lee, *Macromolecular Biosciences*, 2004, **4**, 361.

114. D. G. Abebe and T. Fujiwara, *Biomacromolecules*, 2012, **13**, 1828.

115. S. Li, *Macromolecular Biosciences*, 2003, **3**, 657.

116. S. Li, A. El Ghzaoui and E. Dewinck, *Macromol. Symp.*, 2005, **222**, 23.

117. H. Nouailhas, A. El Ghzaoui, S. Li and J. Coudane, *J. Appl. Polymer Sci.*, 2011, **122**, 1599.

118. H. Nouailhas, F. Li, A. El Ghzaoui, S. Li and J. Coudane, *Polymer Int.*, 2010, **59**, 1077.

119. Y. Hu, Y. Liu, X. Qi, P. Liu, Z. Fan and S. Li, *Polymer Int.*, 2012, **61**, 74.

120. M. N. Mason, A. T. Metters, C. N. Bowman and K. S. Anseth, *Macromolecules*, 2001, **34**, 4630.

121. K. Kondo, T. Kida, Y. Ogawa, Y. Arikawa and M. Akashi, *J. Am. Chem. Soc.*, 2010, **132**, 8236.

122. R. Liu, B. He, D. Li, Y. S. Lai, J. Z. Tang and Z. W. Gu, *Macromol. Rapid Commun.*, 2012, **33**, 1061.

CHAPTER 3

Crystallization of PLA-based Materials

A. J. MÜLLER,*[a,b,c] M. ÁVILA,[a] G. SAENZ[a] AND J. SALAZAR[a]

[a] Grupo de Polímeros USB, Departamento de Ciencia de los Materiales, Universidad Simón Bolívar, Apartado 89000, Caracas 1080A, Venezuela; [b] Institute for Polymer Materials (POLYMAT) and Polymer Science and Technology Department, Faculty of Chemistry, University of the Basque Country (UPV/EHU), Paseo Manuel de Lardizabal 3, 20018 Donostia-San Sebastián, Spain; [c] IKERBASQUE, Basque Foundation for Science, E-48011 Bilbao, Spain
*Email: alejandrojesus.muller@ehu.edu; amuller@usb.ve

3.1 Introduction

A considerable amount of poly(lactic acid) (PLA) properties depend on its degree of crystallinity. In fact, for many applications, the poor dimensional stability as a function of temperature of fully amorphous PLAs is an important limitation, since their glass transition temperatures are around or even below 60 °C. Therefore, the study of the PLAs crystallization has tremendous relevance from both academic and industrial points of view. The applications of PLA and its copolymers range from packaging, environmentally friendly containers and bottles to many biomedical devices, such as controlled release matrices for drug delivery, scaffolds for tissue engineering and biodegradable elastomers for cardiovascular implanted devices amongst many others.

There is a vast literature on the crystallization of PLA and PLA-based copolymers and blends. This chapter is by no means exhaustive but it touches

RSC Polymer Chemistry Series No. 12
Poly(lactic acid) Science and Technology: Processing, Properties, Additives and Applications
Edited by Alfonso Jiménez, Mercedes Peltzer and Roxana Ruseckaite
© The Royal Society of Chemistry 2015
Published by the Royal Society of Chemistry, www.rsc.org

upon most aspects related to polymer crystallization. Some of the subjects not reviewed here are contained in other chapters of this book (like stereo-complexation and nanocomposites). In this chapter, we focus on the morphology, nucleation and crystallization of the different types of PLA and on some specific examples of block copolymers based on PLA. There is particular emphasis on reviewing published works dealing with crystal-lization kinetics.

3.2 Crystal Structure and Single Crystals

L-lactic acid is a monomer that can be obtained from natural sources. It is worth pointing out that (*S,S*)-lactide is referred to as L-lactide (PLLA), (*R,R*)-lactide corresponds to D-lactide (PDLA) and (*S,R*)-lactide is a meso-lactide (PLDA). Different types of PLA crystals can be obtained depending on the crystallization conditions.

The α-form is the crystal type most commonly found for PLA. It was first reported by De Santis *et al.*[1] They determined that the α-form crystal corresponds to a helical conformation by X-ray fibre photographs. The α-form crystal consists of 3.3 monomers per turn and a monomer repeat unit of 27.8 nm. Zhang *et al.*[2] reported another modification called α′-form crystal, a structure similar to the α-form but with less dense packing. The α′-form crystals were obtained at crystallization temperatures (T_c) below 120 °C. Other studies suggest that α′-form crystals form below 100 °C, while in the temperature range of 100–120 °C both α and α′-forms can arise.[3,4]

Eling *et al.*[5] observed β-form crystals, whose melting temperature is 10 °C lower than the melting temperature of the corresponding α-form. Cartier *et al.*[6] reported γ-form crystals, which are obtained by epitaxial crystallization of PLA in hexamethylbenzene. In γ-form, two chains are antiparallel to the crystal unit cell.

Fischer *et al.*[7] prepared lactide copolymers (meso-lactide) single crystals containing 2.75% and 10.2% D-units. Figure 3.1 shows TEM micrographs of these PLDA single crystals obtained by crystallization from dilute xylene solution at an isothermal crystallization temperature of 60 °C. The PLDA single crystals exhibit a more regular terraced morphology for lower D-unit content. However, they were not formed when the content of D-units was greater than 15%.

Kalb *et al.*[8] prepared PLLA single crystals grown from very dilute solutions (0.08% by weight). Figure 3.2 shows representative TEM micrographs. Lamellae were well defined lozenge-shaped at $T_c = 55$ °C, with thicknesses about 10 nm. At shallower undercoolings, *i.e.* $T_c = 90$ °C, lamellae were truncated lozenges with thicknesses about 12 nm. More recently, Vasanthakumari *et al.*[9] obtained PLLA single crystals from the melt. The PLLA samples were crystallized from the melt at 164 °C, and the single crystals were hexagonal-shaped.

An equimolar mixture of PLLA/PDLA was first studied by Okihara *et al.*[10] Their results showed that PLLA/PDLA can co-crystallize into a stereo-complex,

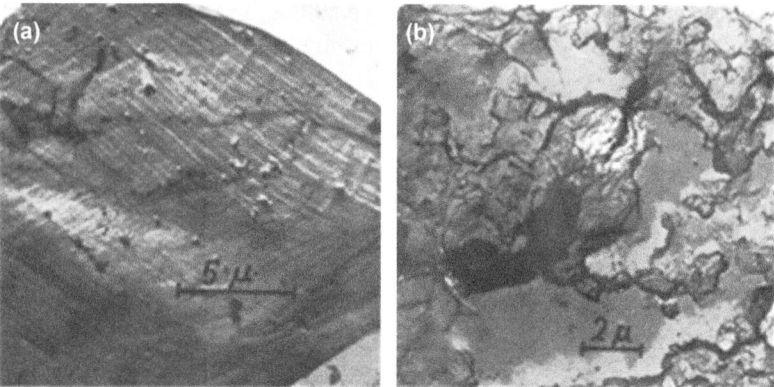

Figure 3.1   TEM micrographs of single crystals of lactide copolymers (meso-lactide) grown from a 0.1 wt% solution in xylene at 60 °C. a) Lactide copolymers containing 2.75% D-units and b) lactide copolymers containing 10.2% D-units.[7]

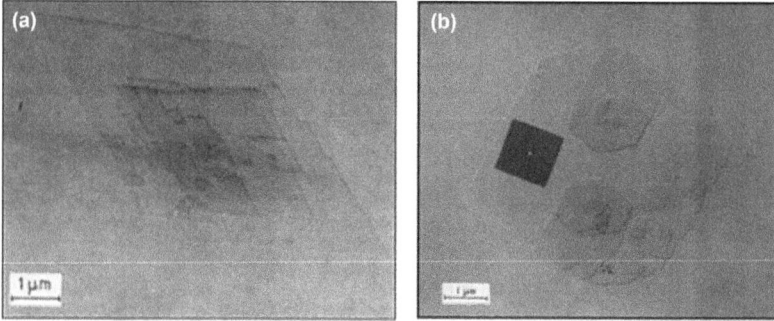

Figure 3.2   TEM micrographs of single crystals of PLLA grown from a 0.08 wt% solution in a) toluene at 55 °C and b) *p*-xylene at 90 °C. The insert shows an electron diffraction pattern of one of the crystals.[8]

whose crystal system is triclinic. Furthermore, Cartier *et al.*[11] proposed a structure wherein the chains of PDLA and PLLA form 3_1 and 3_2 helices, respectively. Chapter 2 of the present book is devoted to stereo-complexation and full information on these observations is given there. A summary of the crystal unit parameters reported for homo- and stereo-PLA crystals is presented in Table 3.1.

3.3   Melting and Glass Transition Temperatures

Molecular weight has a considerable influence on the melting temperature (T_m) of polymeric crystals. Figure 3.3a shows a compilation of literature data that clearly demonstrates how T_m values increase with number average molecular weight (M_n). For $M_n > 100$ kg mol^{-1}, an asymptotical T_m value is achieved. However, T_m is also a function of optical purity of PLA. The higher

Table 3.1 Crystal structure and unit cell parameters of PLA.

Crystal Structure	Crystal system	Chain conformation	Unit cell parameters						Ref.
			a (nm)	b (nm)	c (nm)	α (°)	β (°)	γ (°)	
α	Pseudo-orthorhombic	10_3	1.07	0.645	2.78	90	90	90	1
α	Pseudo-orthorhombic	10_3	1.06	0.61	2.88	90	90	90	12
α	Orthorhombic	10_3	1.078	0.604	2.873	90	90	90	13
α	Orthorhombic	10_3	1.078	0.604	2.873	90	90	90	14
β	Orthorhombic	3_1	1.04	1.82	0.90	90	90	90	12
β	Trigonal	3_1	1.052	1.052	0.88	90	90	120	15
β	Orthorhombic	3_1	1.04	1.77	0.90	90	90	90	16
γ	Orthorhombic	3_1	0.995	0.625	0.88	90	90	90	6
Stereo-complex	Triclinic	3_1	0.916	0.916	0.870	109.2	109.2	109.8	10
	Trigonal	3_1 y 3_2	1.498	1.498	0.87	90	90	120	11
	Triclinic	3_1	0.912	0.913	0.930	110	110	109	17

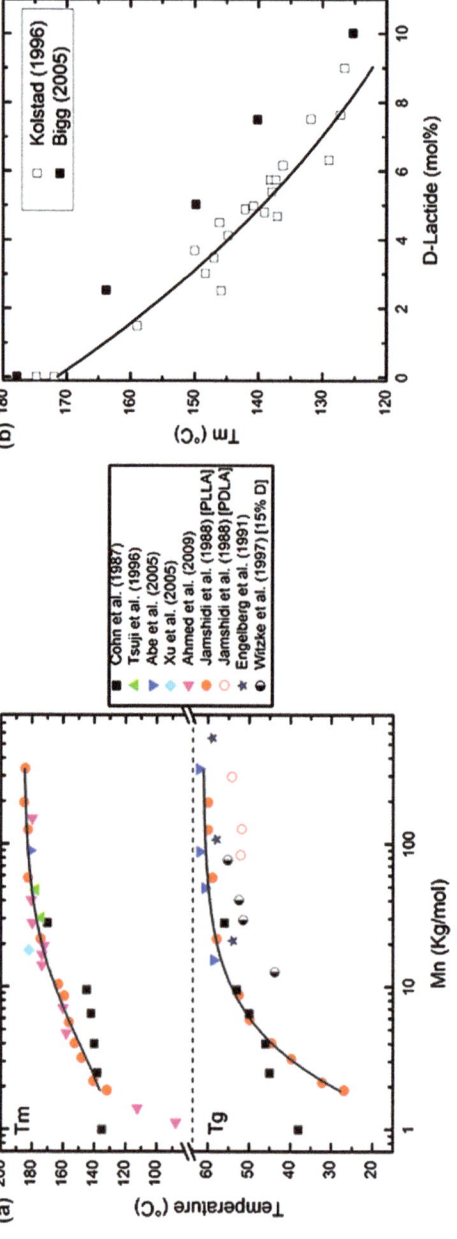

Figure 3.3 a) Melting and glass transition temperature *versus* molecular weight of PLLA,[18–25] b) melting temperature *versus* D-lactide content.[26,27]

the amount of D-lactide incorporated into PLDA structure the lower the T_m obtained, as shown in Figure 3.3b.

The T_m value is proportional to the molecular weight according to the Flory equation[28]

$$\frac{1}{T_m} - \frac{1}{T_m^\infty} = \frac{2RM_O}{\Delta H_{mu}M_n} \tag{1}$$

where T_m^∞ is the melting temperature at infinite molecular weight, R is the gas constant, ΔH_{mu} is the heat of fusion per mole of the repeating unit and M_O is the molecular weight of the repeating unit.[28]

Figure 3.3a shows how T_g values also depend on the molecular weight. The data can be fitted with the Flory–Fox equation[28]:

$$Tg = Tg^* - \frac{K}{M_n} \tag{2}$$

where, T_g^* is the glass transition temperature at infinite molecular weight and K is a constant.[28] According to the recent and comprehensive review of Saeidlou et al.,[29] the value of K increases with the D-lactide concentration (X_D) linearly as:

$$K = 52.23 + 791X_D \tag{3}$$

In addition, the relationship between the T_g^* and the concentration of D-lactide can be predicted with a rational function:

$$Tg^* = \frac{13.36 + 1371.68X_D}{0.22 + 24.3X_D + 0.42X_D^2} \tag{4}$$

where T_g^* and K are expressed in °C and °C kg mol^{-1}, respectively. The parameters T_g^* and K can be used in the Flory–Fox equation to obtain a good approximation of the T_g for PLA.[29] On the other hand, the T_g of PLA can be influenced by chain architecture. Pitet et al.[30] reported a decrease in T_g for hyper-branched PLA, because of the increase in free volume with the number of chain ends.

Celli et al.[31] studied physical ageing of PLA. They found that PLLA samples of different molecular weight experienced physical ageing depending on molecular weight and ageing temperature (at comparable supercooling). The lower the molecular weight the greater the ageing effect, as a consequence of the ability of short polymer chains to relax free volume. In addition, significant ageing effects are produced from room temperature up to temperatures near T_g.

As mentioned above, Figure 3.3b shows the variation of T_m as a function of D-lactide content. As an example of this behaviour, Figure 3.4 shows differential scanning calorimetry (DSC) heating runs for PLA ($M_w = 218$ kg mol^{-1}, %D = 1.4) and PDLA ($M_w = 116$ kg mol^{-1}, %D = 99.9). The T_m of PLA samples shows a difference of 9 °C, and both cold crystallization and glass transition are affected by D-lactide content, as shown in this example.

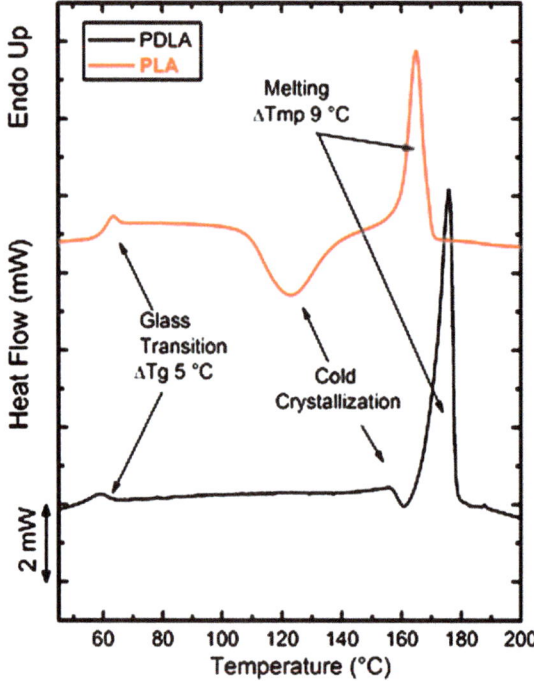

Figure 3.4 Examples of DSC heating runs for two samples of PLA $(M_w =$
218 kg mol^{-1}, %D$= 1.4$) and PDLA $(M_w = 116$ kg mol^{-1}, %D$= 99.9)$ at
10 °C min^{-1}.

The physical, mechanical and barrier properties of PLA are dependent on
the morphology, crystalline form and crystallinity degree. PLA can be either
amorphous or semicrystalline depending on its stereochemical architecture
and thermal history.[3,28,32–35] The degradation behaviour (the rate of hydrolytic or enzymatic degradability) of polylactides is also strongly influenced by
crystallinity.[33,35] One of the most generally employed methods to determine
the crystallinity of PLA is DSC by following eqn (5):

$$\text{crystallinity}(\%) = \frac{\Delta H_m}{\Delta H_m^\circ} \tag{5}$$

where ΔH_m is the enthalpy of fusion of the studied sample and ΔH_m° is the
enthalpy for 100% crystalline PLA samples, assuming no cold crystallization
taking place during the heating run, otherwise the cold-crystallization enthalpy should be subtracted from the melting enthalpy.

The PDLA, PLLA or high D- or L-lactide copolymers have regular structures.
The polylactides are either amorphous or semicrystalline at room temperature, depending on the molecular weight and content of L, D or meso-lactide
in the main chain. PLA can be totally amorphous or up to 40% crystalline.
PLA resins containing more than 93% L-lactic acid can crystallize. However,
high molecular weight can reduce the crystallization rate, and therefore the

crystallinity degree.[3] If the PLA contains less than 93% L units it is usually amorphous.

3.4   Superstructural Morphology

The morphology observed in any given polymer depends on the crystallization temperature. Figure 3.5 shows spherulitic morphologies of PLLA. Typically, PLLA adopts a non-banded spherulitic form when it crystallizes from the melt and exhibits negative birefringence.[8,9,36,37] The material can exhibit banding depending on the thermal history applied to the sample.

Vasanthakumari *et al.*[9] obtained at very high supercooling well-defined spherulites of PLLA. However, as the supercooling was reduced, the observed spherulites were increasingly irregular in form and coarse grained. But, as the supercooling was even smaller, a morphological change from spherulites to axialites was produced in PLLA samples of low molecular weight. This behaviour was also reported by Yuryey *et al.*[38] and by Kalb and Pennings[8] on PLLA crystallized from solutions.

Banding is a common extinction pattern exhibited by almost all chiral polymers or even organic and inorganic compounds with a chiral structure. PLLA is a polymer with main-chain chirality. Nevertheless, the spherulitic morphology of PLLA is widely debated in the literature.[39] In most published scientific research, PLLA has shown non-banded spherulites.[8,9,14,36,37,40] Gazzano *et al.*[41] investigated the spherulitic structure of PLLA by using microfocus X-ray diffraction, and they concluded that no lamellar twisting existed in neat PLLA. However, some authors have obtained banded spherulites of PLLA. Xu *et al.*[21] observed banded spherulites of PLLA (see an example in Figure 3.1) with alternative negative and positive birefringence, when PLLA samples were annealed at 160 °C for 2 hours and cooled slowly at 0.5 °C min^{-1} to T_c. They also demonstrated that thermal history of PLLA before crystallization has a significant role in the formation of regular-banded spherulites. Yasuniwa *et al.*[42] observed circular ring morphology on spherulites grown at $T_c = 121$ °C and 131 °C, *via* isothermal crystallization. Wang *et al.*[39] showed a clear banding-to-non-banding morphological transition in PLLA at a critical crystallization temperature. The PLLA samples of different M_w (151 and 301 kg mol^{-1}) were isothermally crystallized at

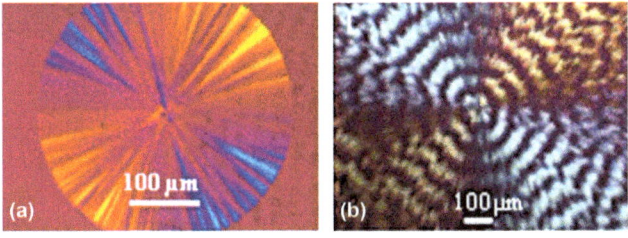

Figure 3.5   Polarized light optical micrographs (PLOM) during isothermal crystallization of PLLA a) after 8 min. at 140 °C (modified from Ref. 37), b) after annealing at 160 °C for 2 h followed by slow cooling to 120 °C (modified from Ref. 21).

different T_c, and exhibited a band spacing pattern depending on M_w. Banded spherulites were also observed in PLLA blends, for instance, when PLLA is mixed with atactic poly(DL-lactide) (PDLLA),[21,43,44] atactic poly(3-hidroxy-butyrate) (a-PHB)[21,45] or poly(ethylene oxide)/poly(vinyl acetate) (POE/PVAc).[46] Xu *et al.*[21] reported that incorporation of a second amorphous polymer induced the lamellar twisting that characterizes banding.

PLLA-*b*-PEG diblock copolymers exhibit banded spherulites when iso-thermally crystallized at $T_c \geq 110$ °C.[47,48] Yang *et al.*[48] proposed that the ex-istence of the PEG block as a diluent played an important role in the formation of ring-banded spherulites. Huang *et al.*[49] studied crystalline morphologies of PLLA-*b*-PEO diblock copolymers. They also observed ring-banded spherulites. Wang and Dong[50] first reported spherulites with ban-ded textures in PCL-*b*-PLLA copolymers. The PLLA blocks within PCL-*b*-PLLA copolymers formed spherulites that gradually changed from ordinary to banded spherulites; this effect could also be attributed to a dilution effect of the molten PCL block on the PLLA block spherulites formed at 120 °C (*i.e.* above the PCL block melting point).

3.5 Crystallization Kinetics

3.5.1 Spherulite Growth Kinetics

Generally, polarized light optical microscopy (PLOM) is used to determine nucleation density as well as spherulite growth rates in isothermal con-ditions.[51–53] First, the polymer film is heated above its melting temperature and quickly cooled to a specific temperature. Then, the size and number of spher-ulites can be evaluated as a function of time. The number of spherulites in the view field can be counted at each crystallization temperature. The relationship between the number of crystallization sites (spherulite density) with crystal-lization temperature for PLLA is reported in Figure 3.6. Several authors[42,43,54–56] observed (for different molecular weights of PLLA) that the spherulite density increases substantially with a decrease in T_c. When spherulite density in-creases, the induction period for primary nucleation decreases. Once the crystallization temperature reaches the melting temperature, the nucleation process becomes difficult because of the low degree of supercooling.[54]

To investigate the morphology and crystal growth in PLLA, atomic force microscopy (AFM) can also be used. This technique can produce high-quality images and more accurate measurements than PLOM, although it is usually only restricted to the sample surface.[21,38,57]

The crystal growth process is normally studied by measuring the spher-ulite radius (r) as a function of the crystallization time (t). The spherulite radius increases linearly with crystallization time at constant temperature unless the molten polymer experiences diffusion problems to the growth front (an unusual occurrence). The spherulite growth rate (G) is simply equal to the slope of the r *vs.* t curve. Therefore, G is a constant value at a given crystallization temperature.[29,42,58] Figure 3.7 shows plots of spherulite

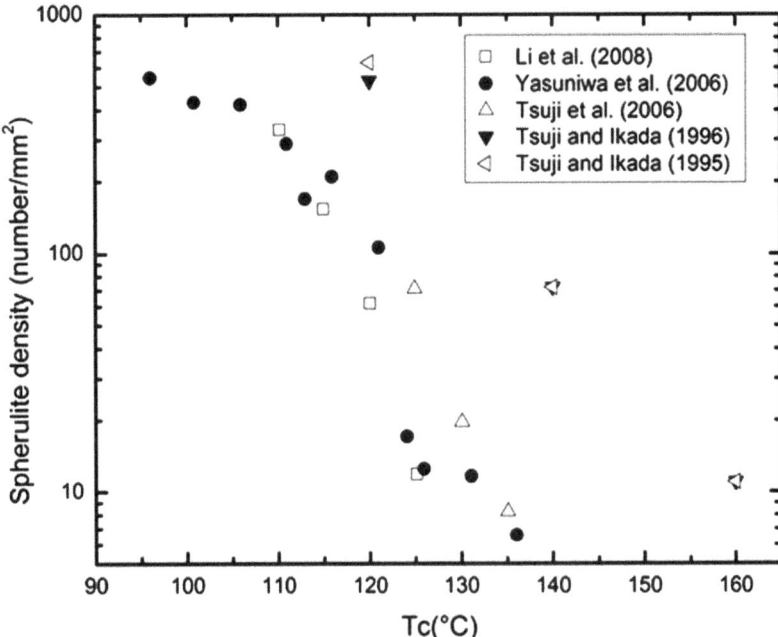

Figure 3.6 Spherulite density *versus* crystallization temperature.[42,43,54–56]

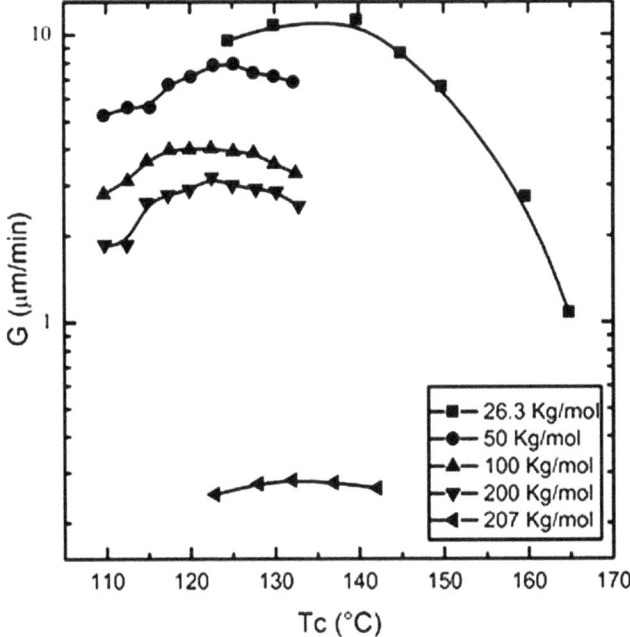

Figure 3.7 Spherulitic growth rate (*G*) as a function of crystallization temperature (*T_c*) for different weight-averaged molecular weight samples of PLLA.[14,37,59]

growth rate as a function of crystallization temperature for various molecular weights of PLLA reported by several authors as indicated in the figure caption.

The spherulite growth rate increases with crystallization temperature (at high supercooling), and a maximum value is reached at around 110–140 °C. After the maximum, G decreases with T_c. The behaviour of G reported in Figure 3.7 corresponds to the well-known bell-shaped trend exhibited by many polymers. As supercooling increases, the thermodynamic driving force for secondary nucleation also increases causing a general improvement in G (right-hand side of the bell-shaped curve). However, as temperature decreases the melt-viscosity also increases limiting chain diffusion to the growth front and causing a decrease in G (left-hand side of the bell-shaped curve). Moreover, the spherulite growth rate decreases with an increase in molecular weight because of more restricted chain mobility.[3,9,14,29,39,42,60] This behaviour has been also reported for several synthetic polymers, such as isotactic polystyrene, nylon-6, poly(oxypropylene) and poly(chlorotrifluoroethylene), among others.[61]

However, it must be pointed out that an unusual behaviour in the spherulite growth rate curve has been reported by several researchers for intermediate molecular weight PLLAs. Figure 3.8 illustrates the anomalous behaviour where instead of displaying one maximum, the G *vs.* T_c curves exhibit two maxima. The lower temperature peak has been correlated to a *Regime* transition in the growth rate of PLLA spherulites, namely a change from *Regime* II to *Regime* III.[3,38,42,60,62,63] In the next section evidence of this *Regime* transition is presented when the data are analyzed with the Lauritzen and Hoffman kinetic theory. Di Lorenzo[62] demonstrated that the discontinuity in the spherulite growth rate is not associated to any change in superstructural morphology. Tsuji *et al.*[63] and Yuryev *et al.*[38] also observed this unusual bimodal crystallization behaviour for pure PLLA, while the normal characteristic bell-shaped spherulite growth rate dependence was seen for poly(L/D-lactide) copolymers.

3.5.2 Application of the Lauritzen and Hoffman Theory to PLLA

The Lauritzen and Hoffman (LH) theory[64] can be used to describe the crystallization kinetics of linear flexible macromolecular chains based on nucleation and growth. Each *Regime* corresponds to the rate balance of chain deposition on the crystal surface. *Regime I* occurs at low supercooling, where chain mobility is high and surface nucleation is low (*i.e.* nucleation rate is much smaller than spreading rate). As temperature decreases *Regime* II is reached, at which chain mobility is diminished while surface nucleation increases (*i.e.* nucleation and spreading rate become comparable). At even larger supercooling, *Regime* III is encountered. In *Regime* III the nucleation rate is much larger than the spreading rate.

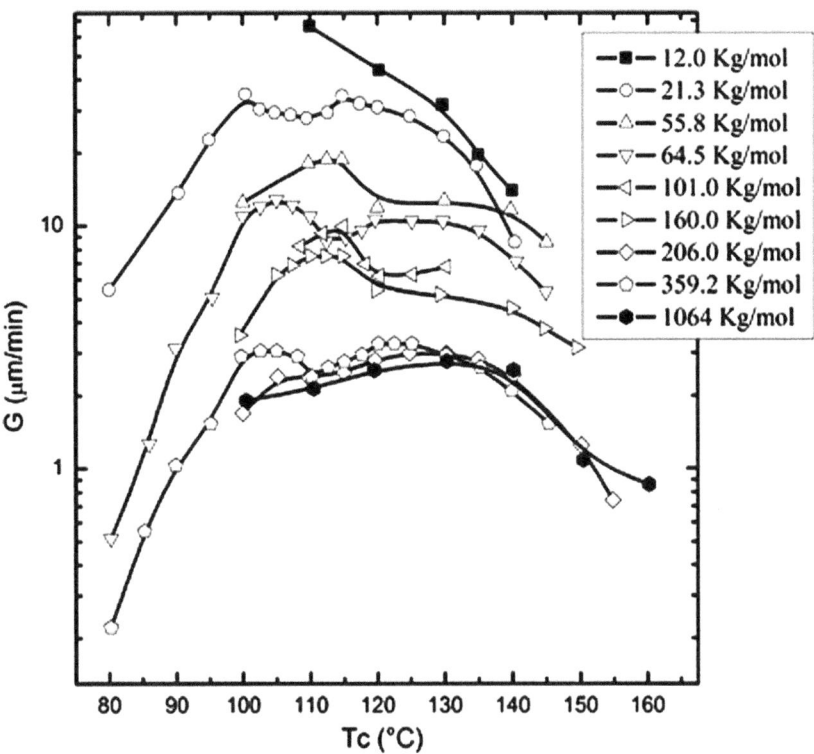

Figure 3.8 Spherulitic growth rate (G) as a function of crystallization temperature (T_c) for different weight-averaged molecular weights samples of PLLA.[3,60,62,63]

According to LH theory, the crystal growth rate (G) of linear flexible chains is given by:

$$G = G_0 \exp\left(-\frac{U^*}{R(T_c - T_\infty)}\right) \exp\left(\frac{-K_g}{T_c \Delta Tf}\right) \quad (6)$$

The first term represents the diffusion of chains to the growth front while the second is related to the secondary nucleation barrier. G_0 represents a pre-exponential factor, U^* is the activation energy for chain mobility, R is the gas constant, T_c is the isothermal crystallization temperature and $\Delta T = T_m^0 - T_c$ is the supercooling (T_m^0 is the equilibrium melting temperature). T_∞ is the temperature where viscous flow ceases ($\Delta T_g - 30K$) and f is a temperature correction factor defined as $2T_c / (T_m^0 + T_c)$, while K_g is the nucleation constant (which is proportional to the energy barrier for secondary nucleation) given by:

$$K_g = \frac{jb\sigma\sigma_e T_m^0}{k_B \Delta H_m} \quad (7)$$

where j is a constant value that depends on the crystallization *Regime* (*i.e.* it is equal to 2 in *Regime* II and 4 for both *Regimes* I and III), b is the layer thickness, σ is the lateral surface free energy, σ_e is the free energy of folding, k_B is the Boltzmann constant and ΔH_m is the melting heat. The values used in the literature to apply the LH theory for PLA crystallization are shown in Table 3.2.

When a plot of $LnG + U^*/R(T_c - T_\infty)$ as a function of $1/T_c\Delta Tf$ is made, linear trends should be obtained with a slope that is indicative of the crystallization *Regime* according to eqns (6) and (7) and the particular value of the j constant. The overall crystallization kinetics of PLA has been investigated by the LH theory and Figure 3.9 shows the LH plots for four different samples.

Vasanthakumari *et al.*[9] showed that at 163 °C, the crystallization behaviour of PLLA ($M_v = 150$ kg mol^{-1}) exhibits a sharp break in the LH plot (see Figure 3.5) that was interpreted as a change from *Regime* I (at low supercooling) to *Regime* II (at high supercooling). Kawai *et al.*[3] studied PLLA samples of $M_w = 206$ kg mol^{-1}, and reported that a change from *Regime* III to *Regime* II was occurring at 120 °C. The estimated values of K_g corresponded to 6.02×10^5 and 3.22×10^5 K^2 for below and above 120 °C, respectively, in accordance with previous results published by Tsuji *et al.*[63] The change in crystallization *Regimes* of PLLA has been found to depend on molecular weight, amount of D co-monomer and tacticity.[63,68] It is worth noting that the transition from *Regime* III to *Regime* II reported at 120 °C occurs without a morphological change, but the transition from *Regime* II to *Regime* I reported at 147 °C[68] occurs with a change in crystal morphology. At $T > 147$ °C hexagonal-shaped single crystals were formed while at $T < 147$ °C only spherulites were observed.

Experimental data of the three crystallization *Regimes* of PLA are summarized in Table 3.3. Di Lorenzo[62] proposed a classification based on three temperature ranges of growth rate data and K_g values of PLLA: (*i*) For 75 °C $\leq T_c \leq 100$ °C: $K_g = 4.38 \times 10^5$ K^2, (*ii*) for 108 °C $\leq T_c \leq 120$ °C: $K_g = 5.97 \times 10^5$ K^2, (*iii*) for $T_c \geq 128$ °C: $K_g = 1.85 \times 10^5$ K^2. However, the K_g^{III}/K_g^{II} ratio (Table 3.3) is not fully consistent with the theoretical value of 2 predicted by the LH model. Taking G data of PLLA samples with 101 kg mol^{-1}, corresponding to ranges (*i*) and (*ii*), the K_g^{III}/K_g^{II} ratios are 2.37 and 3.23, respectively. On the other hand, PLLA samples with twice this molecular weight exhibit a K_g^{III}/K_g^{II} ratio of 1.87 for range (*ii*).

Regardless of the molecular weight, the transition temperature from *Regime* III to *Regime* II appears to be very close to 120 °C.[3,58,62,68] Di Lorenzo suggests that in the interval 75 °C $\leq T_c \leq 120$ °C, the K_g^{III}/K_g^{II} ratio is 3.66. However, the K_g^{III}/K_g^{II} ratio becomes 1.97 using $T_c \geq 128$ °C. In this case, the *Regime* II–III transition temperature can be considered to occur at around 130 °C. Nevertheless, Di Lorenzo[62] considers that LH analysis described above contains uncertainties associated with the *Regimes*, even though there are other variables such as tacticity and molecular weight distribution that can influence the crystallization behaviour.

Table 3.2 Values used in LH theory.

Parameter	Description	Value		Ref.
U^*, J mol^{-1}	Activation energy for local motion	6.27×10^3		9
b, m	Surface nucleus thickness	5.17×10^{-10}		8
σ, J m^{-2}	Lateral surface free energy	12.03×10^{-3}		9
σ_e, J m^{-2}	Fold surface free energy	60.89×10^{-3}		9
ΔH_m^0, J g^{-1}	Melting enthalpy for 100% crystallinity	93	Flory model for solution grown crystals	7
	Flory model	203		19
	Baur	82		23
	Extrapolation to crystal density	135		14
		106		65
		91 ± 3		66
T_m^0, °C	Equilibrium melting temperature	215	$M_v = 550$	8
	Hoffman–Weeks	207	$M_v = 68$	9
		205	$M_w = 13000$	43
		206	$M_w = 640$	67
		224.8	$M_w = 22.1$	68
	Marand	226.6	$M_w = 68.6$	
		227.1	$M_w = 151.3$	
		227.4	$M_w = 701.4$	
	Modified Hoffman–Weeks	215	$M_w = 151.3$	20
	Hoffman–Weeks (linear extrapolation)	180.3	$M_w = 104.3$	69
		175.4	$M_w = 77.8$	
		171.4	$M_w = 36$	
	Hoffman–Weeks (non-linear extrapolation)	201.3	$M_w = 104.3$	
		192.2	$M_w = 77.8$	
		183.5	$M_w = 36$	

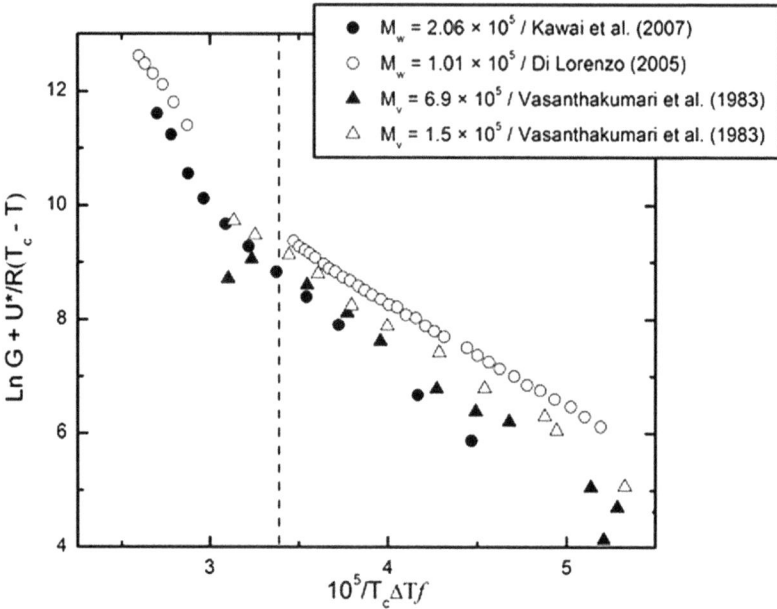

Figure 3.9   Lauritzen–Hoffman plots for the indicated PLLA samples.[3,9,62]

3.5.3   Overall Crystallization Kinetics

From isothermal crystallization experiments performed by DSC, the inverse of the experimental half-crystallization time can be obtained, a quantity that is proportional to the overall crystallization rate since it includes contributions from primary nucleation and growth. Figure 3.10 shows the inverse of half-crystallization time $(\tau_{50\%})$ as a function of crystallization temperature for different PLA samples (different molecular weights and D-isomer concentration). Some of the curves presented in Figure 3.9 also contain the two maxima that have the same origin as in Figure 3.8. Di Lorenzo[62] indicates that PLAs display discontinuity in crystallization rate at around 118 °C.

Generally, the inverse of half-crystallization time decreases with the D-isomer content, as reported by Kolstad,[26] who indicates that the reduction is approximately 45% with increasing 1% in meso-lactide content. On the other hand, the effect of the molecular weight and similar D-isomer concentration shows that the inverse of half-crystallization time decreased with molecular weight as a consequence of the low chain mobility.[28,37,70]

Even though the discontinuity reported in some curves in Figure 3.10 may correspond with the dual peaks reported for the spherulite growth rate plots (Figure 3.8), in some cases the results are not as clear cut. This is probably the result of the influence of nucleation on the overall crystallization rates.

The data obtained during the isothermal crystallization experiments by DSC can be fitted by the Avrami equation,[71,72] which can be expressed as follows:

$$1 - V_c(t - t_0) = \exp(-K(t - t_0)^n) \tag{8}$$

Table 3.3 Reported values of Lauritzen and Hoffman fitting parameters for PLA.

PLA	M (kg mol⁻¹)	T_g (K)	$K_g \times 10^{-5}$ (K²)	K_g^{III}/K_g^{II}	Regime	T_b (K)	G_0 (µm min⁻¹)	T_c range, K	Ref.
	690	333	2.34		II		1.56×10^7	$387 < T_c < 439$	9
	350	331	2.29		II		1.63×10^7		
	260	330	2.37		II	436 (II-I)	2.40×10^7		
	150	328	2.44		II		3.38×10^7		
			4.87		I		2.13×10^{13}		
	206	336	6.02	1.87	III	393 (III-II)		$373 < T_c < 428$	3
			3.22		II				
PLLA	101	336	5.97	3.23	III	393 (III-II)		$348 < T_c < 448$	58,62
			4.38	2.37	II				
			1.85						
	17	332	4.64		II		2.70×10^{10}	$385 < T_c < 435$	68
	49	334	4.79		II	393 (III-II)	2.33×10^{10}		
	89	335	4.97		II		2.21×10^{10}		
	334	335	5.01		II	420 (II-I)	1.53×10^{10}		

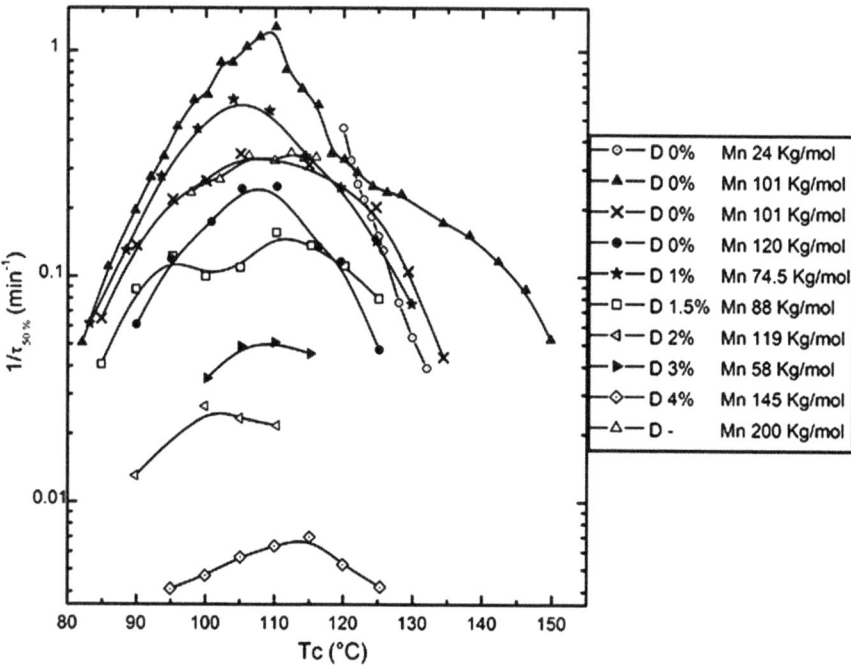

Figure 3.10 Inverse of crystallization half-time as a function of crystallization temperature, for PLAs with different molecular weight and D-isomer concentrations.[47–49,68]

where t is the experimental time, t_0 is the induction time, V_c is the relative volumetric transformed fraction, n is the Avrami index and K is the overall crystallization rate constant. An Avrami exponent value for PLA of around 2.6[70] has been reported. If the value is close to 3, this indicates the instantaneous growth of spherulites, a common occurrence in PLAs.[78] Table 3.4 reports a wider range of values fluctuating from 2–3.5 for different types of PLAs.

Table 3.5 reports the Avrami fitting parameters for PLA-based blends, where the parameters are based on the crystallization of the PLA matrix. Shibata *et al.*[70] studied blends of PLLA with poly(butylene succinate-*co*-l-lactate) (PBSL) and poly(butylene succinate) (PBS) and showed that the addition of small amounts of PBSL enhanced the isothermal and non-isothermal PLLA crystallization, but the addition of PBS was much less effective in modifying PLLA crystallization.

Auliawan *et al.*[79] studied ternary polymer blends of PLLA, poly(methyl methacrylate) (PMMA) and poly(ethylene oxide) (PEO) as matrix for nanocomposites. Their conclusions were that nanoclays enhanced the non-isothermal crystallization of the blend, since this ternary polymer blend hardly crystallized when cooled from the melt. The addition of vermiculite at loadings larger than 1% by weight retarded the crystallization process, while montmorillonite enhanced the crystallization process to some extent before retarding it.

Table 3.4 Reported values of K and n for PLA.

K (s^{-n})	n	%D	M_w (kg mol^{-1})	Supplier	Ref.
2.47×10^{-8}	2.59	1.0	–	Mitsui Chem (Lacea H100)	70
1.97×10^{-9}	2.50	1.3	207	Natureworks (PLA 4032D)	59
1.70×10^{-6}	2.40	1.3	207	Natureworks (PLA 4032D)	73
$(0.52–12.88)\times10^{-9}$	2.76–3.23	–	720	Boegringer Ingelgeim (Resomer L214)	67
2.62×10^{-7}	2.19	–	120	Shimadzu, Inc	74
–	3.5	–	63	–	75
$(0.023–6.25)\times10^{-8}$	2.76–3.11	–	737	Boegringer Ingelgeim (Resomer RS210)	76
$(11.08–4.22)\times10^{-6}$	2.0	–	116	Cargill Dow LLD Chem. Co.	77
$5.53\times10^{-11}–1.13\times10^{-6}$	2.51–2.91	–	221	Lakeshore Biomaterials (Medisorb 100L)	78

Table 3.5 Blends based on PLA.

Blend	ΔT_g (°C)	ΔT_c (°C)	ΔH_c (J g^{-1})	ΔT_m (°C)	ΔH_m (J g^{-1})	K (s^{-n})	n	Ref.
PLLA/PBSL 99/1	−0.9	−4.0	+16.7	−1.7	+18.3	5.38×10^{-8}	2.67	70
PLLA/PBSL 95/5	−2.3	−12.1	+33.2	−2.1	+31.1	2.48×10^{-7}	2.69	70
PLLA/PBSL 90/10	−2.4	−9.0	+28.4	−0.7	+24.6	2.86×10^{-7}	2.81	70
PLLA/PBS 99/1	+0.1	−2.7	+2.5	−0.5	−2.5	2.07×10^{-8}	2.66	70
PLLA/PBS 99/5	−0.6	−0.7	+14.9	0.3	+9.6	1.20×10^{-7}	2.45	70
PLLA/PBS 99/10	−2.4	−7.4	+30.6	−1.7	+26.4	1.77×10^{-7}	2.36	70
PLA/PBAT 80/20	−1.5	–	24.4	+1.4	+20.7	2.58×10^{-9}	2.5	59
PLA/PBAT 60/40	−2.3	–	19.1	+1.9	+13.9	1.28×10^{-8}	2.4	59
PLA/PBAT 75/25	−1.7	−2.0	–	−2.1	–	8.80×10^{-7}	2.4	36
PLA/PBAT 50/50	−0.9	0.0	–	−0.7	–	1.50×10^{-7}	2.9	36
PLA/DHP (98/2)	+2	–	+32	–	–	7.17×10^{-8}	2.65	80
PLA/Starch (98/2)	+2	–	+33	–	–	4.41×10^{-7}	2.47	80
PLA/ 15 TPP	–	–	–	–	–	1.50×10^{-7}	2.6	73
PLA/ 1.2 Talc	–	–	–	–	–	2.00×10^{-5}	2.8	73
PLA/ 15 TPP/ 1.2 Talc	–	–	–	–	–	3.75×10^{-6}	3.0	73

Xiao *et al.*[59] prepared blends of PLA with poly(butylene adipate-*co*-terephthalate) (PBAT), and followed their isothermal crystallization. They showed that the Avrami exponent was almost unchanged. However, the crystallization rate increased with the PBAT content.

Blends of PLA with a dendritic hyperbranched polymer (DHP) and starch were studied by Zhang *et al.*[80] They showed that addition of DHP and starch increased both the cold crystallization ability and the isothermal crystallization rate of PLA.

Talc and triphenyl phosphate (TPP) have been reported as nucleation agents for PLLA. The isothermal kinetics of neat PLA and its blends with TPP and talc were reported by Xiao *et al.*[73] However, the crystallization rate decreased with the incorporation of TPP.

Additionally, Quero *et al.*[36] studied the isothermal cold-crystallization of PLA/PBAT blends with and without acetyl tributyl citrate (ATBC). SEM results showed that the blends exhibited two phases, but partial miscibility: changes in both, T_g and T_m, of the PLA phase, a relationship of the spherulitic growth rate with the blend composition and the occlusion of PBAT droplets within PLA spherulites. ATBC acted like a plasticizer for both phases (neat PLA and PBAT). In the case of blends, ATBC prefers to be included inside the PBAT-rich phase. There was a synergistic effect on the overall crystallization rate of PLA when both ATBC and PBAT were present in the blend.

3.6 Block Copolymers Based on PLA

Table 3.6 presents some representative examples of the wide range of copolymers containing PLA units that have been prepared in recent literature. We will focus in this section on a few selected works that have dealt with the crystallization of PLA-based copolymers. Several reviews on the crystallization of block copolymers have also been published.[81–117]

3.6.1 PLLA–PEO Copolymers

PLLA-*b*-PEO diblock copolymers were initially studied by Younes and Cohn in 1987[118] and are probably one of the most studied systems. In these copolymers, the PEO block is expected to improve the physical properties of PLA, as well as to enhance its degradation resistance and improve drug release properties.[119] The PLA-*b*-PEO-*b*-PLA triblock copolymer has also been much studied.[109,120–129]

These block copolymers are usually miscible or weakly segregated in the melt (depending on their segregation strength, a function of the average molecular weight of each block time the Flory–Huggins interaction parameter).[40,121,122,125,126,130–137] Microphase separation takes place during crystallization.[120,122,125,130] Miscibility in the melt has also been reported for some high-molecular-weight blends of PLLA and PEO.[133] The miscibility has been explored by changes in T_g of one or both blocks by DSC[118,123,126] and DMTA.[120] However, as the T_g of PLA and the T_m of PEO nearly overlap in many cases, it is difficult to evaluate miscibilities from simple T_g measurements. Nevertheless, if data published by different groups are analyzed,[120,125,126,130] both the T_g of PLA and that of PEO blocks are a function of the PEO composition and exhibit trends close to the predictions of the Fox equation for miscible systems.

Table 3.6 Selected research papers about copolymers based on PLA published since 2000.

Copolymer	Authors [Ref]	Year	Applications
PLA copolymer based on lactide acid (LA) and glycolic acid (GA) PLA–*co*–GA, PLA–*b*–GA: The first commercialised biodegradable polymers	Barakat *et al.*[88] Schwach *et al.*[89] Dechy-Cabaret *et al.*[90] Min *et al*[91] Tsuji *et al.*[92]	2001 2002 2004 2007 2012	Medicine and Pharmaceutical Applications
PLLA–*b*–PDLA PLLA–*co*–PDLA	Yuryev *et al.*[93] Tsuji *et al.*[94] Tsuji *et al.*[92]	2008 2010 2012	Nanotechnology and Biomedical Applications
PLLA–*b*–PCL PLLA–*co*–PCL PDLA–*co*–PCL	Stolt *et al.*[95] Broström *et al.*[96] Tsuji *et al.*[97] Hamley *et al.*[98] Hamley *et al.*[99] Declercq *et al.*[100] Müller *et al.*[84] Laredo *et al.*[101] Ajami-Henriquez *et al.*[102] Castillo *et al.*[37] Yilmaz *et al.*[103] Casas *et al.*[104] Purnama *et al.*[105]	2004 2004 2005 2005 2006 2006 2007 2007 2008 2010 2011 2011 2012	
PLLA–*b*–PEO PLDA–b–PEO	Barakat *et al.*[88] Klok *et al.*[106] Deng *et al.*[107] Onyari *et al.*[108] Li *et al.*[109] Yu *et al.*[110] Yang *et al.*[48] Min *et al.*[91] Agrawal *et al.*[111] Huang *et al.*[112] Castillo *et al.*[85] Dimitrov *et al.*[113]	2001 2002 2002 2003 2003 2006 2006 2007 2008 2008 2009 2013	Biomedical, Drug Delivery Applications
PDLA–*b*–PE PLLA–*b*–PE	Müler *et al.*[114] Müler *et al.*[115] Castillo *et al.*[116]	2006 2006 2008	Biomedical Applications
Poly(D– and L–lactide)–*b*– Poly(*N,N*–dimethylamino–2– ethyl methacrylate)	Michell *et al.*[117]	2011	Biomedical Materials Applications

Sun *et al.*[47] studied the crystallization of PLLA-*b*-PEO diblock copolymers with different compositions (*i.e.* 68, 44 and 29 wt% of PEO) by DSC, WAXS, PLOM and AFM. A PEO block length of 5 kg mol^{-1} was constant for all the block copolymers studied. The T_c of the PLLA block increased with its content. However, for contents of PEO lower than 44 wt% T_c values higher than those for the homopolymer were found, probably because of the higher molecular weight of the PLLA blocks (6.3, 12 kg mol^{-1}) as compared to the homopolymer (4.8 kg mol^{-1}). Melting temperatures of the PLLA block were influenced by the plasticizing effect of PEO and the variation in molecular weights. In the case of the PEO block, the previous crystallization of PLLA dominates the superstructural morphology and for low PEO contents fractionated crystallization[47,131] of the minor component revealed inter-lamellar confined crystallization of PEO within the templated PLLA spherulite superstructure. Similar results for PLLA-*b*-PCL have also been reported.[131]

Huang *et al.*[132] studied miscible methoxy poly(ethylene glycol)-*b*-poly(L-lactide) (MPEG-*b*-PLLA) diblock copolymers with 39, 57 and 67 wt% of MPEG. They reported an acceleration of the crystallization kinetics of PLLA with MPEG addition at low PEO contents. At higher MPEG amounts, shifts of the crystallization rate *versus* T_c curves to lower temperatures were reported as a consequence of the decrease in $T_m^{\circ}{}_{PLLA}$ and T_g due to the presence of the MPEG phase and/or decreases in PLLA molecular weight.

The relative length between blocks markedly influences the crystallization behaviour of PLLA-*b*-PEO copolymers.[129,130] PEO cannot crystallize when its content is lower than 20 or 10 wt%.[125–133,129,131] It was already mentioned above that at contents around 30–40% PEO, fractionated crystallization can occur.[131]

Kim *et al.*[127] studied the crystallization kinetics of the PLLA block (denoted L) within L$_{50}$EO$_{50}$ and L$_{10}$EO$_{80}$L$_{10}$ (M_n of the PLLA block was 5.2 kg mol^{-1} and the subscripts denote the composition in wt%) by employing real time WAXS at 40 °C (a temperature too high for the PEO block to crystallize). They found that the PLLA block crystallization was slowed down as compared to the neat homo-PLLA by the presence of the molten covalently bonded PEO block.

The morphology of block copolymers that are based on PEO and PLLA has been reported in several studies.[47,48,107,122,131] It is common to observe well-developed spherulites by PLOM. Both PLLA and PEO homopolymers crystallize in negative spherulites with well-defined Maltese cross patterns. However, in block copolymers with PEO contents of 29 to 68 wt%, banded spherulites have been reported during the crystallization of the PLLA block. Similar morphologies for diblock copolymers with 14 and 24 wt% PEO were reported by Huang *et al.*[120] In the case of diblock and triblock copolymers with PEO contents of 10 and 20 wt%, spherulites with Maltese cross extinction patterns have been observed but without banding.[131] Shin *et al.*[122] also reported that a PLLA-*b*-PEO-*b*-PLLA triblock copolymer and a homopolymer blend of equal composition formed banded spherulites at the same crystallization temperature range. Finally, at high supercoolings, at which the PEO block can also crystallize, the morphology of the banded spherulites does not change on the microscopic length scale.[47] As a consequence, it is accepted that the crystalline lamellae of PLLA previously formed strongly templates the crystallization of the PEO chains.

3.6.2 Copolymers *versus* Blends of PLLA and PCL

In this section, we have chosen an interesting example, where blends and block copolymers of poly(L-lactic acid) (PLLA) and poly(ε-caprolactone) will be presented. The PLLA and PCL blends are immiscible and, concomitantly, they exhibit two crystallization and melting temperatures that are almost identical to those obtained for the parent homopolymers irrespective of the composition. Figure 3.11 shows additional evidence of blend immiscibility by PLOM. At 120 °C, see Figure 3.11 (top), only the PLLA blend component can crystallize. The PLLA in the blend forms well-developed negative spherulites that engulf dark isotropic regions (one of which has been marked by a white ellipse) where the PCL blend component is segregated and molten. These dark PCL molten regions (without birefringence) are clearly filled with small PCL spherulites when the sample is quenched to 35 °C, a temperature at which PCL can crystallize (see Figure 3.11 (bottom)).

Figure 3.12 (left) shows how the crystallization and melting temperature of the PLLA block within PLLA-*b*-PCL diblock copolymers is depressed when the PCL content increases. As opposed to the blends, the PLLA-*b*-PCL diblock copolymer systems are partially miscible or weakly segregated. Therefore, a

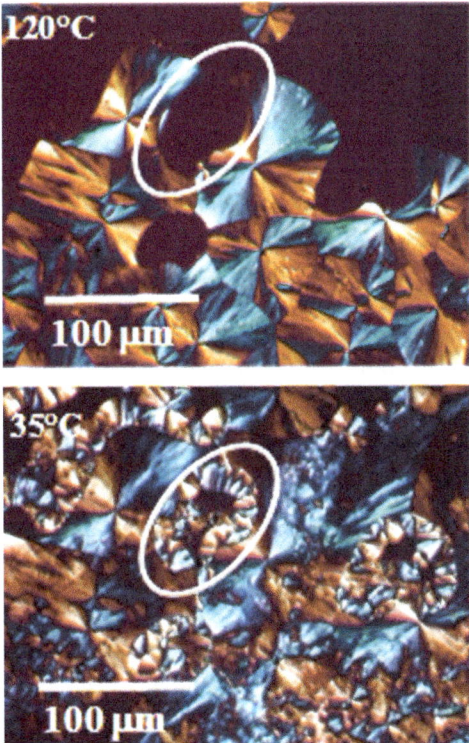

Figure 3.11 PLOM micrographs of a 32/68 PLLA/PCL blend at the indicated iso-thermal crystallization temperatures (modified from Ref. 37).

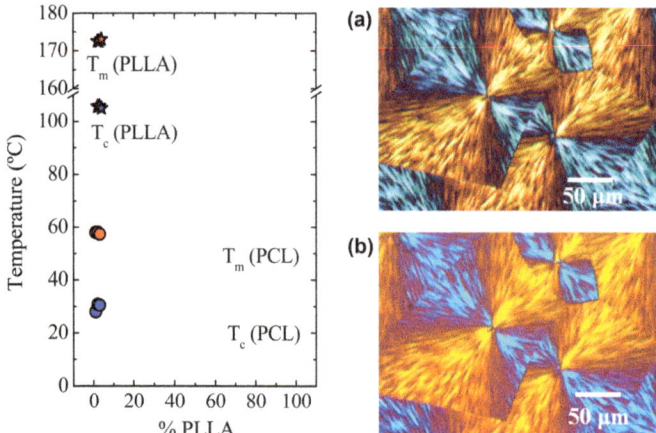

Figure 3.12   Left: Peak crystallization and melting temperatures for the PLLA and
PCL block for a wide range of PLLA-*b*-PCL diblock copolymers *versus*
PLLA content. Right: PLOM micrographs for a) $L_{32}{}^{7}C_{68}{}^{15}$ after 30 min. at
122 °C, b) the same sample after 15 min. additional time at 42 °C
(modified from Ref. 107).

certain level of thermodynamic interactions between the blocks is expected
and that allows PCL to act as a plasticizer for PLLA.

Figure 3.12 (right) shows PLOM micrographs of one PLLA-*b*-PCL diblock
copolymer at different crystallization temperature. The diblock copolymer
$L_{32}{}^{7}C_{68}{}^{15}$ (subscripts indicate the composition in weight percent and
superscripts the number average molecular weight in kg mol^{-1} of each
block) was first crystallized at 122 °C, a temperature at which the PCL chains
are molten. The PLOM field was filled with PLLA spherulites as shown in
Figure 3.12 (right). These spherulites are well developed and exhibit clear
Maltese crosses and some irregular banding pattern. They contain within the
amorphous interlamellar regions molten PCL block chains mixed together
with those PLLA block chains that remained amorphous. Figure 3.12b (right)
shows that once the sample is cooled to 42 °C, the PCL block crystallizes
within the interlamellar regions, simply filling the template spherulitic
superstructure formed by the PLLA block previous crystallization.
Such mixed spherulitic structures are composed of interdigitized PLLA
and PCL lamellae. Both micrographs in Figure 3.12 (right) are very similar
except for the fact that once the PCL crystallized at 42 °C, the birefringence
value changed (it became more negative), thereby causing a brighter
micrograph.

Figure 3.13 shows a complete series of mostly previously unpublished
micrographs of the remarkable ability of the PLLA block (and homo-PLLA) to
form spherulites at temperatures where the PCL block is molten, thus
templating the spherulitic morphology that will characterize the sample
upon cooling, since PCL crystallizes within the interlamellar region of
the previously formed spherulites.[99–101] The PLLA block crystallized in a

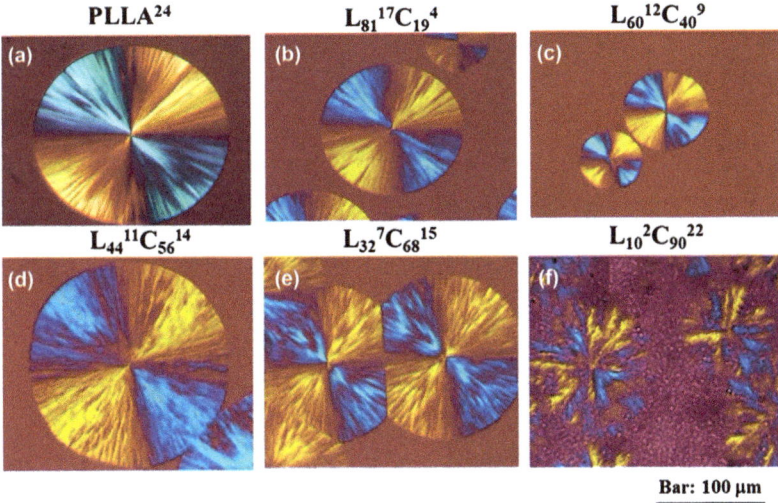

Figure 3.13 Polarized light optical micrographs during isothermal crystallization of a copolymer PLLA-PCL: a) PLLA[24] at 122 °C, b) $L_{81}^{17}C_{19}^{4}$ at 120 °C, c) $L_{60}^{12}C_{40}^{9}$ at 120 °C, d) $L_{44}^{11}C_{56}^{14}$ at 122 °C, e) $L_{32}^{7}C_{68}^{15}$ at 120 °C, f) $L_{10}^{2}C_{90}^{22}$ at 115 °C.

well-defined spherulitic morphology regardless of composition[37,99–101] except in the extreme case of the $L_{10}C_{90}^{24}$ diblock copolymer. In this case, only 10% of the sample is composed of PLLA chains and they still try to fill up the entire field of view with axialite-like superstructural crystal aggregates that are formed above the PCL melting point. In fact, the PLLA block can only achieve a 50% crystallinity (but the sample only contains 10% of PLLA) and therefore 95% of the sample remains in the melt (*i.e.* 50% of the 10% PLLA block plus 90% of the PCL block).

3.6.3 PE-*b*-PLA Block Copolymers

The last example that we will present is that of strongly segregated PE-*b*-PLA diblock copolymers.[114–116] Because of the strong segregation neither PLLA nor PE blocks are expected to form spherulites because of confined crystallization within the melt microdomains. Nevertheless, Ring *et al.*[134] reported the synthesis of PE-*b*-PLLA block copolymers of 12–29 kg mol^{-1} and 89–96 wt% of PLLA, combining catalytic ethylene oligomerization with "coordinate-insertion" ring-opening polymerization. They observed microphase separation by AFM even though spherulites were detected by PLOM during isothermal crystallization after erasing thermal history at 200 °C for 3 min. In this particular case, the PLLA content in the diblock copolymer is very high (and this may have triggered a break-out process in spite of the strong segregation) or some PLLA degradation may have occurred at 200 °C making the thermodynamic segregation weaker.

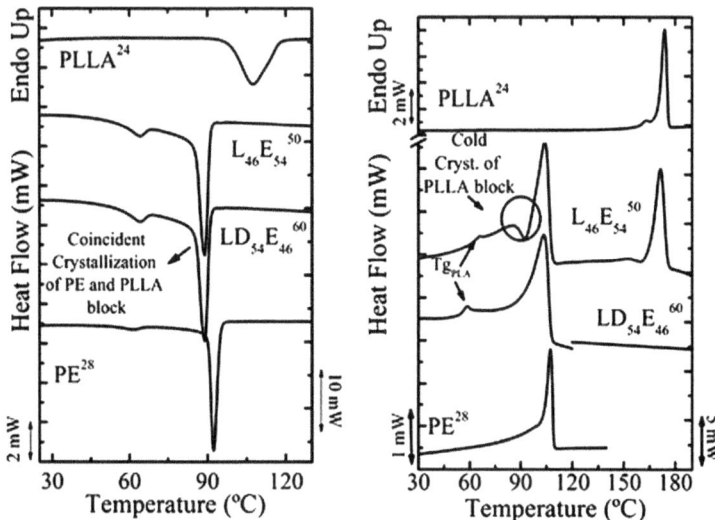

Figure 3.14 Left: curves of DSC cooling scans at 10 °C min⁻¹. Right: subsequent heating scans (modified from Refs. 122 and 123).

Müller *et al.*[114,115] and Castillo *et al.*[116] have studied in detail double crystalline PLLA-*b*-PE and amorphous-crystalline PLDA-*b*-PE diblock copolymers by DSC, PLOM, TEM, SAXS and WAXS. Figure 3.14 shows DSC cooling and subsequent heating scans at 10 °C min⁻¹ for PLLA-*b*-PE, PLDA-*b*-PE and homo-PLLA and homo-PE. The copolymers are nearly symmetric in composition and exhibit well-defined lamellar microstructure that was characterized by SAXS and TEM. The PLLA-*b*-PE diblock displays a single coincident exotherm upon cooling from the melt (as indicated with an arrow and text in Figure 3.14), even though during the subsequent heating run, two well-defined melting endotherms can be seen. Müller *et al.*[114,115] demonstrated, by using multiple experimental techniques, that the crystallization kinetics of the PLLA block is slowed down because of the covalently bonded and molten PE block. Therefore, during cooling from the melt at 10 °C min⁻¹ the PLLA block delayed crystallization process overlaps with that of the PE block which starts at lower temperatures. In fact, the crystallization process of each block can be separated in temperatures by the use of a slower cooling rate[114] or by self-nucleating the PLLA block[116] in a similar way to that reported previously for PPDX-*b*-PCL diblock copolymers.[135]

Figure 3.14 (right) shows that the melting trace of the PLLA-*b*-PE diblock copolymer shows a bimodal endotherm at lower temperatures (in the temperature range corresponding to the melting of the PE block crystals). This is caused by the interference of the PLLA block cold crystallization exotherm that occurs at temperatures above T_g. This can be easily demonstrated by especially designed DSC experiments.

The PLLA block within PLLA-*b*-PE does not form spherulites at any examined temperature, as expected for a strongly segregated system, since

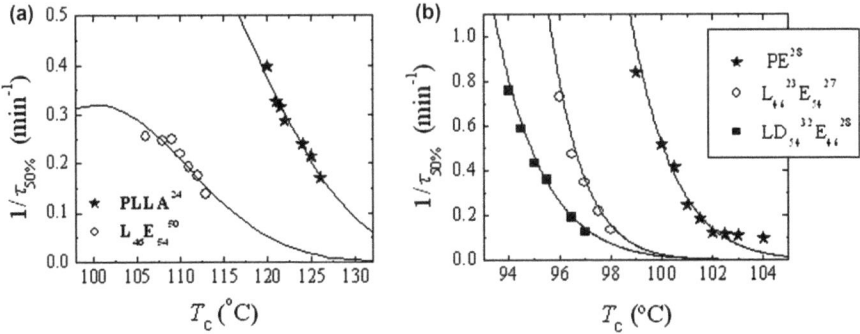

Figure 3.15 a) The inverse of the half-crystallization time *versus* isothermal crystallization temperature for PLLA and the PLLA block of the indicated diblock copolymer. b) Inverse of the half-crystallization time *versus* isothermal crystallization temperature for the PE block within the indicated copolymers and for PE homopolymer (modified from Refs. 122 and 123).

the crystallization of each component takes place in the confined space defined by the lamellar microdomains, *i.e.* no break-out can occur.[114,116]

Figure 3.15a shows the values of the overall crystallization rates, expressed as $1/\tau_{50\%}$ as a function of T_c for a PLLA-*b*-PE and a corresponding homo-PLLA.[114,115] The PLLA block within the copolymer clearly crystallizes at much slower rates than homo-PLLA when similar crystallization temperatures are considered by extrapolation. Such a decrease in the overall crystallization rate of the PLLA block within the copolymer together with the higher supercooling needed for crystallization is responsible for the coincident crystallization effect that can be observed when the PLLA-*b*-PE diblock copolymer is cooled down from the melt at rates larger than 2 °C min⁻¹ (see Figure 3.15a).[114–116]

Figure 3.15b shows the overall crystallization kinetics data for the PE block within PLLA-*b*-PE (after the PLLA block was crystallized until saturation), PDLA-*b*-PE and also for a corresponding homo-PE. The crystallization rate of the PE block is reduced, as compared to homo-PE, regardless of whether it is covalently linked to amorphous PLDA or to semicrystalline PLLA. However, in the case of PLLA-*b*-PE, since the PLLA block was crystallized to saturation first, a nucleation effect of the PLLA crystals on the PE block was observed. This is obvious if we compare the PE crystallization in both PLDA-*b*-PE and PLLA-*b*-PE block copolymers. As the result of this nucleating effect, even though the crystallization rate of the PE block is depressed in the PLLA-*b*-PE diblock copolymer, the nucleation effect compensates this rate reduction and, in the end, the PE block attached to the semicrystalline PLLA can crystallize faster than that attached to amorphous PLDA.[136]

It is noteworthy that in weakly segregated diblock copolymers like PPDX-*b*-PCL, the rubbery PCL block has a dramatic effect depressing PPDX crystallization rate by at least one order of magnitude.[135,137]

Castillo *et al.*[116] demonstrated by SAXS that the crystallization of the PE block within PLLA-*b*-PE and PDLA-*b*-PE is strictly confined within the pre-existing lamellar microdomains in the melt regardless of whether $T_{c\ PE} > T_{g\ PLDA,PLLA}$ (soft confinement) or $T_{c\ PE} < T_{g\ PLDA,PLLA}$ (hard confinement).

3.7   Conclusions and Outlook

The morphology and crystallization behaviour of PLA have been extensively studied by WAXS/SAXS, TEM, PLOM, AFM and DSC. Literature results indicate some general features:

1) Single crystals of PLA can be formed in different crystalline forms: α-, α'-, β- and γ-forms, although the most common is the α-form. Additionally, single crystals containing 2.75% and 10.2% D-units can still be formed. However, if the content of D-units is higher than 15%, single crystals are not viable.

2) The glass and melting temperatures of PLA are a function of the chain lengths and of the D-lactide content incorporated into the repeating units.

3) PLLA generally form non-banded spherulites from the melt that exhibit negative birefringence. Banding can appear after special annealing conditions or in blends and block copolymers.

4) Crystallization rate increases with decreasing molecular weight of PLA in the range of molecular weights commonly reported in the literature.

5) The spherulite growth rate (and overall crystallization rate in some cases) of PLLA exhibits a discontinuity from the usual bell-shaped curve of polymer crystal growth (for intermediate molecular weight samples), when it is examined in a large supercooling range. This behaviour has been correlated to a transition in crystallization *Regimes* based on the Lauritzen and Hoffman theory. However, the origin of this discontinuity is still controversial.

6) The crystallization rate of PLAs can be tailored by blending with either plasticizers or miscible polymers with low T_g values (*e.g.* PEO) and by copolymerization. Block copolymers can provide extremely complex morphologies and properties depending on whether the neighbouring block can crystallize or not. So, in block copolymers the number of variables that can be tuned to tailor PLAs properties is very large: composition, molecular weight, segregation strength, capability of the second block to crystallize or to nucleate the first block, *etc.*

If we consider the large number of possible applications of PLAs in the biomedical field and in biodegradable packaging applications, to name just two of the most important areas, the study of their crystallization will always be decisive to tailor their properties. Additionally, the production of a high-impact PLA with improved thermal stability and processability is still under investigation. Crystallization is likely to play a vital role in the development of future PLA-based materials.

References

1. P. De Santis and A. J. Kovacs, *Biopolymers*, 1968, **6**, 299.
2. J. Zhang, Y. Duan, H. Sato, H. Tsuji, I. Noda, S. Yan and Y. Ozaki, *Macromolecules*, 2005, **38**, 8012.
3. T. Kawai, N. Rahman, G. Matsuba, K. Nishida, T. Kanaya, M. Nakano, H. Okamoto, J. Kawada, A. Usuki, N. Honma, K. Nakajima and M. Matsuda, *Macromolecules*, 2007, **40**, 9463.
4. J. Zhang, K. Tashiro, H. Tsuji and A. J. Domb, *Macromolecules*, 2008, **41**, 1352.
5. B. Eling, S. Gogolewski and A. J. Pennings, *Polymer.*, 1982, **23**, 1587.
6. L. Cartier, T. Okihara, Y. Ikada, H. Tsuji, J. Puiggali and B. Lotz, *Polymer*, 2000, **41**, 8909.
7. E. Fischer, H. Sterzel and G. Wegner, *Kolloid Z. Z. Polym*, 1973, **251**, 980.
8. B. Kalb and A. J. Pennings, *Polymer*, 1980, **21**, 607.
9. R. Vasanthakumari and A. Pennings, *Polymer*, 1983, **24**, 175.
10. T. Okihara, M. Tsuji, A. Kawaguchi, I. Katayama, H. Tsuji, S. Hyon and Y. Ikada, *J. Polymer Sci. Polymer Phys.*, 1991, **30**, 119.
11. L. Cartier, T. Okihara and B. Lotz, *Macromolecules*, 1997, **30**, 6313.
12. W. Hoogsteen, A. R. Postema, A. J. Pennings, G. Ten Brinke and P. Zugenmaier, *Macromolecules*, 1990, **23**, 634.
13. J. Kobayashi, T. Asahi, M. Ichiki, A. Oikawa, H. Suzuki, T. Watanabe, E. Fukada and Y. Shikinami, *J. Appl. Phys.*, 1995, **77**, 2957.
14. T. Miyata and T. Masuko, *Polymer*, 1998, **39**, 5515.
15. J. Puiggali, Y. Ikada, H. Tsuji, L. Cartier, T. Okihara and B. Lotz, *Polymer*, 2000, **41**, 8921.
16. D. Sawai, K. Takahashi, T. Imamura, K. Nakamura, T. Kanamoto and S. Hyon, *J. Polymer Sci. Polymer Phys.*, 2002, **40**, 95.
17. D. Brizzolara, H. J. Cantow, K. Diederichs, E. Keller and A. J. Domb, *Macromolecules*, 1996, **29**, 191.
18. D. Cohn, H. Younes and G. Marom, *Polymer*, 1987, **28**, 2018.
19. H. Tsuji and Y. Ikada, *Macromol. Chem. Phys.*, 1996, **197**, 3483.
20. H. Abe, M. Harigaya, Y. Kikkawa, T. Tsuge and Y. Doi, *Biomacromolecules*, 2005, **6**, 457.
21. J. Xu, B. H. Guo, J. J. Zhou, L. Li, J. Wu and M. Kowalczuk, *Polymer*, 2005, **46**, 9176.
22. J. Ahmed, J. Zhang, Z. Song and S. Varshney, *J. Ther. Anal. Calorim.*, 2009, **95**, 957.
23. K. Jamshidi and Y. Ikada, *Polymer*, 1988, **29**, 2229.
24. I. Engelberg and J. Kohn, *Biomaterials*, 1991, **12**, 292.
25. D. Witzke, PhD Thesis, Department of Chemical Engineering, MI Michigan State University, 1997.
26. J. J. Kolstad, *J. Appl. Polymer Sci.*, 1996, **62**, 1079.

27. D. M. Bigg, *Adv. Polymer Tech.*, 2005, **24**, 69.
28. D. Garlotta, *J. Polymer. Environ.*, 2001, **9**, 63.
29. S. Saeidlou, M. A. Huneault, H. Li and C. B. Park, *Progr. Polymer Sci.*, 2012, **37**, 1657.
30. L. Pitet, S. Hait, T. Lanyk and D. Knauss, *Macromolecules*, 2007, **40**, 2327.
31. A. Celli, M. Scandolat, C. G. Ciamician, C. Studio and V. Selrni, *Polymer*, 1992, **33**, 2699.
32. L. T. Lim, R. Auras and M. Rubino, *Progr. Polymer Sci.*, 2008, **33**, 820.
33. R. Auras, B. Harte and S. Selke, *Macromolecular Biosciences*, 2004, **4**, 835.
34. J. Sarasua, L. Arraiza, P. Balerdi and I. Maiza, *J. Mater. Sci.*, 2005, **40**, 1855.
35. G. Kortaberria, C. Marieta, A. Jimeno, P. Arruti and I. Mondragon, *J. Microsc.*, 2006, **224**, 277.
36. E. Quero, A. Müller, F. Signori, M. Coltelli and S. Bronco, *Macromol. Chem. Phys.*, 2012, **213**, 36.
37. R. V. Castillo, A. J. Müller, J. M. Raquez and P. Dubois, *Macromolecules*, 2010, **43**, 4149.
38. Y. Yuryev, P. Wood-Adams, M. C. Heuzey, C. Dubois and J. Brisson, *Polymer*, 2008, **49**, 2306.
39. Y. Wang and J. Mano, *Appl. Polymer Sci.*, 2007, **105**, 3500.
40. Y. Wang, J. Gómez, M. Salmerón and J. Mano, *Macromolecules*, 2005, **38**, 4712.
41. M. Gazzano, M. L. Focarete, C. Riekel and M. Scandola, *Biomacromolecules*, 2004, **5**, 553.
42. M. Yasuniwa, S. Tsubakihara, K. Iura, Y. Ono, Y. Dan and K. Takahashi, *Polymer*, 2006, **47**, 7554.
43. H. Tsuji and Y. Ikada, *Polymer*, 1996, **37**, 595.
44. H. Tsuji, K. Tashiro, L. Bouapao and M. Hanesaka, *Polymer*, 2012, **53**, 747.
45. M. L. Focarete, G. Ceccorulli, M. Scandola and M. Kowalczuk, *Macromolecules*, 1998, **31**, 8485.
46. K. S. Kim, I. J. Chin, J. S. Yoon, H. J. Choi, D. C. Lee and K. H. Lee, *J. Appl. Polymer Sci.*, 2001, **82**, 3618.
47. J. Sun, Z. Hong, L. Yang, Z. Tang, X. Chen and X. Jing, *Polymer*, 2004, **45**, 5969.
48. J. Yang, T. Zhao, L. Liu, Y. Zhou, G. Li, E. Zhou and X. Chen, *Polymer J.*, 2006, **38**, 1251.
49. S. Huang, S. Jiang, L. An and X. Chen, *J. Polymer Sci. Polymer Phys.*, 2008, **46**, 1400.
50. J. L. Wang and C. M. Dong, *Macromol. Chem. Phys.*, 2006, **207**, 554.
51. U. Gedde, *Polymer Physics*, Chapman & Hall, London, 1995, ch. 7, pp. 151–155.
52. B. Wunderlich, *Macromolecular Physics*, Academic Press, New York, 1976, vol. 1, ch. 7, p. 325.

53. J. Schultz, *Polymer Materials Science*, Prentice-Hall, Englewood Cliffs, NJ, USA, 1974, ch. 3, pp. 141–209.
54. X. J. Li, Z. M. Li, G. J. Zhong and L. B. Li, *J. Macromol. Sci. Phys.*, 2008, **47**, 511.
55. H. Tsuji, H. Takai and S. K. Saha, *Polymer*, 2006, **47**, 3826.
56. H. Tsuji and Y. Ikada, *Polymer*, 1995, **36**, 2709.
57. Y. Kikkawa, H. Abe, M. Fujita, T. Iwata, Y. Inoue and Y. Doi, *Macromol. Chem. Phys.*, 2003, **204**, 1822.
58. M. L. Di Lorenzo, *Polymer*, 2001, **42**, 9441.
59. H. Xiao, W. Lu and J. Yeh, *J. Appl. Polymer Sci.*, 2009, **112**, 3754.
60. P. Pan, W. Kai, B. Zhu, T. Dong and Y. Inoue, *Macromolecules*, 2007, **40**, 6898.
61. J. Hoffman, G. T. Davis and J. Lauritzen, *Treatise on Solid State Chemistry*, Springer, New York, 1976, vol. 3, ch. 7, p. 497.
62. M. L. Di Lorenzo, *Eur. Polym. J.*, 2005, **41**, 569.
63. H. Tsuji, Y. Tezuka, S. K. Saha, M. Suzuki and S. Itsuno, *Polymer*, 2005, **46**, 4917.
64. J. D. Hoffman and J. I. Lauritzen, Jr., *J. Res. NBS A*, 1961, **65**, 297.
65. J. Sarasua, R. E. Prud, M. Wisniewski, A. Le Borgne and N. Spassky, *Macromolecules*, 1998, **31**, 3895.
66. M. Pyda, R. C. Bopp and B. Wunderlich, *J. Chem. Therm.*, 2004, **36**, 731.
67. S. Iannace and L. Nicolais, *J. Appl. Polymer Sci*, 1997, **64**, 911.
68. H. Abe, Y. Kikkawa, Y. Inoue and Y. Doi, *Biomacromolecules*, 2001, **2**, 1007.
69. Y. He, Z. Fan, Y. Hu, T. Wu, J. Wei and S. Li, *Eur. Polym. J.*, 2007, **43**, 4431.
70. M. Shibata, Y. Inoue and M. Miyoshi, *Polymer*, 2006, **47**, 3557.
71. M. Avrami, *J. Chem. Phys.*, 1941, **9**, 177.
72. A. T. Lorenzo, M. L. Arnal, J. Albuerne and A. J. Müller, *Polymer Test*, 2007, **26**, 222.
73. H. W. Xiao, P. Li, X. Ren, T. Jiang and J. T. Yeh, *J. Appl. Polymer Sci.*, 2010, **118**, 3558.
74. T. Ke and X. Sun, *J. Appl. Polymer Sci.*, 2003, **89**, 1203.
75. Y. Li, C. Chen, J. Li and X. S. Sun, *J. Appl. Polymer Sci.*, 2012, **124**, 2968.
76. E. Urbanovici, H. A. Schneider, D. Brizzolara and H. J. Cantow, *J. Therm. Anal.*, 1996, **47**, 931.
77. Y. Shieh, Y. Twu, C. Su, R. Lin and G. Liu, *J. Polymer Sci. Polymer Phys.*, 2010, **48**, 983.
78. W. Y. Zhou, B. Duan, M. Wang and W. L. Cheung, *J. Appl. Polymer Sci.*, 2009, **113**, 4100.
79. A. Auliawan and E. M. Woo, *J. Appl. Polymer Sci.*, 2012, **125**, E444.
80. J. Zhang and X. Sun, *Polymer Int.*, 2004, **53**, 716.
81. Y. Loo and A. Register, in *Developments in Block Copolymer Science and Technology*, ed. I. W. Hamley, Wiley, New York, 2004, pp. 213–243.
82. A. J. Müller, V. Balsamo and M. L. Arnal, *Adv. Polymer Sci*, 2005, **190**, 1.

83. B. Nadan, J. Hsy and H. Chen, *J. Macromol. Sci. Polymer Rev*, 2006, **46**, 143.
84. A. J. Müller, V. Balsamo and M. L. Arnal, in *Lecture Notes in Physics: Progress in Understanding of Polymer Crystallization*, ed. G. Reiter and G. Strobl, Springer, Berlin, 2007, vol. 714, p. 229.
85. R. V. Castillo and A. J. Müller, *Progr. Polymer Sci.*, 2009, **34**, 516.
86. A. J. Müller, M. L. Arnal and A. T. Lorenzo, in *Handbook of Polymer Crystallization*, ed. E. Piorkowska and G. Rutledge, Wiley, New York, 2013, pp. 347–378.
87. R. M. Michell, I. Blaszczyk-Lezak, C. Mijangos and A. J. Müller, *Polymer*, 2013, **54**, 4059.
88. I. Barakat, Ph. Dubois, Ch. Grandfils and R. Jérôme, *J. Polymer Sci. Polymer Chem*, 2001, **39**, 294.
89. G. Schwach, J. Coudane, R. Engel and M. Vert, *Biomaterials*, 2002, **23**, 993.
90. O. Dechy-Cabaret, B. Martin-Vaca and D. Bourissou, *Chem. Rev.*, 2004, **104**, 6147.
91. C. Min, W. Cui, J. Bei and S. Wang, *Polymer. Adv. Tech.*, 2007, **18**, 299.
92. H. Tsuji, K. Tashiro, L. Bouapao and M. Hanesaka, *Macromol. Chem. Phys.*, 2012, **213**, 2099.
93. Y. Yuryev, P. Wood-Adams, M.-C. Heuzey, C. Dubois and J. Brisson, *Polymer*, 2008, **49**, 2306.
94. H. Tsuji, T. Wada, Y. Sakamoto and Y. Sugiura, *Polymer*, 2010, **51**, 4937.
95. M. Stolt, M. Viljanmaa, A. Södergård and P. Törmälä, *J. Appl. Polymer Sci.*, 2004, **91**, 196.
96. J. Broström, A. Boss and I. S. Chronakis, *Biomacromolecules*, 2004, **5**, 1124.
97. H. Tsuji and Y. Tezuka, *Macromolecular Biosciences*, 2005, **5**, 135.
98. W. Hamley, V. Castelletto, R. V. Castillo, A. J. Müller, C. M. Martin, E. Pollet and Ph. Dubois, *Macromolecules*, 2005, **38**, 4631.
99. W. Hamley, P. Parras, V. Castelletto, R. V. Castillo, A. J. Müller, E. Pollet, Ph. Dubois and C. M. Martin, *Macromol. Chem. Phys.*, 2006, **207**, 941.
100. H. A. Declercq, M. J. Cornelissen, T. L. Gorskiy and E. H. Schacht, *J. Mater. Sci.*, 2006, **17**, 113.
101. E. Laredo, N. Prutsky, A. Bello, M. Grimau, R. V. Castillo, A. J. Müller and Ph. Dubois, *Eur. Phys. J.*, 2007, **23**, 295.
102. D. Ajami-Henriquez, M. Rodríguez, M. Sabino, V. Castillo, A. J. Müller, A. Boschetti-de-Fierro, C. Abetz, V. Abetz and Ph. Dubois, *J. Biomed. Mater. Res.*, 2008, **87A**, 405.
103. M. Yilmaz, S. Eğri, N. Yildiz, A. Çalimli and E. Pişkin, *Polymer J.*, 2011, **43**, 785.
104. M. T. Casas, J. Puiggalí, J.-M. Raquez, P. Dubois, M. E. Córdova and A. J. Müller, *Polymer*, 2011, **52**, 5166.
105. P. Purnama, Y. Jung and S. H. Kim, *Macromolecules*, 2012, **45**, 4012.

106. H.-A. Klok, J. J. Hwang, S. N. Iyer and S. I. Stupp, *Macromolecules*, 2002, **35**, 746.
107. X. Deng, Y. Liu, M. Yuan, X. Li, L. Liu and W. X. Jia, *J. Appl. Polymer Sci.*, 2002, **86**, 2557.
108. J. M. Onyari and S. J. Huang, *Macromol. Symp.*, 2003, **193**, 143.
109. S. Li and M. Vert, *Macromolecules*, 2003, **36**, 8008.
110. G. Yu, J. Ji and J. Shen, *J. Mater. Sci.*, 2006, **17**, 899.
111. S. K. Agrawal, N. Sanabria-DeLong, G. N. Tew and S. R. Bhatia, *Macromolecules*, 2008, **41**, 1774.
112. S. Huang, S. Jiang, L. An and X. Chen, *J. Polymer Sci. Polymer Phys.*, 2008, **46**, 1400.
113. I. V. Dimitrov, I. V. Berlinova and V. I. Michailova, *Polymer J.*, 2013, **45**, 457.
114. A. J. Müller, R. V. Castillo and M. Hillmyer, *Macromol. Symp.*, 2006, **242**, 174.
115. A. J. Müller, A. T. Lorenzo, R. V. Castillo, M. L. Arnal, A. Boschetti-de-Fierro and V. Abetz, *Macromol. Symp.*, 2006, **245–246**, 154.
116. R. V. Castillo, A. J. Müller, M.-C. Lin, H. Chen, U. Jeng and M. A. Hillmyer, *Macromolecules*, 2008, **41**, 6154.
117. R. M. Michell, A. J. Müller, M. Spasova, P. Dubois, S. Burattini, B. W. Greenland, I. W. Hamley, D. Hermida-Merino, N. Cheval and A. Fahmi, *J. Polymer Sci. Polymer Phys.*, 2011, **49**, 1397.
118. H. Younes and D. Cohn, *J. Biomed. Mater. Res.*, 1987, **21**, 1301.
119. Q. Cai, J. Bei and S. Wang, *Polymer*, 2002, **43**, 3585.
120. G. Maglio, A. Migliozzi and R. Palumbo, *Polymer*, 2003, **44**, 369.
121. C. G. Mothé, W. S. Drumond and S. H. Wang, *Thermochim. Acta*, 2006, **445**, 61.
122. D. Shin, K. Shin, K. A. Aame, G. N. Tew, T. P. Russell, J. H. Lee and J. Y. Jho, *Macromolecules*, 2005, **38**, 104.
123. K. E. Uhrich, S. M. Cannizzaro, R. S. Langer and K. M. Shakesheff, *Chem. Rev.*, 1999, **99**, 3181.
124. Y. K. Choi, Y. H. Bae and S. W. Kim, *Macromolecules*, 1998, **31**, 8766.
125. D. Lim and T. G. Park, *J. Appl. Polymer Sci.*, 2000, **75**, 1615.
126. D. Kubies, F. Rypáček, J. Kovářová and F. Lednický, *Biomaterials*, 2000, **21**, 529.
127. K. S. Kim, S. Chung, I. J. Chin, M. N. Kim and J. S. Yoon, *J. Appl. Polymer Sci*, 1999, **72**, 341.
128. H. S. Choi, T. Ooya, S. Sasaki, N. Yui, Y. Ohya, T. Nakai and T. Ouchi, *Macromolecules*, 2003, **36**, 9313.
129. S. M. Li, I. Rashkov, J. L. Espartero, N. Manolova and M. Vert, *Macromolecules*, 1996, **29**, 57.
130. I. Rashkov, N. Manolova, S. M. Li, J. L. Espartero and M. Vert, *Macromolecules*, 1996, **29**, 50.
131. C. Cai, L. Wang and C. Dong, *J. Polymer Sci. Polymer Chem.*, 2006, **44**, 2034.
132. C. Huang, S.-H. Tsai and C. M. Chen, *J. Polymer Sci. Polymer Phys.*, 2006, **44**, 2438.

133. A. J. Nijenhuis, E. Colstee, D. W. Grijpma and A. J. Pennings, *Polymer*, 1996, **37**, 5849.
134. J. O. Ring, R. Thomann, R. Mülhaupt, J.-M. Raquez, P. Degée and Ph. Dubois, *Macromol. Chem. Phys.*, 2007, **208**, 896.
135. A. J. Müller, J. Albuerne, L. Márquez, J. M. Raquez, P. Degée, Ph. Dubois, J. Hobbs and I. W. Hamley, *Faraday Discuss.*, 2005, **128**, 231.
136. R. V. Castillo and A. J. Müller, unpublished results.
137. J. Albuerne, L. Marquez, A. J. Müller, J. M. Raquez, P. Degée, Ph. Dubois, V. Castelletto and I. W. Hamley, *Macromolecules*, 2003, **36**, 1633.

Processing, Characterization and Physical Properties of PLA

CHAPTER 4

Reactive Extrusion of PLA-based Materials: from Synthesis to Reactive Melt-blending

JEAN-MARIE RAQUEZ,* RINDRA RAMY-RATIARISON, MARIUS MURARIU AND PHILIPPE DUBOIS

Laboratory of Polymeric and Composite Materials, Center of Innovation and Research in Materials and Polymers (CIRMAP), University of Mons & Materia Nova Materials R&D Centre, Place du Parc 20, 7000, Mons, Belgium
*Email: jean-marie.raquez@umons.ac.be

4.1 Introduction

This chapter emphasizes the synthesis and modification of polylactide-based materials by means of reactive extrusion (REX) technology. Polylactide (PLA) represents the most investigated bio-based polymer due to its attractive mechanical properties, renewability, biodegradability and relatively low production costs.[1] Due to these interesting features, PLA-based materials have been considered for a large set of applications, from packaging to electronic devices. However, industrial implementation of PLA-based materials is impeded due to its low thermal stability, sensitivity to hydrolysis, low crystallization rate and high brittleness. A range of modifications and processes including chemical modifications and blending have therefore been regarded for PLA-based materials, and several reviews in the

RSC Polymer Chemistry Series No. 12
Poly(lactic acid) Science and Technology: Processing, Properties, Additives and Applications
Edited by Alfonso Jiménez, Mercedes Peltzer and Roxana Ruseckaite
Published by the Royal Society of Chemistry, www.rsc.org

realm have been published. Moreover, some of us featured a general review on REX processing of biodegradable polymer-based compositions.[2] This chapter hence emphasizes an up-to-date review only about the chemical aspects of REX processes, *i.e.* for the synthesis and modification of PLA-based materials. These different REX processes of PLA-based materials are discussed here in terms of their scope, limitations and specificity.

4.2 Polylactide (PLA)-based Materials: General Aspects and Reactive Extrusion (REX) Processing

Belonging to the family of aliphatic polyesters,[3] poly(lactic acid) or polylactide (PLA) is composed of lactic acid repetitive units, which is the simplest α-hydroxy acid with an asymmetric carbon atom.[4] Interestingly, the L-lactic acid monomer, and more recently the D-lactic acid monomer, can be straightforwardly obtained by bacterial fermentation from renewable resources (namely starch), making both monomers and therefore the resulting polymers environmentally friendly.[1,5,6] Polycondensation of lactic acid and ring-opening polymerization (ROP) of lactide (LA), *i.e.* cyclic diesters of lactic acid, are currently used to prepare PLA polymers (Scheme 4.1).

However, ROP promoted by organometallics like tin(II) octoate (Sn(Oct)$_2$) represents the most privileged synthetic pathway as it overcomes the major drawbacks of the polycondensation process in terms of release of by-products and use of high temperatures, and can be carried out in solution as well as in bulk, that is in the absence of solvents.[7] Due to the presence of two chiral centres, LA exists as optically active D,D-lactide (D-LA) or L,L-lactide (L-LA)

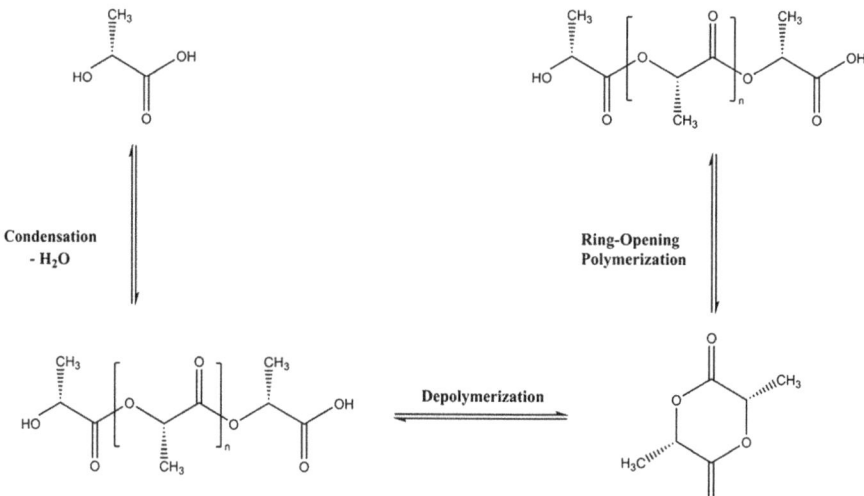

Scheme 4.1   PLA production *via* polycondensation and ring-opening polymerization (reproduced with John Wiley and Sons' permission from Ref. 2).

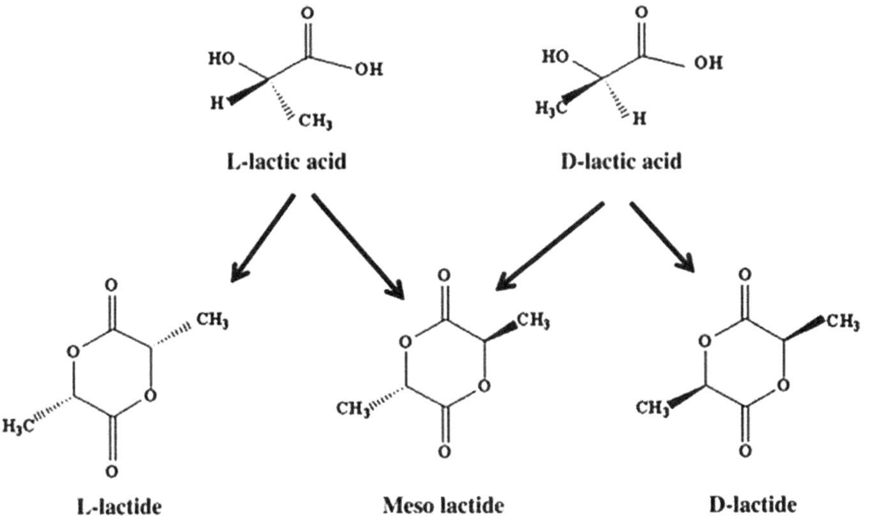

Figure 4.1 Stereoisomers of lactide monomers (reproduced with Elsevier's permission from Ref. 9).

monomers (Figure 4.1) as well as optically inactive meso-lactide (meso-LA) monomer or the racemic mixture of L-LA and D-LA (*i.e.* rac-LA).[8,9]

PLA-based materials span a large set of properties such as biodegradability, renewability, biocompatibility, transparency (in most cases), broad possibility of processing using standard equipment, high stiffness (similar to PS), *etc.*[10] However the molar content of both enantiomers within the polymeric backbone can alter, to some extent, some properties of PLA-based materials.[11] For instance, PLAs fully derived from the optically active D-monomers are semicrystalline materials with a melting temperature (T_m) of *ca.* 175 °C and a glass transition temperature (T_g) of *ca.* 60 °C and exhibit mechanical properties with a Young's elastic modulus of *ca.* 3000 MPa that are equivalent to or in some cases even better than many petroleum-based polymers, like PET.[12] In contrast, when PLA contains more than 10 wt% optical impurities, the resulting materials are amorphous, displaying a T_g of around 55–60 °C. Both thermal parameters, *i.e.* T_g and T_m, are strongly dependent on the polymer molar mass and optical purity.

The first commercial successes of PLA were achieved as biocompatible and resorbable materials in particular for the development of improved multifilament sutures in the 1980s and drug-delivery devices in the 1990s.[1,13] In the 2000s, end-life concerns about the wastes issued from conventional/petroleum-based plastics merely pushed forward the utilization of biodegradable and renewable PLA-based materials in short-time applications, such as packaging (cups, bottles, films and containers) and in textiles.[6,14–16] However, the real breakthrough of PLA-based materials thrived when NatureWorks Company (formally Cargill-Dow) set up a large-scale process with a production capacity of *ca.* 140 000 ton year^{-1} of PLA.[17]

This process involves the continuous production of the LA monomer, starting from the fermentation of corn glucose into L-lactic acid, followed by its batch ROP promoted by Sn(Oct)$_2$ as catalyst. Recently PLA, blended or not-blended with petroleum-based polymers, has been prospecting new directions into higher added-value applications (electronics, automotive, high-performance components, *etc.*) as renewable materials. However, even if PLA exhibits good properties there are still some shortcomings to be addressed in these applications, such as low ductility, heat-resistance and crystallization rate, relatively high sensitivity to moisture and low impact resistance. Melt-blending PLA with selected additives represents the most popular strategy to partly or totally implement these desired properties.[18] A variety of materials (plasticizers, impact modifiers, (nano)fillers, reinforcing fibres, flame retardants, enantiomeric PLA for stereo-complexation and other polymeric partners) has been blended with PLA.[19,20] The properties of the resulting PLA blends are highly dependent on the nature of partners as well as the final composition. For instance, adding plasticizers, such as glycerol, into PLA enables the melt-processing of PLA-based materials to be improved as well as increasing their flexibility and ductility.[21,22] However, plasticizers tend to migrate out due to their high mobility through the PLA matrix.[23] The development of PLA-based melt-blends/composites with satisfactory overall thermo-mechanical properties requires controlling the interfacial energy, generating dispersed phases of limited size and strong interfacial adhesion and improving the stress transfer between the component phases.[24,25] Using a reactive process such as the reactive extrusion (REX), has proven to be an elegant and efficient way that strengthens the interface between the different partners, even in a continuous manner.[26]

In the case of REX processing, single-screw extruders and especially twin-screw extruders operate as a solvent-free and continuous "chemical reactor", where a large set of chemical modifications can proceed such as continuous polymerization, free-radical grafting and reactive blending.[27] REX processing indeed shows to be relevant in order to tackle high-viscosity materials and resolve the heat- and mass-transfer issues arising from batch reactions.[28] Variation in screw design and barrel provides a good control over the residence time and offers many opportunities for adding (or removing) additives, such as micro- and nanofillers, plasticizers, injection of gases, *etc.* during processing. Of course, the successful achievement of these reactive processes depends on the residence time, which should be usually between 5 and 10 minutes. REX can therefore enhance the commercial viability and cost-competitiveness of these materials as already featured about biodegradable polymers like PLA.[29,30] In this regard, REX processing of PLA-based materials has been investigated to conduct a large set of reactive modifications in order to make the properties of PLA-based materials suitable from short-time to long-time and durable high-performance properties. In addition to its polymerization (i), most investigations take advantage of the (relatively) high reactivity of PLA chains to conduct (ii) coupling reactions *via* its hydroxyl functionalities (and eventually carboxylic functionalities for, *e.g.* PLA polycondensates), (iii) exchange reactions *via* its ester functionalities and (iv) free-radical grafting reactions *via* its methine chemical units using REX processing.[29] Several reviews on the chemical modification of PLA-based materials including a few instances related to REX have been already provided.[2,29,31] The following sections will discuss the main up-to-date features and the new prospects of using melt-processing, in particular REX processing, for the synthesis and chemical modification of PLA-based materials including:

- REX synthesis of PLA-based materials *via* ring-opening polymerization (ROP);
- REX coupling reactions from PLA precursors;
- REX free-radical grafting reactions of PLA chains;
- REX transesterification (exchange) reactions.

4.3 REX Synthesis of PLA-based Materials *via* Ring-opening Polymerization

Ring-opening polymerization (ROP) of lactide (LA) represents an efficient synthetic route to the production of high-molar-mass PLA in solution and in bulk (absence of solvent).[32–34] In this regard, aluminium and tin(II) alkoxides have shown to be the most efficient initiators by favouring the propagation rate as well as preventing inter- and intramolecular transesterification reactions as side-reactions in ROP of LA.[35] As a consequence, good control over dispersity, end-functionality and molar masses can be obtained. These metallic alkoxides propagate through a coordination-insertion mechanism during ROP of LA, involving the formation of covalently bound propagating species between the initiators and the monomer. In addition to these metallic alkoxides, carboxylate salts, in particular tin(II) octoate ($Sn(Oct)_2$), are employed as ROP catalysts in both academic and industrial purposes.[36,37] From the industrial viewpoint, $Sn(Oct)_2$ offers several advantages including both easy handling and purification, while it is noteworthy to mention its acceptance as a food additive. Mechanistically, like in the case of

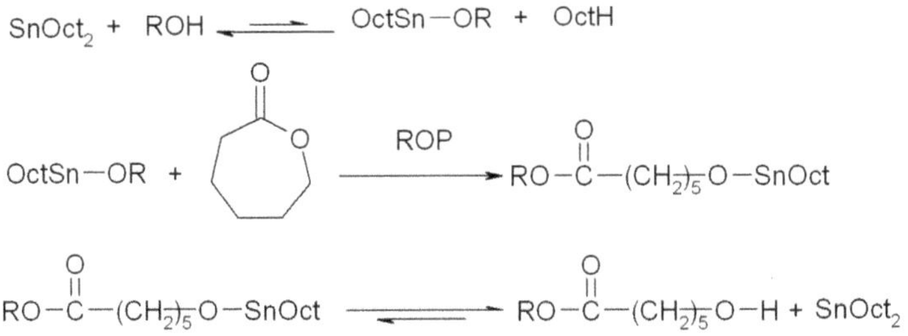

Scheme 4.2 Proposed activation mechanism for the ROP of ε-caprolactone promoted by $Sn(Oct)_2$ (only tin monoalkoxide, Oct-Sn-OR, is shown even though the formation of tin dialkoxide cannot be ruled out. Reproduced with Elsevier permission from Ref. 1).

ε-caprolactone (Scheme 4.2), Sn(Oct)$_2$-promoted ROP of LA proceeds *via* a coordination-insertion mechanism after the *in situ* formation of Sn-alkoxide moieties at the chain extremity as active centres. These active centres are derived from a rapid exchange reaction of protic compounds (ROH) with tin alkoxides.

Sn(Oct)$_2$ has shown to be an efficient catalyst for REX polymerization of LA as reported by some of us.[38–41] However, to get the best processing conditions, an equimolar amount of triphenyl-phosphine (P(C$_6$H$_5$)$_3$) must be added to Sn(Oct)$_2$ in order to promote a fast polymerization of LA as well as to suppress (or at least delay) degradation reactions, such as transesterification reactions, during extrusion processing. From this catalytic complex and using a closely intermeshing co-rotating twin-screw extruder (with a suitable processing and screw concept), PLA molar masses of 70 000 g mol^{-1} can be achieved within a residence time of *ca.* 7 min. at high monomer conversions (*ca.* 98%) and at temperatures of 180–185 °C. The intentional addition of protic compounds (ROH) can also adjust the molar masses of the as-recovered PLA. However, an optimal [LA]$_0$/[Sn]$_0$ molar ratio is required for the REX polymerization of LA in order to recover PLA with good melt-stability when any post-processing, such as melt-spinning, is subsequently carried out. For more specific applications, the melt-stability of the resulting PLA can be further enhanced after adding thermal stabilizers such as Ultranox® 626 (that is, bis-(2,4-di-*t*-butylphenyl) pentraerythritol diphosphite) during REX, without influencing the polymerization reaction itself. Interestingly REX polymerization of LA is under industrial implementation[1] from Futerro™, a joint venture between Galactic and Total Petrochemicals in Belgium.

Recently, Gallos *et al.* have reported on the synthesis of PDLA-*b*-PLLA diblock copolymers able to form PLA stereo-complexes (PLA$_{sc}$) by REX.[42] In long-lasting applications, PLA$_{sc}$ have been viewed as an efficient solution to improve PLA properties, in particular its poor thermal-resistance (increased HDT), crystallization rate and resistance to hydrolysis.[43–47] In general, PLA stereo-complexation can occur when L-chains and D-chains are present in the system as long as both polymer chains contain at least seven successive repeating units of L-lactide or D-lactide.[48] An increase of the crystallization rate and T_m to a maximum of 230 °C is noticed for the resulting PLA$_{sc}$ with respect to PLA homopolymers. Unfortunately, PLA$_{sc}$ can be merely recovered at high yields using solvent-based processes as they can readily undergo undesirable and irreversible thermal degradation reactions during melt-processing, that is, when heated up to or even below their T_m (>230 °C). In the work reported by Gallos *et al.*, a two-stage ROP promoted by Sn(Oct)$_2$ complexed with triphenyl-phosphine (initial molar [LA]/[Sn] ratio of 5000) was carried out using a DSM micro-extruder, to polymerize first L-lactide, followed by D-lactide.[42] Accordingly, stereo-complexed PDLA-*b*-PLLA diblock copolymers with a maximum T_m of 220 °C were produced. However, when heated, DSC and ^{13}C NMR analyses of stereo-complexed PLA-based materials revealed some degradation reactions including unzipping depolymerization.

In order to achieve enhanced intumescent properties of stereo-complexed PLA, the *in situ* REX synthesis of PLA was similarly reported for PLA homopolymers in the presence of nanoclays and carbon nanotubes[49] and for stereo-complexed PDLA-*b*-PLLA diblock copolymers in the presence of a mixture of ammonium polyphosphate, melamine and nanoclay.[50]

We must mention that intensive research is on-going in the design of metal-free catalysts for the REX ROP of lactide.[34] By comparison with metal-based catalysts like Sn(Oct)$_2$, the advantages of these metal-free catalysts are that they are less harmful for biomedical applications and readily removed during, *e.g.*, extrusion processes. Among the simplest ones, some selected enzymes, tertiary amines, urea- and iodine-based compounds, phosphines and carbenes have proven efficient metal-free catalysts. However, most of these investigations have shown that these metal-free catalysts are less active and may yield intensive side-reactions, such as epimerization, when applied during bulk REX ROP of LA.[51]

4.4 REX Coupling Reactions from PLA Precursors

Although ROP of LA is the most widely investigated way to produce high-molar-mass PLA chains, it is necessary to develop other (complementary) synthetic methods, such as solid-state polymerization and chain-extension. Depending on the synthetic pathway (polycondensation or ROP), PLA may possess either two hydroxyl functional end-extremities or two distinct functional end-extremities, *i.e.* hydroxyl and carboxyl. These end-groups can further participate to selective functionalizations,[52] in particular coupling reactions between hydroxyl/isocyanate,[53] hydroxyl/anhydride[54,55] and carboxyl/epoxy[56,57] during REX melt reactions in order to, *e.g.*, compatibilize PLA-based blends as well as to modify the structure of PLA chains (branching).[58] Another interest in these coupling agents is to improve the PLA thermal stability upon thermal processing, predominantly implying random main-chain scission and unzipping depolymerization reactions. Almost all the active chain end-groups, residual catalysts, residual monomers and other impurities indeed affect the PLA thermal degradation. As a consequence, an undesired molar mass reduction and weight loss occur from 180 °C to 220 °C. In this regard, adding coupling agents into PLA yields long and branched structures as well as limits the occurrence of the thermal degradation of PLA-based materials. In fact, many authors have shown that coupling agents are able to relink cleaved chains, therefore increasing the polymer molar mass and strength of melt and they can also be successfully used as reactive compatibilizers in blends. Coupling agents do not release any by-products during melt-processing, while being generally multifunctional oligomers (epoxides, diisocyanates, dianhydrides, bisoxazolines, tris(nonyl-phenyl) phosphate and polycarbodiimides), thermally stable and largely available.[27] The most prominent examples of REX coupling reactions will be outlined, *i.e.* epoxy- and isocyanate-based functionalizations.

4.4.1   REX Isocyanate-based Coupling Reactions

Isocyanate moiety is highly reactive towards, *e.g.*, hydroxyl end-groups of PLA chains, yielding urethane and eventually allophanate linkages when an excess of isocyanate is used.[59] There is a large set of diisocyanate compounds that can react towards hydroxyl moieties including hexamethylene diisocyanate, methylene bis(4-cyclohexylisocyanate), toluene diisocyanate, 1,4-butane diisocyanate, lysine diisocyanate, *etc*. When reacted with α,ω-dihydroxy PLA oligomers,[60] it resulted in the formation of poly(ester-urethane)s (PEUs) with enhanced thermal properties. These PEUs are of segmental structure that can be readily controlled upon the length of PLA, the type of diisocyanate, the -OH/-NCO ratio and the processing conditions.[61] To extend the interactions between PEU segments, short chain-extenders such as 1,4-butanediol may be added, *e.g.* to bring more stiffness to the chains.[62] Recently, ROP of LA combined with a coupling process has been investigated using two-stage REX processing in order to elaborate PEUs.[63] Both processes were catalyzed by the same catalytic system used for REX ROP of LA, that is $Sn(Oct)_2$ complexed with triphenyl-phosphine. The first step was to synthesize α,ω-dihydroxy PLA macrodiols in batch reactions and then react them using REX processing with 4,4-diisophenylmethane diisocyanate and 4,4-diaminodiphenylmethane, as coupling-agent and chain-extender, respectively (Figure 4.2).

Interestingly, the resulting PLA-based PEUs were still semicrystalline, but their ability to crystallize was slightly hindered by the presence of urethane moieties along their backbone. Moreover, the straightforward and continuous production of melt-processable PLA-based stereo-complexes was also successfully demonstrated through two successive reactions: coupling and stereo-complexation starting from PLA oligomers of D- and L-configuration. As aforementioned, the coupling reactions are considered as an efficient way to enhance the thermal stability of PLA_{sc} *via* the synthesis of PEUs. However, combining coupling reactions and stereo-complexation remains difficult as they compete during melt-processing, yielding incomplete extent of both reactions. In this regard, a novel T-shaped REX processing was implemented on the synthesis of melt-processable PLA-based stereo-complexes in our research group (Figure 4.3). Interestingly, the technique was so versatile that the combination of ROP of D- and L-Lactide, chain-extension and stereo-complexation could be attempted using this T-shaped REX processing as well. In the latter, the formation of PLA-based stereo-complexes with a T_m of 205 °C (melting enthalpy of 15.6 J g^{-1}) was successfully achieved.

Isocyanate-based coupling reactions encompass other issues for PLA-based materials, that is, by toughening PLA in REX blends with poly(ester-amide),[64] thermoplastic polyester elastomer,[65] starch,[66] polyurethane[67] and poly(ε-caprolactone) (PCL).[68] Harada *et al.* elaborated PLA/PCL blends compatibilized with four different isocyanates (lysine triisocyanate (LTI), lysine diisocyanate, 1,3,5-tris(6-isocyanatohexyl)-1,3,5-triazinane-2,4,6-trione and 1,3,5-tris(6-isocyanatohexyl)biuret) and with industrial epoxide-trimethylol-propane triglycidyl ether using an internal mixer and REX processing.

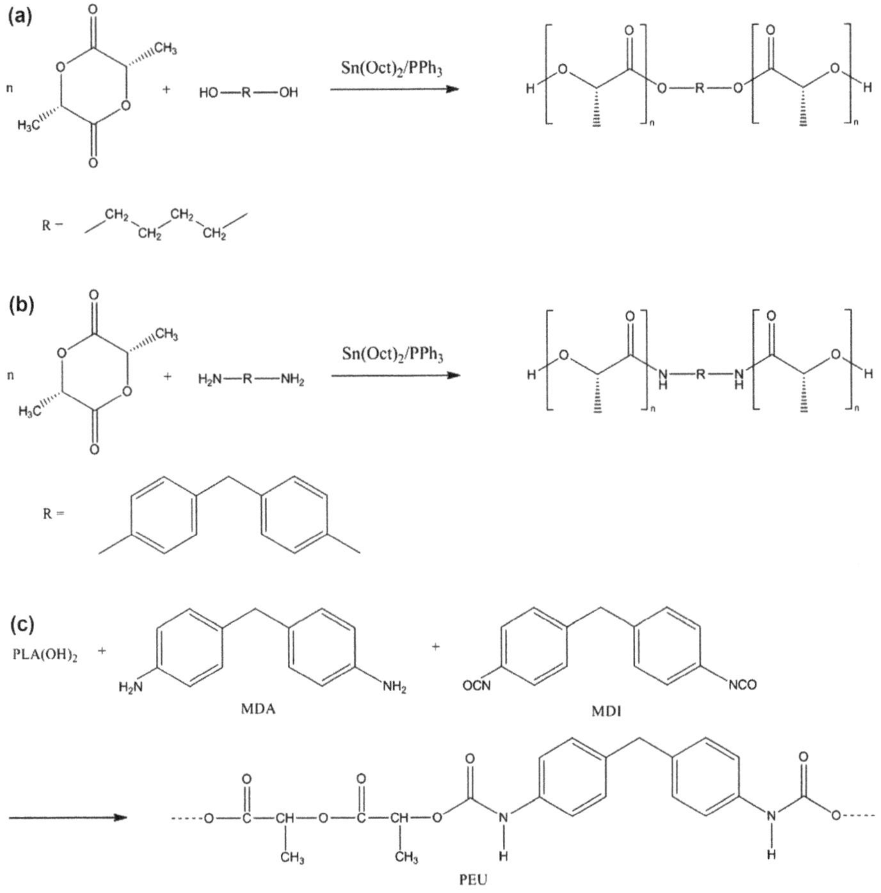

Figure 4.2 Synthesis of α,ω-hydroxyl polylactide pre-polymer initiated by 1,4-butanediol (a) or 4,4'-diaminodiphenylmethane (b) and synthesis of poly(ester-urethane)s by chain extension reaction (c) (adapted and used with Elsevier's permission from Ref. 63).

The result was that PLA/PCL blended in the presence of LTI had the highest torque from experiments carried out in an internal mixer, and LTI was then selected as coupling agent during REX processing. Different PCL mass ratios were thereby evaluated to assess the mechanical performance of these blends. The result was that both impact strength and nominal tensile strain increased considerably for blends containing more than 20 wt% PCL. These results indicated that the compatibility of PLA/PCL blends was improved after adding LTI by REX, *via* possible reactions between LTI and terminal hydroxyl or carboxylic groups of PLA chains.

Nyambo *et al.*[64] investigated the combination of two potential toughening agents for PLA, *i.e.* hydroxyl-terminated hyperbranched poly(ester amide) and isocyanate-terminated prepolymer of butadiene (ITPB). Synergistic effects in impact strength were observed in PLA ternary blends containing

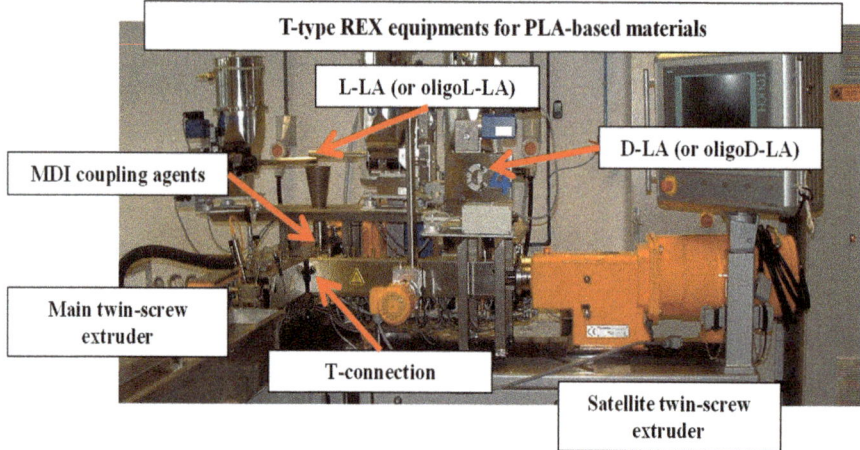

Figure 4.3   T-shaped REX technology.

hyperbranched polymer (HBP) and ITPB. For instance, the impact strength of the PLA/HBP/ITPB ternary blend was improved over 86%, while the elongation at break was increased by over 100%. The authors stated that some physical and chemical interactions between the hydroxyl-terminated HBP and the ITPB might be responsible for these improvements in impact strength. Tensile, flexural, thermal and thermo-mechanical properties of the PLA/HBP blends with varying amounts of the ITPB were studied as well. The results revealed that the fracture behaviour of PLA changed from brittle to ductile in the PLA/HBP/ITPB ternary blends as highlighted by SEM analyses.

Polycarbodiimide (CDI) was employed to enhance the thermal stability of PLA on melt-processing.[69] The polycarbodiimide backbone can be viewed as a hybrid between polyisocyanates (a dynamic helical polymer) and poly-isocyanides (a static helical polymer) and used as an effective crosslinker for carboxyl functional polymers to improve toughness, thermal stability, solvent- and wear-resistance of PET, polyamides and polyurethanes. From this work, the results showed that an addition of CDI in an amount of 0.1–0.7 wt% with respect to PLA led to the polymer stabilization even at 210 °C for up to 30 min. in a kneading mixer. This was evidenced by small changes in molar mass, melt viscosity, tensile strength and elongation at break compared to PLA samples. The authors reported a possible reaction mechanism between CDI and residual moisture, reducing the extent of thermal degradation and hydrolysis of PLA during melt-processing.

4.4.2   REX Epoxy-based Coupling Reactions

The use of epoxy-based coupling agents is well known to yield linear chain extension and/or branched structures (in function of the level/amount of epoxy functionalities), to improve the material properties and to widen the processing window of PLA.[70,71] The main advantages of epoxy-based coupling agents over isocyanate-based coupling agents are that they are less volatile and toxic. The reaction of epoxy-based coupling agents generally occurs with carboxylic-ended PLA chains in short reaction times in the melt as, for the first time, reported by Zhou *et al.*[72] To highlight this reaction, dicarboxylated PLA was first synthesized by reacting succinic anhydride with the PLA prepolymer (as prepared by melt polycondensation) and subsequently reacted with diglycidyl ether of bisphenol A (DGEBA) copolymers as epoxy-based coupling agents. Two types of coupling reactions were thereby reported (Figure 4.4). In reaction I, the linear PLA chains can be obtained, while the ester formed by the epoxy-acid reaction can react with the OH- groups through a transesterification mechanism in reaction II. Reaction II leads to chain branching and gel formation.

Multifunctional epoxy compounds have been extensively used as coupling agents for branching PLA with improved processability using REX processing.[73,74] Mallet *et al.* developed various formulations of PLA with multifunctionalized epoxy derivatives, nucleating agents and plasticizers in order to improve the processability for PLA-based blown films.[73] In this work, they utilized Joncryl® (JON) as multifunctional epoxy compound for its efficiency to create branched structures. Joncryl® (BASF producer) is a glycidyl methacrylate-based copolymer generally obtained by free-radical copolymerization. After adding JON, the shear viscosity increased with its content independently of the shear rate, indicating the significant occurrence of coupling reactions and/or branching mechanisms. A synergistic effect of both nucleating agent and plasticizer was also noticed on their

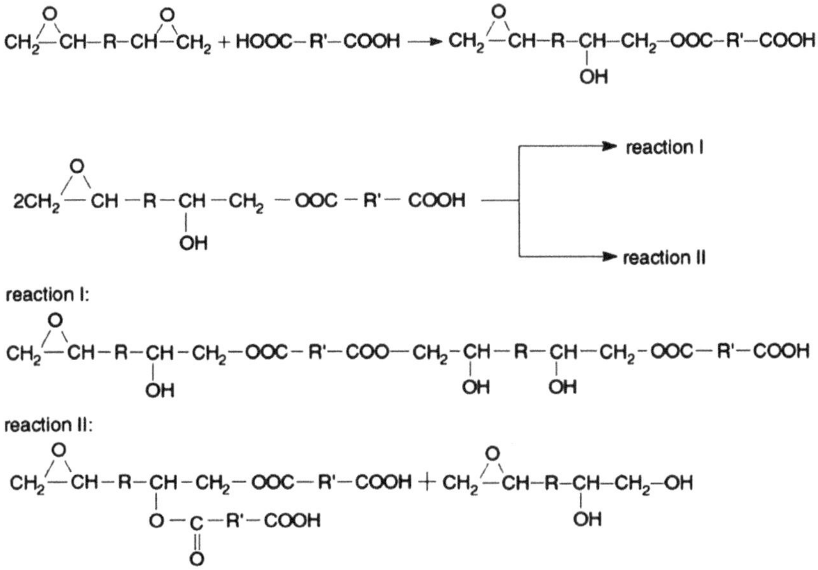

Figure 4.4 Reaction mechanism of dicarboxylated PLA reacting with DGEBA (reproduced with Express Polymer Letters' permission from Ref. 72).

blowing ability and properties. For instance, the thermo-mechanical properties of PLA and the optimized PLA blends confirmed this improvement, as highlighted by the recovery of flexible and ductile PLA-based blends in contrast to the brittle starting materials. Similar results were reported on PLA foaming[75] using Joncryl® as coupling agent.

Epoxy-based coupling agents were also employed to strengthen the compatibility of immiscible blends from PLA/poly(butylene adipate-*co*-terephthalate) (PBAT)[76] and PLA/thermoplastic starch.[77] For instance, blending PLA with PBAT was investigated in order to elaborate flexible and thermally stable PLA-based films. PBAT is a petroleum-based biodegradable aliphatic-aromatic copolyester used in film applications to impart ductility and processability to PLA. Blending PLA with PBAT represents a promising strategy to achieve a toughened multiphase material, but the lack of compatibility between partners affects their performance. REX processing was here employed to compatibilize and blend PLA and PBAT in the presence of JON as chain-extender/branching agent. The authors reported that both coupling and branching reactions could occur in function of the amount of the additive (JON) with glycidyl functions and blending time because of the competition between branching and degradation reactions during processing. Under controlled conditions by favouring the synthesis of branched structures, it was possible to increase the intrinsic viscosity, shear thinning, elasticity and thermal stability of compatibilized blends during melt processing. For instance, compared to unmodified blends, the tensile modulus (from 820 MPa (PLA/PBAT blends) to 1095 MPa (PLA/PBAT/JON)) and the elongation at break (from 50% (PLA/PBAT blends) to 135% (PLA/PBAT/JON)) were improved. Such behaviour was explained by the formation of ester linkages between PLA, PBAT and JON, enhancing the compatibility between polymeric partners as confirmed by transmission electron microscopy (TEM) observations.

Blends made from PLA, epoxidized soybean oil and maleated starch were also processed in a lab-scale extruder in order to elaborate PLA/starch composites toughened with epoxidized soybean oil.[78] The maleic anhydride (MA) grafted onto the starch granules acted as catalyst for the reaction of epoxy groups from epoxidized soybean oil towards carboxylic acid endgroups of PLA, yielding a highly compatible blend. As a consequence both elongation at break and impact strength of PLA/epoxidized soybean oil/starch increased markedly compared to the non-compatibilized composites.

Recently, some authors have reported super-tough PLA ternary blends,[56,57,79,80] *e.g.* consisting of PLA, a glycidyl-based elastomer and a zinc ionomer of ethylene/methacrylic acid-based copolymer (EMAA-Zn). Unlike in other super-toughened PLA blends, zinc ionomer promoted crosslinking of glycidyl elastomers that can further participate in the interfacial compatibilization, in addition to the compatibilization reaction between glycidyl moieties and PLA in this ternary system. Indeed, previous studies demonstrated that EMAA-Zn played an important role in the toughening effect on the ternary blends and showed that both zinc ions and free-carboxylic groups of methacrylic acid in the ionomer actively participated in these dual

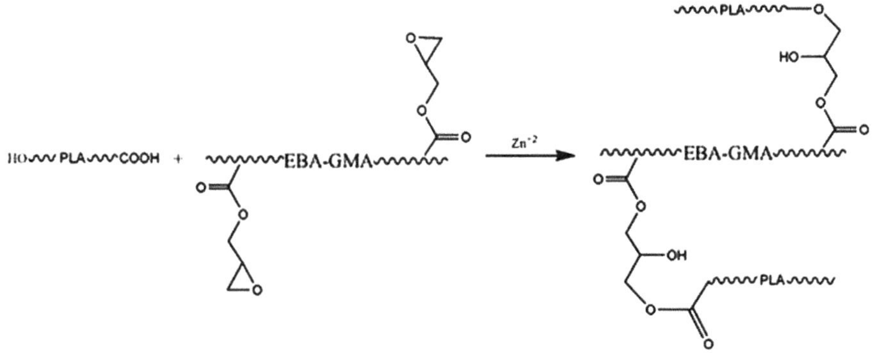

Figure 4.5   Interfacial compatibilization reaction catalyzed by Zn^{2+} present in the ionomer (reproduced with Elsevier permission from Ref. 79).

reactions involved. In a similar work, Song *et al.* reported on the reactive blending PLA with ethylene/*n*-butyl acrylate/glycidyl methacrylate (EBA-GMA) terpolymer and zinc ion-containing ionomer.[79]

The ionomer was prepared by neutralizing the ethylene/methacrylic acid copolymer, *i.e.* ionomer precursor, with ZnO-based solution. The reactive interfacial compatibilization between PLA and EBA-GMA (Figure 4.5) and the crosslinking of EBA-GMA during blending were both tuned upon the degree of neutralization (DN) of the ionomer and the methacrylic acid (MAA) content of ionomer precursors. Interestingly, the ionomers derived from precursor of high MAA content and/or having high DN tended to yield superior impact strength of the PLA blends.

4.5   REX Free-radical Grafting Reactions of PLA Chains

Free-radical reactions of PLA in the presence of, *e.g.*, organic peroxides are very simple to conduct within REX equipment for controlled degradation,[81] branching[82] and functionalization to the main polymeric backbone.[83] The earliest example of free-radical modification of PLA was reported on free-radical branching reactions.[84] It is well known that PLA chains have a low degree of shear-thinning and weakly pronounced strain-hardening, hampering the PLA processability, particularly in extrusion or elongation processing.[85] To resolve these issues, Carlson *et al.* investigated the synthesis of branched PLA by using REX technology in the presence of 2,5-dimethyl-2,5-di-(*t*-butylperoxide) as free-radical initiators in the range of 160 to 200 °C.[84] Depending on the extrusion conditions (temperature and relative content of peroxide), there are three main processes participating, *i.e.*, branching, crosslinking (gelation) and thermal degradation. In the case of branching and crosslinking, free-radicals from PLA are generated after methine H-radical abstraction along the PLA backbone (Figure 4.6). If the peroxide concentration is inside the range 0.1–0.25 wt%, branching is favoured at around 170–180 °C, while highly branched systems (possible microgelation)

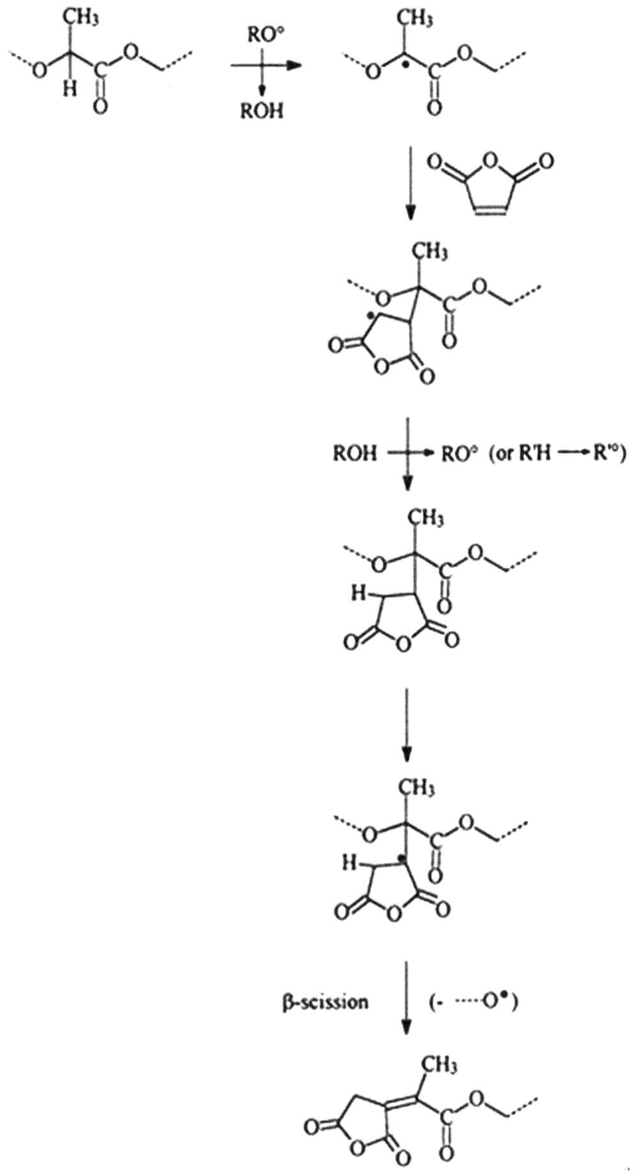

Figure 4.6 Methine H-radical abstraction and maleation process of PLA (reproduced with John Wiley and Sons' permission from Ref. 84).

are favoured at initiator concentrations higher than 0.5 wt% within the same
temperature range. However, on increasing the processing temperature (or
decreasing the peroxide concentration), chain-scissions could be promoted
via β-scissions, intramolecular transesterification and thermohydrolysis, not
all following a free-radical process. A more comprehensive investigation was
also made on the free-radical efficiency and hydrogen radical abstraction

ability of PLA-based radicals upon the type of peroxides during REX free-radical crosslinking of PLA.[81] In addition to the synthesis of branched PLA, adding free-radical initiators has shown to enhance the compatibility of immiscible polymeric partners in the case of blends between PLA and PBAT obtained after internal mixing.[86] In this regard, 2,5-dimethyl-2,5-di(t-butyl-peroxy)hexane was purposely added into these blends as a way to generate PLA/PBAT graft polymers *via* intermolecular free-radical coupling reactions. Interestingly, compatibilization between both partners was noticed, in particular in the presence of atmospheric oxygen due to the increase of the free-radical reactivity. The authors reported that PBAT tended to crosslink faster than PLA at low peroxide contents.

However, much attention has been paid to free-radical grafting of unsaturated monomers, such as maleic anhydride (MA), onto the PLA backbone using REX processing.[87–93] The most investigated unsaturated monomer is MA as moiety, which is able to participate in further compatibilization reactions with other polymeric partners or fillers.[29] Another reason is that MA does not readily polymerize under the conditions used during grafting reactions, leading to high extent of MA grafting. Although there are numerous methods to graft MA, such as melt-grafting and solid-state grafting, REX processing remains the most utilized method because the diffusion of monomers within polymeric matrix in bulk is the highest. Currently, REX grafting of MA to PLA chains is performed in the presence of peroxide such as 2,5-dimethyl-2,5-di-(t-butylperoxide)hexane within a temperature range of 180–200 °C. High extent of MA-grafting at *ca.* 0.7 wt% can be achieved, together with some β-scissions as reported by Mani *et al.*[94]

Maleated PLA can be further utilized as compatibilization agent by using REX technology in PLA-based blends/composites with hydrophilic (nano)fillers such as starch,[83,95] wood,[96] wheat straw,[97,98] soy protein concentrate[99] and cellulose nanocrystals,[95] with biodegradable polymers, such as PBAT[100] and thermoplastic starch.[101] For instance, Yuan *et al.*[100] highlighted that the use of maleated PLA enhanced the compatibility of PLA with PBAT and improved the foaming ability of the resulting blends. In the case of foams, the cell structure of foamed PLA/PBAT blend was much larger and the distribution of cells was more homogenous after adding maleated PLA. Other authors attempted to maleate other polymers, such as polyethylene rather than PLA, giving similar results of enhanced compatibility between PLA and PE.[102]

In addition, plasticized PLA-based blends were investigated in the presence of maleated PLA. Plasticization of PLA is generally conducted to enhance the flexibility of PLA due to its elevated brittleness, related with a T_g above ambient temperature.[22] Plasticizers can be divided into low-molar-mass plasticizers, including diethyl-bishydroxymethyl malonate, glucose monoesters, citrate esters, oligomeric lactic acid, glycerol, *etc.*, and high-molar-mass ones, including poly(ethylene glycol) (PEG), polyester diols, poly(ethylene adipate) and oligoesteramides. However, low-molar-mass plasticizers tend to migrate at high extent due to their high mobility where a reduction of T_g is noticed, while high-molar-mass plasticizers can be readily phase-separated, *e.g.* under certain processing conditions, consequently reducing the mechanical properties of plasticized PLA materials. In order to reduce their tendency to migration and to enhance the compatibility between PLA and plasticizers, grafting reactions between maleated PLA with hydroxyl-functionalized plasticizers,[103] *i.e.* tributyl citrate (TbC), and PEG were carried out through REX. For instance, in the case of α,ω-dihydroxyl poly(ethylene glycol),[103] a clear effect on the thermal properties of plasticized PLA was noticed, where the shift in T_g was even more pronounced after grafting reactions of maleated PLA with α,ω-dihydroxyl poly(ethylene glycol).

In addition to MA, other unsaturated monomers have been free-radically grafted onto the PLA backbone for toughening and plasticization[103] of the PLA-based materials. For instance, PLA was reactively blended with poly(acrylic acid) (PAA) and PEG.[104] While PEG was chosen as plasticizer, PAA was used as stiffening agent as well as to promote some degradation of PLA, through its acid moieties, in order to control the melt-viscosity of PLA during blending. Acrylic acid was first grafted onto the PLA backbone using a free-radical solution process, and was further mixed with PEG to conduct reactive blending between as-grafted PLA and PEG and to characterize their thermal and mechanical performances using tensile testing, DSC and DMTA analyses for the blown films derived from these compositions. Results revealed a ten-fold increase in toughness compared to neat PLA with a limited decrease in tensile strength and modulus.

Recently, Choi *et al.* have reported on free-radical grafting of acrylated poly(ethylene glycol) (Acryl-PEG) using REX to improve the plasticization extent of PLA-based materials (Figure 4.7).[105] In this case, reactive blending was carried out using an internal mixer in order to plasticize PLA at different amounts of Acryl-PEG (from 10 to 40 wt%) in the presence of 0.8 wt% dicumyl peroxide as free-radical initiator at 180 °C. Acryl-PEG was partially grafted onto the PLA backbone and the amount increased with decreasing Acryl-PEG within PLA blend as confirmed by Soxhlet extractions. The most interesting feature in this study was that the T_g values of as-grafted PLAs (PLEAs in Figure 4.7) significantly decreased by more than 20 °C with respect to neat PLA. This also had some influence on the Young's modulus of PLA by lowering it to 66%, that is from 1.2 GPa to 0.4 GPa, while the elongation at break of PLA was significantly increased. The authors reported that the grafted PEG chains also facilitated the hydrolytic degradation, especially when the Acryl-PEG (PEGA, in Figure 4.7) content was above 20%. We also developed a similar procedure exclusively on the basis of REX processing. In this process, polyethylene glycol methyl ether methacrylate (MAPEG) and Acryl-PEG were melt-mixed and extruded with PLA in the absence and in the presence of a free-radical di-tertiary alkyl peroxide, *i.e.* 2,5-dimethyl-2,5-di-(t-butylperoxy)hexane (Luperox101 or L101) used as initiator.[106] Molecular characterization revealed that in the case of PLA/MAPEG/L101 blends (79.5/20/0.5 wt%), about 20% of the initially introduced MAPEG was grafted onto the PLA chains. The remaining fraction (80%) of the plasticizer was a mixture of unreacted/monomeric and "homo-oligomerized" MAPEG. As a result,

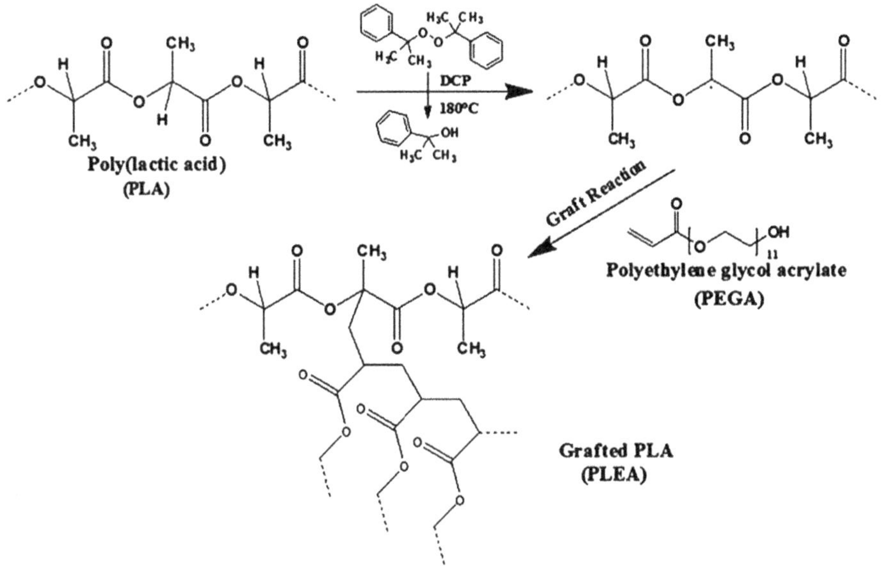

Figure 4.7 Reactive grafting of polyethylene glycol acrylate (PEGA) onto PLA (reproduced with Elsevier's permission from Ref. 105).

an efficient plasticization effect was evidenced by a significant decrease in T_g and storage modulus (E′) as well as by a drastic increase in the tensile elongation at break (approximately 70 times compared to neat PLA). More interestingly, in the case of PLA/Acryl-PEG/L101 (79.5/20/0.5 wt%), up to 65% of the initially introduced Acryl-PEG reacted and was grafted onto the PLA chains. The remaining non-grafted Acryl-PEG was completely homo-oligomerized. As a result, an efficient toughening effect of the resulting materials was reached. This was especially marked by a drastic enhancement of the impact strength, ~ 36 times, and by a significant improvement of the elongation at break, ~ 63 times.

4.6   REX Transesterification (Exchange) Reactions

Among the various methods investigated for PLA modification, such as the copolymerization of LA with other co-monomers, blending PLA with immiscible or miscible polymers is a more practical and economical way of toughening. However, most of the polymer blends are immiscible and the multiphase blends show poor mechanical performance because of the low interfacial adhesion between the polymer phases. In order to overcome the problem of immiscibility, compatibilizing agents, such as (i) *in situ* generated block or graft copolymers made of miscible segments with the blend components or (ii) polymers with complementary reactive groups that can link the matrix with the dispersed phase *via* covalent bonds formed *in situ* during melt-blending process, are used to reduce the interfacial tension between the immiscible phases. In the case of PLA, graft copolymers can be generated using interchange reactions between other functional polymer partners. Indeed, belonging to the family of aliphatic polyesters, PLA contains ester moieties able to undergo dynamic exchanges in the presence of other ester functionalities or protic species, such as alcohol compounds.[107,108] These reshuffling reactions require esterification catalysts and high temperatures.[109–111] For instance, Sadik *et al.* reported the transesterification reactions of PLA with poly[ethylene-*co*-(vinyl alcohol)] (EVOH) in internal mixer and REX processing with different catalysts, in particular 1,5,7-triazabicyclo[4.4.0]dec-5-ene and $Sn(Oct)_2$.[112] The authors reported the formulation of PLA-based blends with good barrier properties with EVOH. Indeed, EVOH is a semicrystalline copolymer widely used in food packaging and produced by co-extrusion principally as a barrier to oxygen and aromatic compounds. In this work, grafting reactions were first conducted in bulk using an internal mixer to optimize the reactive extrusion parameters. It was shown that only a few minutes were necessary to graft PLA onto the hydroxyl groups of EVOH, in particular with $Sn(Oct)_2$ as esterification catalyst. REX processing was further employed to elaborate these as-grafted copolymers for determining their mechanical performances. However, it revealed that the extrusion temperature had a great influence over the grafting extent with a maximum of 34% obtained at 230 °C by using $Sn(Oct)_2$. As supported by similar investigations from Signori *et al.*[113] using an internal mixer, it was indeed shown that PLA was susceptible to thermal degradation more than other biodegradable polyesters, such as PBAT, in melt-processing conditions, particularly using a temperature range between 150 and 200 °C, and to a lesser extent, other relevant processing parameters (moisture and nitrogen atmosphere). They reported the presence of PLA oligomers, so that the melt viscosity ratio between PLA and PBAT related with processing temperatures had a great influence on the extent of the PLA degradation. For instance, under strong degrading conditions, that is, using high processing temperatures, highly random PLA/PBAT copolymers could be obtained through intermolecular chain reactions. When re-processed within a blend containing 75 wt% PLA, some improvement in compatibilization between both polymeric partners was noticed as was shown by thermo-mechanical and morphological testing.

In addition, REX processing was used to chemically modify PLA *via* an alcoholysis mechanism for blown film applications. In this case, intensive alcoholysis reactions from a diol solution or from functionalized alcohol promoted by titanium tetrapropoxide were carried out to tune the PLA chain length and rheology.[48] It was found that under certain alcoholysis conditions, the resulting PLA blends were suitable for melt-blown non-woven processing.

4.7   Conclusions

Due to its renewability, thermo-mechanical properties and biodegradability, PLA is the front-runner of biopolymers for a large range of short-term and long-term applications. However, there are some shortcomings to be addressed, including its low crystallization, requiring applying for reactive melt-modification of PLA. The scope, advantages, limitations and specificity of various processes for the chemical transformation of PLA by reactive extrusion (REX) processing have been reviewed as a sustainable and efficient way to modify PLA covering ROP of lactide, free-radical grafting onto the PLA backbone, coupling reactions (epoxy- and isocyanate-based chain extenders) and exchange reactions of PLA chains. It has been highlighted that REX processing affords a flexible, low-cost process, with the correct selection of chemistry, enabling PLA-based materials to be produced at high-throughput processing with high conversions and few by-products, making PLA-based materials more sustainable and interesting for durable applications.

Acknowledgements

The authors are grateful to the "Région Wallonne" and European Community (FEDER, FSE) in the frame of the "Pôle d'Excellence Materia Nova" and Excellence Program OPTI^{2}MAT for their financial support. CIRMAP thanks the BELSPO for general support in the frame of the Program IUAP-6/27. J.-M. Raquez is "Chercheur Qualifié" by the F.R.S.-FNRS (Belgium).

References

1. J. M. Raquez, R. Mincheva, O. Coulembier and P. Dubois, in *Polymer Science: A Comprehensive Reference*, ed. M. Krzysztof and M. Martin, Elsevier, Amsterdam, 2012, pp. 761–778.
2. J. M. Raquez, R. Narayan and P. Dubois, *Macromol. Mater. Eng.*, 2008, **293**, 447.
3. R. Babu, K. O'Connor and R. Seeram, *Progr. Biomater.*, 2013, **2**, 1.
4. L. Avérous, in *Monomers, Polymers and Composites from Renewable Resources*, ed. B. Mohamed Naceur and G. Alessandro, Elsevier, Amsterdam, 2008, pp. 433–450.
5. J. Ahmed and S. K. Varshney, *Int. J. Food Prop.*, 2011, **14**, 37.
6. S. Bo, T. Vasily and W. James, in *Renewable and Sustainable Polymers*, American Chemical Society, New York, 2011, vol. 1063, pp. 117–132.
7. K. Pang, R. Kotek and A. Tonelli, *Progr. Polym. Sci.*, 2006, **31**, 1009.
8. X. Pang, X. Zhuang, Z. Tang and X. Chen, *Biotechnol. J.*, 2010, **5**, 1125.
9. K. Madhavan Nampoothiri, N. R. Nair and R. P. John, *Bioresour. Technol.*, 2010, **101**, 8493.
10. G. Luckachan and C. K. S. Pillai, *J. Polym. Environ.*, 2011, **19**, 637.
11. A. Södergård and M. Stolt, *Progr. Polym. Sci.*, 2002, **27**, 1123.
12. J. R. Dorgan, B. Braun, J. R. Wegner and D. M. Knauss, in *Degradable Polymers and Materials*, American Chemical Society, New York, 2006, vol. 939, pp. 102–125.
13. S. Suzuki and Y. Ikada, in *Poly(Lactic Acid)*, John Wiley & Sons, Inc., Hoboken, NJ, 2010, pp. 443–456.

14. S. Joseph, M. John, L. Pothen and S. Thomas, in *Polymers – Opportunities and Risks II*, eds. P. Eyerer, M. Weller and C. Hübner, Springer, Berlin-Heidelberg, 2010, pp. 55–80.
15. R. Narayan, in *Degradable Polymers and Materials*, American Chemical Society, New York, 2006, vol. 939, pp. 282–306.
16. S. Obuchi and S. Ogawa, in *Poly(Lactic Acid)*, John Wiley & Sons, Inc., Hoboken, NJ, 2010, pp. 457–467.
17. M. Mochizuki, in *Poly(Lactic Acid)*, John Wiley & Sons, Inc., Hoboken, NJ, 2010, pp. 469–476.
18. P. M. Visakh, A. Mathew and S. Thomas, in *Advances in Natural Polymers*, ed. S. Thomas, P. M. Visakh and A. P. Mathew, Springer, Berlin-Heidelberg, 2013, vol. 18, pp. 1–20.
19. L. T. Lim, R. Auras and M. Rubino, *Progr. Polym. Sci.*, 2008, **33**, 820.
20. L. Yu, K. Dean and L. Li, *Progr. Polym. Sci.*, 2006, **31**, 576.
21. R. M. Rasal, A. V. Janorkar and D. E. Hirt, *Progr. Polym. Sci.*, 2010, **35**, 338.
22. H. Liu and J. Zhang, *J. Polym. Sci. Polym. Phys.*, 2011, **49**, 1051.
23. S. Jacobsen and H. G. Fritz, *Polym. Eng. Sci.*, 1999, **39**, 1303.
24. J. K. Fink, in *Reactive Polymers Fundamentals and Applications (Second Edition)*, William Andrew Publishing, Oxford, 2013, pp. 373–409.
25. B. Imre and B. Pukánszky, *Eur. Polym. J.*, 2013, **49**, 1215.
26. J. K. Fink, in *Reactive Polymers Fundamentals and Applications (Second Edition)*, William Andrew Publishing, Oxford, 2013, pp. 339–371.
27. G. Moad, *Progr. Polym. Sci.*, 2011, **36**, 218.
28. V. P. Eduard and N. Z. Alexandr, *Russ. Chem. Rev.*, 2001, **70**, 65.
29. J. M. Raquez, P. Degée, Y. Nabar, R. Narayan and P. Dubois, *Compt. Rendus Chim.*, 2006, **9**, 1370.
30. E. D. Weil, *J. Polym. Sci. Polym. Chem.*, 1993, **31**, 1343.
31. J. K. Fink, in *Reactive Polymers Fundamentals and Applications*, William Andrew Publishing, Norwich, NY, 2005, pp. 507–530.
32. T. Biela, A. Kowalski, J. Libiszowski, A. Duda and S. Penczek, *Macromol. Symp.*, 2006, **240**, 47.
33. B. G. G. Lohmeijer, R. C. Pratt, F. Leibfarth, J. W. Logan, D. A. Long, A. P. Dove, F. Nederberg, J. Choi, C. Wade, R. M. Waymouth and J. L. Hedrick, *Macromolecules*, 2006, **39**, 8574.
34. A. P. Dove, in *Handbook of Ring-Opening Polymerization*, Wiley-VCH Verlag GmbH & Co., Weinheim, 2009, pp. 357–378.
35. Y. Miao and P. Zinck, *Polym. Chem.*, 2012, **3**, 1119.
36. A. Duda and A. Kowalski, in *Handbook of Ring-Opening Polymerization*, Wiley-VCH Verlag GmbH & Co., Weinheim, 2009, pp. 1–51.
37. O. Dechy-Cabaret, B. Martin-Vaca and D. Bourissou, in *Handbook of Ring-Opening Polymerization*, Wiley-VCH Verlag GmbH & Co., Weinheim, 2009, pp. 255–286.
38. P. Degee, P. Dubois, S. Jacobsen, H. G. Fritz and R. Jerome, *J. Polym. Sci. Polym. Chem.*, 1999, **37**, 2413.
39. S. Jacobsen, P. Degee, H. G. Fritz, P. Dubois and R. Jerome, *Polym. Eng. Sci.*, 1999, **39**, 1311.

40. S. Jacobsen, H. G. Fritz, P. Degee, P. Dubois and R. Jerome, *Polymer*, 2000, **41**, 3395.
41. S. Jacobsen, H. G. Fritz, P. Degee, P. Dubois and R. Jerome, *Ind. Crop. Prod.*, 2000, **11**, 265.
42. A. Gallos, G. Fontaine and S. Bourbigot, *Macromol. Mater. Eng.*, 2013, **298**, 1016.
43. S. R. Andersson, M. Hakkarainen and A. C. Albertsson, *Polymer*, 2013, **54**, 4105.
44. Y. Liu, J. Sun, X. Bian, L. Feng, S. Xiang, B. Sun, Z. Chen, G. Li and X. Chen, *Polym. Degrad. Stabil.*, 2013, **98**, 844.
45. J. M. Raquez, Y. Habibi, M. Murariu and P. Dubois, *Progr. Polym. Sci.*, 2009, **34**, 479.
46. S. Saeidlou, M. A. Huneault, H. Li and C. B. Park, *Progr. Polym. Sci.*, 2012, **37**, 1657.
47. J. Sun, J. Shao, S. Huang, B. Zhang, G. Li, X. Wang and X. Chen, *Mater. Lett.*, 2012, **89**, 169.
48. M. Nishida, T. Tanaka, T. Yamaguchi, K. Suzuki and W. Kanematsu, *J. Appl. Polym. Sci.*, 2012, **125**, E681.
49. S. Bourbigot, G. Fontaine, A. Gallos and S. Bellayer, *Polym. Adv. Tech.*, 2011, **22**, 30.
50. A. Gallos, G. Fontaine and S. Bourbigot, *Polym. Adv. Tech.*, 2013, **24**, 130.
51. O. Coulembier, S. Moins, J. M. Raquez, F. Meyer, L. Mespouille, E. Duquesne and P. Dubois, *Polym. Degrad. Stabil.*, 2011, **96**, 739.
52. T. Tarvainen, T. Karjalainen, M. Malin, S. Pohjolainen, J. Tuominen, J. Seppälä and K. Järvinen, *J. Contr. Release*, 2002, **81**, 251.
53. R. Bhardwaj and A. K. Mohanty, *Biomacromolecules*, 2007, **8**, 2476.
54. A. C. Fowlks and R. Narayan, *J. Appl. Polym. Sci.*, 2010, **118**, 2810.
55. A. Salminen, A. Nykänen, J. Ruokolainen and J. Seppälä, *Eur. Polym. J.*, 2009, **45**, 107.
56. H. Liu, X. Guo, W. Song and J. Zhang, *Ind. Eng. Chem. Res.*, 2011, **52**, 4787.
57. H. Z. Liu, F. Chen, B. Liu, G. Estep and J. W. Zhang, *Macromolecules*, 2010, **43**, 6058.
58. B. Liu, L. Jiang, H. Liu and J. Zhang, *Ind. Eng. Chem. Res.*, 2010, **49**, 6399.
59. J. Kylmä, J. Tuominen, A. Helminen and J. Seppälä, *Polymer*, 2001, **42**, 3333.
60. J. Tuominen, J. Kylmä and J. Seppälä, *Polymer*, 2002, **43**, 3.
61. M. Furukawa, T. Shiiba and S. Murata, *Polymer*, 1999, **40**, 1791.
62. M. Szycher, *Szycher's Handbook of Polyurethanes*, CRC Press, Boca Ratón, FL, 1999.
63. R. M. Michell, A. J. Muller, A. Boschetti-de-Fierro, D. Fierro, V. Lison, J. M. Raquez and P. Dubois, *Polymer*, 2012, **53**, 5657.
64. C. Nyambo, M. Misra and A. Mohanty, *J. Mater. Sci.*, 2012, **47**, 5158.
65. H. Zaman, J. Song, L. S. Park, I. K. Kang, S. Y. Park, G. Kwak, B. S. Park and K. B. Yoon, *Polym. Bull.*, 2011, **67**, 187.

66. Z. Xiong, L. Zhang, S. Ma, Y. Yang, C. Zhang, Z. Tang and J. Zhu, *Carbohydr. Polym.*, 2013, **94**, 235.
67. B. Imre, D. Bedő, A. Domján, P. Schön, G. J. Vancso and B. Pukánszky, *Eur. Polym. J.*, 2013, **49**, 3104.
68. M. Harada, K. Iida, K. Okamoto, H. Hayashi and K. Hirano, *Polym. Eng. Sci.*, 2008, **48**, 1359.
69. L. Yang, X. Chen and X. Jing, *Polym. Degrad. Stabil.*, 2008, **93**, 1923.
70. Y. M. Corre, J. Duchet, J. Reignier and A. Maazouz, *Rheol. Acta*, 2011, **50**, 613.
71. H. B. Li and M. A. Huneault, *J. Appl. Polym. Sci.*, 2011, **122**, 134.
72. Z. F. Zhou, G. Q. Huang, W. B. Xu and F. M. Ren, *Express Polym. Lett.*, 2007, **1**, 734.
73. B. Mallet, K. Lamnawar and A. Maazouz, *Polym. Eng. Sci.*, 2013, **54**, 840.
74. J. Liu, L. Lou, W. Yu, R. Liao, R. Li and C. Zhou, *Polymer*, 2010, **51**, 5186.
75. S. Pilla, S. G. Kim, G. K. Auer, S. Q. Gong and C. B. Park, *Polym. Eng. Sci.*, 2009, **49**, 1653.
76. R. Al-Itry, K. Lamnawar and A. Maazouz, *Polym. Degrad. Stabil.*, 2012, **97**, 1898.
77. Y. C. Zhang, X. Yuan, Q. Liu and A. Hrymak, *J. Polym. Environ.*, 2012, **20**, 315.
78. Z. Xiong, Y. Yang, J. Feng, X. Zhang, C. Zhang, Z. Tang and J. Zhu, *Carbohydr. Polym.*, 2013, **92**, 810.
79. W. Song, H. Liu, F. Chen and J. Zhang, *Polymer*, 2012, **53**, 2476.
80. H. Liu, W. Song, F. Chen, L. Guo and J. Zhang, *Macromolecules*, 2011, **44**, 1513.
81. M. Takamura, T. Nakamura, T. Takahashi and K. Koyama, *Polym. Degrad. Stabil.*, 2008, **93**, 1909.
82. K. M. Dean, E. Petinakis, S. Meure, L. Yu and A. Chryss, *J. Polym. Environ.*, 2012, **20**, 741.
83. P. Dubois and R. Narayan, *Macromol. Symp.*, 2003, **198**, 233.
84. D. Carlson, P. Dubois, L. Nie and R. Narayan, *Polym. Eng. Sci.*, 1998, **38**, 311.
85. H. Fang, Y. Zhang, J. Bai, Z. Wang and Z. Wang, *RSC Advances*, 2013, **3**, 8783.
86. M. B. Coltelli, S. Bronco and C. Chinea, *Polym. Degrad. Stabil.*, 2010, **95**, 332.
87. D. Carlson, L. Nie, R. Narayan and P. Dubois, *J. Appl. Polym. Sci.*, 1999, **72**, 477.
88. E. Chiellini, P. Cinelli, F. Chiellini and S. H. Imam, *Macromolecular Bioscience*, 2004, **4**, 218.
89. K. Kelar and B. Jurkowski, *Polymer*, 2000, **41**, 1055.
90. S. S. Pesetskii, B. Jurkowski, Y. M. Krivoguz and K. Kelar, *Polymer*, 2001, **42**, 469.
91. D. Shi, J. Yang, Z. Yao, Y. Wang, H. Huang, W. Jing, J. Yin and G. Costa, *Polymer*, 2001, **42**, 5549.

92. Q. Shi, L. C. Zhu, C.-L. Cai, J. H. Yin and G. Costa, *Polymer*, 2006, **47**, 1979.
93. S. W. Hwang, J. K. Shim, S. Selke, H. Soto-Valdez, M. Rubino and R. Auras, *Macromol. Mater. Eng.*, 2013, **298**, 624.
94. R. Mani, M. Bhattacharya and J. Tang, *J. Polym. Sci. Polym. Chem.*, 1999, **37**, 1693.
95. C. Zhou, Q. Shi, W. Guo, L. Terrell, A. T. Qureshi, D. J. Hayes and Q. Wu, *ACS Appl. Mater. Interfaces*, 2012, **5**, 3847.
96. D. Plackett, *J. Polym. Environ.*, 2004, **12**, 131.
97. C. Nyambo, A. K. Mohanty and M. Misra, *Biomacromolecules*, 2010, **11**, 1654.
98. C. Nyambo, A. K. Mohanty and M. Misra, *Macromol. Mater. Eng.*, 2011, **296**, 710.
99. R. Zhu, H. Liu and J. Zhang, *Ind. Eng. Chem. Res.*, 2012, **51**, 7786.
100. H. Yuan, Z. Liu and J. Ren, *Polym. Eng. Sci.*, 2009, **49**, 1004.
101. N. Wang, J. G. Yu and X. F. Ma, *Polym. Int.*, 2007, **56**, 1440.
102. R. Gallego, S. López-Quintana, F. Basurto, K. Núñez, N. Villarreal and J. C. Merino, *Polym. Eng. Sci.*, 2013, **54**, 522.
103. F. Hassouna, J. M. Raquez, F. Addiego, V. Toniazzo, P. Dubois and D. Ruch, *Eur. Polym. J.*, 2012, **48**, 404.
104. R. M. Rasal and D. E. Hirt, *Macromol. Mater. Eng.*, 2010, **295**, 204.
105. K. M. Choi, M. C. Choi, D. H. Han, T. S. Park and C. S. Ha, *Eur. Polym. J.*, 2013, **49**, 2356.
106. G. Kfoury, F. Hassouna, J. M. Raquez, V. Toniazzo, D. Ruch and P. Dubois, *Macromol. Mater. Eng.*, 2013, **299**, 583.
107. I. Moura, R. Nogueira, V. Bounor-Legare and A. V. Machado, *React. Funct. Polym.*, 2011, **71**, 694.
108. L. Wang, W. Ma, R. A. Gross and S. P. McCarthy, *Polym. Degrad. Stabil.*, 1998, **59**, 161.
109. I. Moura, R. Nogueira, V. Bounor-Legare and A. V. Machado, *Mater. Chem. Phys.*, 2012, **134**, 103.
110. J. Ding, S. C. Chen, X. L. Wang and Y. Z. Wang, *Ind. Eng. Chem. Res.*, 2008, **48**, 788.
111. J. Ding, S. C. Chen, X. L. Wang and Y. Z. Wang, *Ind. Eng. Chem. Res.*, 2011, **50**, 9123.
112. T. Sadik, F. Becquart, J. C. Majesté and M. Taha, *Mater. Chem. Phys.*, 2013, **140**, 559.
113. F. Signori, M. B. Coltelli and S. Bronco, *Polym. Degrad. Stabil.*, 2009, **94**, 74.

CHAPTER 5
Plasticization of Poly(lactide)

ALEXANDRE RUELLAN,[a,b] VIOLETTE DUCRUET[a,b] AND
SANDRA DOMENEK*[a,b]

[a] AgroParisTech, UMR 1145 Ingénierie Procédés Aliments, F-91300, Massy,
France; [b] INRA, UMR 1145 Ingénierie Procédés Aliments, F-91300, Massy,
France
*Email: sandra.domenek@agroparistech.fr

5.1   Introduction

Poly(lactic acid) (PLA) offers the opportunity to replace petroleum-based
commodity plastics in a cost-effective way because of its numerous advantages, such as ease of processing, satisfying mechanical properties, glass
transition higher than room temperature, ability for heat sealing, high
transparency, gloss, printing ability, *etc.*[1-3] Examples of these advantages
and applications are given throughout the different chapters of this book.

PLA is, however, a rigid and brittle polymer with medium tensile strength
but low deformation at break. Table 5.1 gives main characteristics of PLA
compared to petrochemical commodity plastics. Furthermore, PLA exhibits
slow crystallization rate.[4] Plasticization is a widely used technique to increase polymer ductility. Furthermore, it can also improve processability, for
example by melting point depression, improving crystallization rates and
change service applications by a decrease of the glass transition temperature. It affects moreover other properties, such as optical clarity, electric
conductivity or resistance to abiotic or biological degradation. In consequence a large literature addresses the plasticizing of PLA with a number
of different molecules and processes used.

RSC Polymer Chemistry Series No. 12
Poly(lactic acid) Science and Technology: Processing, Properties, Additives and Applications
Edited by Alfonso Jiménez, Mercedes Peltzer and Roxana Ruseckaite
© The Royal Society of Chemistry 2015
Published by the Royal Society of Chemistry, www.rsc.org

Table 5.1 Mechanical properties of PLA and main commodity plastics.

	T_g (°C)	E (GPa)	σ (MPa)	ε (%)
PET	70–80	2.8–4.1	275	60–165
PS	105	2.9	45	3
HDPE	− 125	0.4–0.95	10–60	400–1800
LDPE	− 130	0.1–0.9	9–15	100–800
PP	− 20- − 5	0.9	31–45	20
PLA	58–60	2–3	40–70	10

Plasticizers can be classified as internal or external plasticizers: internal plasticizers can be co-polymerized into the polymer structure or react with the original polymer; external plasticizers are not chemically attached to the polymer chain by covalent bonds. They are typically high boiling point liquids having a molecular mass lower than 600 g mol^{-1}.

This chapter treats the external plasticizing of PLA, while Chapter 4 "Reactive extrusion" discusses internal plasticizers. This chapter seeks to give an overview of recent results in order to provide the reader with some simple guidelines to successfully plasticizing PLA with external plasticizers.

5.2   Principles of Plasticizing

Plasticization can be explained by several theories[5] which are initially based on physical observations such as the decrease of elastic moduli, viscosity or glass transition temperature. Since the beginning of the 1940s, scientists have elaborated different theories which all tried to clarify critical phenomena observed in plasticized polymer materials.

First works[6] modelled the small plasticizer molecules as being partly attached to the macromolecules with a pending portion acting as lubricants. Thereby, plasticizer molecules should have a specific chemical structure in order to create points of attraction with the polymer chains and leave an unattached portion. Then, this lubrication model was extended[7] with the notion of voids filled by plasticizer between the macromolecules, establishing a structure of parallel alternate layers of polymers lubricated by strata made of plasticizers, which break intermolecular bonds between polymer chains. Furthermore, for Houwink[8] the quantity of broken links between macromolecules was dependent on the swelling rate induced by the dissolving ability of the plasticizer, which is higher when the respective polarities of polymer and plasticizer are similar.

Afterwards, based on the hypothesis that the elasticity development in plasticized PVC was due to the enhancement of micro-Brownian motions of the polymer,[9,10] Doolittle,[11–14] Stickney and Cheyney[15] and Alfrey *et al.*[16] developed the Gel Theory. They considered that the mechanical properties of a polymer were due to the elastic resistance of entangled segments of the macromolecules structured as a three-dimensional network. The plasticizer molecules which are dissolved into the polymer matrix break some points of attachment between macromolecules leading to an easier glide. Moreover, due to the dynamic equilibrium between plasticizers and polymer chains, phenomena of aggregation and disaggregation of cohesion links between macromolecules occur. However, due to the too high cohesion forces existing inside the polymer crystallites, these events only happen in the amorphous regions of the polymer network.[17]

The main effect of the addition of plasticizers to a polymer can also be rationalized as an increase in free volume of the system. The Free Volume Theory relies on the assumption that the free volume (*i.e.* not occupied space) of a polymer system can be decomposed into interstitial free volume, distributed continuously in space and a discontinuous distribution of holes. The free volume theory has been principally developed to describe the transport of small molecules in solid matrices. The Cohen–Turnbull–Fujita model[18,19] describes successfully the dependence of diffusivity in swollen polymer systems in assuming that the free volume of the binary system is the sum of the fractional free volume of the polymer and of the solvent. The solvent has thus a large contribution to the free volume of the system. The main action of plasticizers is to increase the free volume of the polymer/plasticizer system by augmenting the space between polymer chains in spreading them apart. They form secondary bonds with the polymer and reduce secondary bonds between polymer chains. Fox and Flory postulated the Free Volume Theory for plasticizing after they studied viscosity, thermal expansion coefficients and specific volume of polymers, as a function of temperature,[20,21] and showed that viscosity at the glass transition temperature was similar for all polymers.[22] To explain this phenomenon, Williams related the viscosity to the volume between macromolecules.[23] He concluded that for the glass transition to take place, a proper volume between macromolecules, regardless of their chemical structure, has to be reached. It was later estimated for all polymers to be 0.0646 cm^3 g^{-1}.[24] Thereby, above this volume, enough energy and space are available for configurational changes, and below T_g the amorphous structure is frozen. Increasing free volume allows for more mobility of the macromolecular chain and affects in consequence the relaxation properties of the polymer in lowering the energy needed (*i.e.* the glass transition temperature) to enable the motion of the macromolecules segments. It explains typical plasticizer effects, such as increased ductility, lowered glass transition temperature, raised crystallization rates, changed optical and conductivity properties, *etc.* A plasticizer needs thus to be bulky to develop large free volume, molecules should be miscible to permit introduction into the system and to avoid phase separation, and have preferentially low volatility to prevent loss during ageing.

The inverse effect of plasticizing is sometimes observed when a small quantity of plasticizer is added to the polymer not allowing for sufficient swelling of the amorphous phase. In this case, the free volume increase allows the amorphous phase to gain more order and to enhance the size and the number of the hypothetical crystallites.[25] This results in stronger elastic moduli and higher tensile strength,[25] opposite effects to the expected ones, being called antiplasticization.

The choice of a plasticizer is mainly ruled by its compatibility with the polymer material involved. Indeed, as explained by the plasticization theories, plasticizer molecules have to deeply penetrate into the macromolecules network and remain stable inside. Thereby, the initial emulsion during the mixing process using an extruder or roll mixers requires a thermodynamically favourable plasticizer/polymer pair. In order to quantify the compatibility, several approaches have been developed.[5] The most complex ones are the QSAR (Quantitative Structure-Activity Relationship)[26] or UNIFAP (Universal Functional Activity coefficient for Polymers),[27] which are highly effective, but need extensive amounts of data and consequently might be uncomfortable to employ in a first approach.

Another way is the determination of the Flory–Huggins interaction parameter[28] noted χ. It is a dimensionless semi-empirical constant which can depict the compatibility level between a plasticizer and a polymer. The lower is the χ value, the better will be the compatibility. The highest value for a plasticizer to be considered as compatible enough with the polymer is 0.5.

This theory is based on the Gibbs free energy of mixing equation[29–31] and can be expressed as:

$$\Delta G_{mix} = \Delta H_{mix} - T\Delta S_{mix} \tag{1}$$

$$\frac{\Delta G_{mix}}{RT} = n_a \ln \phi_a + n_b \ln \phi_b + \chi\phi_a\phi_b \left(n_a + n_b \frac{V_b}{V_a} \right) \tag{2}$$

$$\chi = \frac{z\Delta\varepsilon_{ab}}{RT} \tag{3}$$

$$\Delta\varepsilon_{ab} = \left(\frac{\varepsilon_{aa} + \varepsilon_{bb}}{2} \right) - \varepsilon_{ab}, \tag{4}$$

where ΔG_{mix} is the Gibbs free energy of mixing, n_a and n_b are respectively the mole fractions of plasticizer and polymer, ϕ_a and ϕ_b are respectively the volume fractions of plasticizer and polymer, χ is the Flory–Huggins interaction parameter, z is the coordination number and ε_{aa}, ε_{bb}, ε_{ab} are respectively the net energy associated to the contacts established between pure plasticizer molecules, pure polymer molecules and their mixture.

Another useful tool is the Hildebrand solubility theory, which is applicable to apolar and moderately polar systems. For strongly polar systems, it is unable to correctly qualify the compatibility between components.[32] However, the massive amount of interaction parameters data obtained in recent decades, and mainly Small's method,[33–35] allowing to assess them, make this method quite efficient and readily applicable. The Hildebrand solubility parameter, δ, can be defined as the square root of the cohesive energy density (CED) and it is measured in $(MJ\ m^{-3})^{0.5}$. This parameter indicates the polarity level of the component and goes from 12 $(MJ\ m^{-3})^{0.5}$ for nonpolar components to 23 $(MJ\ m^{-3})^{0.5}$ for water. The larger the difference between the plasticizer and the polymer solubility parameters is, the lower is their compatibility.

$$\delta = \sqrt{\frac{\Delta E}{V_m}} = \sqrt{\frac{\Delta H - RT}{V_m}}, \tag{5}$$

where ΔE is the cohesive energy, V_m is the molar volume and ΔH is the heat of vaporization.

Another interesting way to predict compatibility, and probably the most useful theory for PLA, has been developed by Hansen.[36,37] He assumed that the cohesive energy (derived from the free energies of vaporization of solvents as the Flory–Huggins parameter χ and Hildebrand's δ) could be split into three terms, each representing a kind of contribution: the dispersion energy (E_D), the polar cohesion energy (E_P) and the hydrogen bonding energy (E_H). Then, divided by the molar volume, the cohesive energy becomes:

$$\frac{\Delta E}{V_m} = \delta^2 = \delta_D^2 + \delta_P^2 + \delta_H^2 \tag{6}$$

The strength of this theory lies in the three-dimensional interpretation related to this method. Indeed, each of the three contribution parameters can be considered as a coordinate in space. Since the plasticizer and the polymer both have some specific contribution parameters, it is therefore possible to calculate the spatial distance between them and so determine their thermodynamical similarity degree. Thereby, the acceptance degree of the polymer, which partly depends on its molecular weight or its crystallinity ratio, can be defined as a sphere radius. The plasticizers whose coordinates are located inside the sphere are dissolving or swelling the polymer whereas those outside are not miscible within the polymer matrix. Another interesting point is that compounds with identical solubility parameters according to Hildebrand's theory are now able to be distinguished.[38] Finally, once again, the closer both plasticizer and polymer coordinates are, the better would be the compatibility:

$$\text{distance} = \left[4\left(\delta_{D_{Plast}} - \delta_{D_{Pol}}\right)^2 + \left(\delta_{P_{Plast}} - \delta_{P_{Pol}}\right)^2 + \left(\delta_{H_{Plast}} - \delta_{H_{Pol}}\right)^2 \right]^{1/2} \tag{7}$$

Note that the factor "4" is used to obtain a convenient fit to experimental data. Moreover, as for Hildebrand's solubility parameter, it is possible to calculate the Hansen parameters using a similar increment method as Small's approach.

Some solubility parameters and molar volume values[37] for the main plasticizers chemical families are given in Table 5.2. Regarding the Hansen parameters data table, the relative energy difference (RED) between PLA and molecules has been calculated by dividing the distance by the PLA solubility sphere radius.[38] Therefore, the lower the difference value is, the better is the miscibility. Furthermore, a distance value higher than 1 means that the molecule is out of the PLA solubility sphere and it will not be miscible within the PLA matrix. On the other hand, the molar volume informs on the diffusion ability of the molecule getting into the polymer network.[38] The smaller it is, the faster the molecule is likely to diffuse. Thereby, a molecule having a low molar volume and a high distance value will exude out of PLA and induce poor ageing properties. Moreover, according to the free volume theory, a plasticizer with a high molar volume would be more efficient. Taking into account these indications, molecules such as di(ethyl-hexyl phthalate (DEHP), tricresyl phosphate or triisooctyl trimellitate would potentially be good plasticizers for PLA. Finally, comparing the solubility values obtained for Hildebrand and Hansen theories, respectively, some differences can be noticed. For example, according to Hildebrand, tributyl citrate (TBC) and butyl benzoate would have nearly the same compatibility with PLA, whereas Hansen theory says that butyl benzoate would be almost "twice as compatible" as TBC. Nevertheless, the Hansen theory usually provides more truthful indications for PLA compatibilities than Hildebrand's.[38]

5.3 Plasticizer Permanence, Migration and Interaction with Contact Media

Miscibility of plasticizers with the polymer matrix and their diffusion rate inside the structure are important for their permanence. Migration means removing the plasticizer from the material, being extracted into a neighbouring gas (evaporation), liquid or solid phase (extraction), and it is one of the toughest challenges in the choice or development of an external plasticizer. The physicochemical mechanism behind the migration is the transport (diffusion) of molecules in solid phases and the partitioning with the neighbouring phase. Therefore, the plasticizer should have low vapour pressure and diffusion rate in the polymer. Plastic materials are often in contact with stationary or flowing liquids or with solid materials to where plasticizers can be leached out. Besides the technological problems raised by leaching out of the plasticizer, there are evident potential risks for health and the environment. PLA is furthermore a biodegrading polymer, and consequently the plasticizer becomes inevitably freed in end-of-life treatments, such as composting. These problems are addressed in PLA plasticizing by using plasticizers admitted for food contact materials and, preferentially, biodegradable by themselves. To decrease migration problems and increase permanence, the rate of transport of the molecules inside the polymer matrix has to be slowed down. This can be achieved by using high molecular mass plasticizers, as diffusivity decreases with increasing molecular mass.[46–49]

In the case of small plasticizers, the considered molecules need to be miscible, *i.e.* to present strong intermolecular forces with the polymer. In the case of partial miscibility, phase separation occurs when the miscibility limit of the plasticizer is exceeded and pure plasticizer phase and enriched

Table 5.2 Solubility parameters of monomeric and oligomeric plasticizers.

Chemical family	Name	Abbreviation	Cas N°	δ (MJ m^{-3})$^{0.5}$	Difference
		Reference		Hildebrand solubility parameter	
	Polylactide	PLA	33135-50-1	21.9	–
Citrates	Triethyl citrate	TEC	77-93-0	20.4	1.5
	Tributyl citrate	TBC	77-94-1	19.8	2.1
	Acetyl triethyl citrate	ATEC	77-89-4	19.0	2.9
	Acetyl tributyl citrate	ATBC	77-90-7	18.4	3.5
Polycitrates	Tri(tributyl citrate)	TBC-3	–	18.6[41]	1.5[41]
	Hepta(tributyl citrate)	TBC-7	–	18.6[41]	1.5[41]
Adipates	Dioctyl adipate	DOA	123-79-5	17.6	4.3
	Bis(2-ethylhexyl) adipate	DEHA	103-23-1	17.6	4.3
	Diisobutyl adipate	DiBA	141-04-8	18.0	3.9
	Diisononyl adipate	DiNA	33703-08-1	17.0	4.9
Polyadipates	Poly(ethylene adipate)	PEA	–	18.9[42]	0.3[42]
	Poly(butylene adipate)	PBA	–	18.1[42]	1.1[42]
	Poly(hexalethylene adipate)	PHA	–	17.6[42]	1.6[42]
	Poly(diethylene adipate)	PDEA	–	18.1[42]	1.1[42]
	Glyplast® 2006/2		–	21.9[43]	2[43]
	Glyplast® 206/7		–	22.9[43]	3[43]
Phthalates	Dioctyl phthalate	DOP	117-84-0	18.3	3.6
	Ditridecyl phthalate	DtDP	119-06-2	17.6	4.4
	Dimethyl phthalate	DMP	131-11-3	22.1	0.1
Sebacates	Dimethyl sebacate	DMS	106-79-6	18.1	3.8
	Dibutyl sebacate	DBS	109-43-3	17.8	4.1
				17.7[44]	5.4[44]
	Di-(2-ethyl hexyl) sebacate	DEHS	122-62-3	17.5	4.4
Phosphates	Tributyl phosphate	TBPA	126-73-8	18.0	3.9
	Tricresyl phosphate	TCPA	1330-78-5	23.1	1.2
	Trioctyl phosphate	TOPA	1806-54-8	17.7	4.2
	Triphenyl phosphate	TPPA	115-86-6	22.2	0.3
Maleates	Dibutyl maleate	DBM	105-76-0	19.0	2.9
Azelates	Di-(2-Ethyl Hexyl) Azelate	DEHAz	103-24-2	17.4	4.5
Trimellitates	Triisooctyl trimellitate	TiOT	27251-75-8	17.8	4.1
	Triisononyl trimellitate	TiNT	53894-23-8	17.7	4.2
Stearates	Butyl stearate	BS	123-95-5	15.4	6.5
Cinnamates	Ethyl cinnamate	EC	103-36-6	20.6	1.4
Benzoates	Benzyl benzoate	BzB	120-51-4	21.3	0.6
	Butyl benzoate	BuB	136-60-7	19.9	2.0
Acetates	Glycerol triacetate (triacetin)	GTA	102-76-1	19.4	2.5
	Diethylene glycol butyl ether acetate	DEGBEA	124-17-4	18.4	3.5
	2-Ethyl hexyl acetate	EHA	103-09-3	16.9	5.1
Polymalonates	Diethyl bishydro-xymethyl malonate	DBM	20605-01-0	20.2[45]	0.1[45]
	DBM/adipoyl dichloride copolymer	DBM-A-8	–	18.7[45]	1.4[45]
	DBM/adipoyl dichloride copolymer	DBM-A-18	–	18.7[45]	1.4[45]
	DBM/succinyl dichloride copolymer	DBM-S-4	–	18.4[45]	1.7[45]
	DBM/succinyl dichloride copolymer	DBM-S-7	–	18.4[45]	1.7[45]

δ_D (MJ m^{-3})$^{0.5}$	δ_P (MJ m^{-3})$^{0.5}$	δ_H (MJ m^{-3})$^{0.5}$	Distance	Relative Energy Difference (Distance/radius)	Molecular weight (g mol^{-1})	Molar volume (cm^3 mol^{-1})[37]	Boiling point at 760 mmHg (°C)	Approval	Specific migration limit (mg kg^{-1})
18.6[38]	9.9[38]	6.0[38]	–	(radius 10.7)[38]	–	–	–	Yes	60
16.5	4.9	12.0	8.9	0.83	276	243	294	Yes	60
16.6	3.8	10.1	8.4	0.78	360	346	325	No	–
16.6	3.5	8.6	8.0	0.75	318	280	327	No	–
16.7	2.5	7.4	8.4	0.79	402	384	441	Yes	60
–	–	–	–	–	980	–	–	No	–
–	–	–	–	–	2240	–	–	No	–
16.7	2.0	5.1	8.8	0.82	370	400	398	No	–
16.7	2.0	5.1	8.8	0.82	370	400	374	Yes	18
16.7	2.5	6.2	8.3	0.78	258	270	293	No	–
16.2	1.8	4.9	9.5	0.89	398	434	405	No	–
–	–	–	–	–	2000	–	–	No	–
–	–	–	–	–	2000	–	–	No	–
–	–	–	–	–	2000	–	–	No	–
–	–	–	–	–	2000	–	–	No	–
–	–	–	–	–	$M_n = 1532$	–	–	No	–
–	–	–	–	–	$M_n = 2565$	–	–	No	–
16.6	7.0	3.1	5.7	0.54	390	377	230	No	–
16.6	5.4	1.9	7.3	0.68	530	558	508	No	–
18.6	10.8	4.9	1.4	0.13	194	163	284	No	–
16.6	2.9	6.7	8.1	0.76	230	233	288	No	–
16.7	4.5	4.1	6.9	0.64	314	339	345	Yes	60
16.8	1.0	4.7	9.7	0.91	424	469	551	No	–
16.3	6.3	4.3	6.1	0.57	266	274	289	No	–
19.0	12.3	4.5	2.9	0.27	368	316	438	No	–
16.2	5.9	4.2	6.5	0.61	434	470	414	No	–
20.1	6.4	6.8	4.7	0.44	326	272	447.0	No	–
16.5	6.1	7.2	5.8	0.54	228	230	280	No	–
16.7	1.4	4.8	9.4	0.88	412	450	417	No	–
16.6	6.0	2.5	6.6	0.62	547	553	585	No	–
16.6	5.7	2.2	6.9	0.65	588	603	617	No	–
14.5	3.7	3.5	10.6	0.99	340	382	343	Yes	60
18.4	8.2	4.1	2.6	0.2	176	167	270	No	–
20.0	5.1	5.2	5.6	0.52	212	191	324	No	–
18.3	5.6	5.5	4.4	0.41	178	178	250	Yes	60
16.5	4.5	9.1	7.5	0.70	218	188	258	No	–
16.0	4.1	8.2	8.1	0.76	204	208	240	No	–
15.8	2.9	5.1	9.0	0.84	172	196	200	No	–
–	–	–	–	–	220	180	350	No	–
–	–	–	–	–	$M_n = 2500$ $M_w = 4200$	–	–	No	–
–	–	–	–	–	$M_n = 5300$ $M_w = 8900$	–	–	No	–
–	–	–	–	–	$M_n = 1300$ $M_w = 1800$	–	–	No	–
–	–	–	–	–	$M_n = 2100$ $M_w = 3500$	–	–	No	–

Hansen solubility parameters[37] | Physical properties[39] | Food contact safety[40]

Table 5.2 (*Continued*)

	Reference			Hildebrand solubility parameter	
Poly(ethylene glycol)	Ethylene glycol	EG	107-21-1	33.0	11.0
	Diethylene glycol	DEG	111-46-6	29.1	7.2
	Triethylene glycol	TEG	112-27-6	27.5	5.6
	Tetraethylene glycol	PEG 200	112-60-7	24.3	2.4
				23.5^{44}	0.4^{44}
	Poly(ethylene glycol) 400	PEG 400	25322-68-3	22.5^{44}	0.6^{44}
	Poly(ethylene glycol) 800	PEG > 800	25322-68-3	20.6	1.3
	Poly(ethylene glycol) 1000	PEG 1000	25322-68-3	21.9^{44}	1.2^{44}
	Poly(ethylene glycol)	PEG	25322-68-3	22.1	0.2
	Acetyl glycerol monolaurate	AGM	–	18.5^{44}	4.6^{44}
	Poly(1,3-butanediol)	PBOH	–	21.3^{44}	1.8^{44}
Bio-based	Limonene	Li	6876-12-6	17.8	4.1
	Glycerol	Gly	56-81-5	36.2	14.3
	Oleic acid	OA	112-80-1	17.4	4.5
	Castor oil	CO	8001-79-4	18.2	3.7
	Linseed oil	LO	8001-26-1	14.4	7.5
	Lactic acid	LA	50-21-5	34.1	12.21

polymer phase are observed. Thus, two glass transitions can be observed, one of the plasticizer enriched or pure plasticizer phase and one of the polymer containing the plasticizer at its miscibility limit. This can give rise to the presence of plasticizer droplets on the surface especially during ageing. Polymeric plasticizers, possessing the advantage of low volatility and high permanence, have, however, the drawback of low miscibility and reduced efficiency. Compared to monomeric plasticizers, they have fewer chain ends per mass of plasticizer and develop lower free volume. Branched structures will bring more free volume to the system and will be often more efficient in increasing polymer mobility inside the material than linear molecules of same molecular weight. An interesting result was obtained recently by Fang *et al.*,[48] who showed that the introduction of flexible chemical structures near the centre of mass of a molecule can have a tremendous impact on the diffusion coefficient. In the case of biphenyls, it was found that the addition of one -CH$_2$ group between benzene rings diminished the diffusion coefficient by one order of magnitude. However, compatibility with the target polymer needs to be ensured for example by using those methods of solubility prediction discussed earlier.

The compatibility of the plasticizer in the PLA matrix, the morphological stability of the plasticized material and the prevention of the plasticizer migration from the material bulk should be optimized as leaching out of additives from materials could impact the media in contact. This is of particular importance where food packaging applications are concerned, because food safety has to be ensured and contamination risks should be minimized. As with all substances intentionally added to the packaging polymer, the choice of plasticizer must comply with the European Commission regulation EU 10/2011 on plastic materials intended to come into

Hansen solubility parameters[37]					Physical properties[39]			Food contact safety[40]	
17.0	11.0	26.0	20.3	1.90	62	56	197	Yes	30
16.6	12.0	20.7	15.4	1.44	106	95	245	Yes	30
16.0	12.5	18.6	13.9	1.30	150	114	286	Yes	60
16.6	5.7	16.8	12.3	1.15	194	175	314	Yes	60
–	–	–	–	–	400	–	–	Yes	60
16.6	7.3	9.8	6.1	0.57	>800	–	–	Yes	60
–	–	–	–	–	1000	–	–	Yes	60
17.0	11.0	8.9	4.5	0.42	>10000	–	–	Yes	60
–	–	–	–	–	358	–	–	No	–
–	–	–	–	–	$M_w = 2100$	–	–	No	–
17.2	1.8	4.3	8.7	0.82	136	–	75	No	–
17.4	12.1	29.3	23.5	2.20	92	73	290	Yes	60
16.0	2.8	6.2	8.8	0.82	282	317	390	Yes	60
13.6	6.0	10.5	11.6	1.09	–	–	–	Yes	60
13.5	3.5	3.7	12.3	1.15	–	–	–	No	–
17.0	8.3	28.4	22.7	2.1	90	72	227	Yes	60

contact with food. This regulation establishes a positive list of these compounds authorized for use in plastic formulations and manufacturing and provides migration limits for quite a number of molecules. Table 5.2 gives information on the regulatory status of the different plasticizers described.

In comparison with other additives used in packaging materials, plasticizers focus attention as they need to be used in high amounts in glassy polymers, up to 50 wt% in PVC and up to 30 wt% in PLA to reach the expected mechanical properties.

With respect to migration during food contact, obviously cyclic lactide monomers, oligolactic acid additives and linear oligolactic acid could be considered as ideal plasticizers in comparison to traditional approaches, as they have a similar structure to the polylactide chain and they are, thus, not expected to cause new toxic migrants.[50] On the contrary, these low-molecular-mass molecules can diffuse rapidly in the bulk of the polymer and cause contamination problems to the media in contact with the polymer. As an example, lactide is an environmentally degradable, non-toxic additive that is well known as a good plasticizing agent for PLA, but its rather fast migration rate results in a stiff material with time and it can easily contaminate the processing equipment.[51]

Thus, controlling the degradation rate, the release of degradation products and other migrants are key issues during the design of plasticized polymers. Few studies related the influence of plasticizer structure on the migration (*i.e.* leach out) phenomena in contact with food or simulants (according to the regulation EU 10/2011). In general, large leaching out of plasticizer is expected when phase separation occurs and that plasticizer diffuses in the bulk to reach the material's surface, as was shown for PEG, for example.[52]

Non-intentional plasticization could occur during food storage. In fact, polymers used in packaging are not inert and mass transfers occur between them and foodstuff, even in the case of glassy polymers like PET or PVC.[53] On one hand, sorption of organic compounds into packaging such as aroma compounds could impact food quality. On the other hand, a number of studies showed that the sorption of flavour molecules can change the properties of the packaging material by swelling and/or plasticization. Plasticization and solvent-induced crystallization of PLA by ethyl acetate were observed using high ethyl acetate activities 0.5 and 0.2, respectively.[54,55] For example, Colomines *et al.*[54] found that T_g decreased from 50.9 °C to 16.9 °C, after 3 days of contact at 0.5 activity. Recently Salazar *et al.*[56] showed that a small plasticization of PLA occurs even at very low concentration (<200 ppm) in vapour contact with aroma compounds like ethyl esters. Benzaldehyde, an aromatic molecule, showed high interaction with PLA compared to ethyl esters, although the latter have a more similar chemical structure. Furthermore, this aldehyde played a synergy role towards the low interacting molecules present in the mixture and augmented their solubility in the polymer. The good solubility of benzaldehyde in PLA was in accordance with the prediction of the Hansen solubility (high δ_D) parameters and could be compared with the good miscibility of PLA with aromatic solvents.

5.4   PLA Plasticizers: Properties, Effects and Processability

The first goal of plasticizing a polymer is, as discussed, the improvement in the mechanical properties, mainly the increase in the elongation at break, and the decrease of T_g to render the polymer more flexible at service conditions. In this respect, many different molecules have been tested as PLA plasticizers. But there have been only a few reviews on the subject, such as Saeidlou *et al.*[4] and Liu and Zhang.[2] To get a clear picture on the main findings and advances in PLA plasticizing, we choose to separate the discussion of literature results in systems using monomeric and polymeric molecules. The discussion is supported by tables gathering main literature results for either monomeric (Table 5.3) or polymeric (Table 5.4) plasticizers. In the aim of creating bio-based polymer materials, the use of additives derived from renewable resources, often being co-products of agriculture or food industry, has conceptual importance. To highlight these novel approaches, we voluntarily separated these novel plasticizers into a specific chapter.

5.4.1   Plasticizer Impact on Mechanical Properties

5.4.1.1   *Monomeric Plasticizers*

Table 5.3 gathers the main literature results on PLA plasticized with monomeric external plasticizers. It shows furthermore that different

Table 5.3 Glass transition and mechanical properties of PLA plasticized with monomeric external plasticizers.

Plasticizer	Process	M (g mol^{-1})	Content (%)	Tg (°C) DSC	σ (MPa)	E (MPa)	ε (%)	Ref.
LA	–	144	1.3	–	51.7	1993	3	57
			17.3	–	15.8	820	288	
			19.2	32–40	29.2	658	536	
			25.5	–	16.8	232	546	
PLA (blank)	MM + E	M_w = 137000						58
TEC		276	0	59.1	51.7	–	7	
			10	42.1	28.1	–	21.3	
			20	32.6	12.6	–	382	
TBC		360	10	40.4	22.4	–	6.2	
			20	17.6	7.1	–	350	
ATEC		318	10	50.8	34.5	–	10	
			20	30	9.6	–	320	
ATBC		402	10	25.4	17.7	–	2.3	
			20	17	9.2	–	420	
PLA	TS	M_n = 74000	0	54	57	3750	5	51
Loxiol GMS 95		–	2.5		52	3400	15	
			5		48	3200	7	
			10	45	45	3000	8	
Dehydat VPA 1726 (GME)		–	2.5		53	3300	5	
			5		47	3000	6	
			10	40	38	2500	13	
PLA	IM + E	M_w = 49000	0	58	–	2050 ± 44	9 ± 2	59
Glycerol		92	10	54	–	–	–	
			20	53	–	–	–	
TBC		360	10	51	–	–	–	
			20	46	–	–	–	
PLA	TS	–	0	60 (onset)	66	3300	2	60
ATBC		402	5	49	53	3200	5	
			10	41	50	2900	7	
			12.5	26	18	100	218	
			15	27	21	100	299	
			20	24	23	100	298	
PLA (blank)	IM	M_w = 100000	0	54 (DMA)	–	–	–	61
TBC		360	15	29	–	–	–	
			20	16	–	–	–	
			25	1	–	–	–	

Table 5.3 (*Continued*)

Plasticizer	Process	M (g mol^{-1})	Content (%)	Tg (°C) DSC	σ (MPa)	E (MPa)	ε (%)	Ref.
GTA		218	15	29	–	–	–	62
			20	21	–	–	–	
			25	10	–	–	–	
			30	0	–	–	–	
PLA (blank)	TS	$M_w = 100000$	0	54	–	–	6	62
GTA		218	11	29			355	
TBC		360	12	29			350	
PLA	TS	$M_w = 100000$	0	52	–	–	–	63
TBC		360	15	25				
DBM		220	15	29				
PLA	IM	$M_w = 74000$	0	59.2	64 ± 1.5	2840 ± 50	3 ± 0.3	44
DBS		314	10	39.9	39.2 ± 4	2000 ± 80	2.3 ± 0.2	
			20	− 66.9/26.1	23.1 ± 0.9	430 ± 50	269 ± 6	
AGM		358	10	45.8	52.1 ± 4	2240 ± 100	32 ± 2.1	
			20	− 65.8/24.3	27.1 ± 3.1	35 ± 5	335 ± 2.3	
PLA	C	$M_n = 63000$	0	58.2	–	–	–	64
DOA		371	10	40.8				
			20	40.1				
PLA	TS + I	–	0	–	69.2 ± 0.41	3680 ± 130	6.0 ± 0.29	65
ATEC		318	2	–	64.3 ± 0.68	3490 ± 70	8.8 ± 1.78	
			5	–	57.8 ± 1.32	3460 ± 50	5.2 ± 0.34	
			10	–	50.1 ± 0.28	3150 ± 80	6.1 ± 1.30	
PLA	IM	$M_n = 74500$	0	62	66 ± 2	1020 ± 100	11 ± 3	66
ATBC		402	10	44	51 ± 1	970 ± 70	11 ± 4	
			15	32	37 ± 1	590 ± 50	221 ± 8	
			20	38	30 ± 1	270 ± 20	317 ± 4	
DOA		371	5	49	–	–	–	
			10	45	29 ± 2	720 ± 90	36 ± 5	
			15	45	22 ± 1	710 ± 50	77 ± 44	
			20	45	21 ± 1	670 ± 120	78 ± 33	

Sample	Method	M	Plasticizer content					Ref.
GTA		218	10	48	38 ± 3	760 ± 140	8 ± 2	67
			15	45	31 ± 5	590 ± 110	223 ± 19	
			20	29	24 ± 1	10 ± 3	443 ± 13	
PLA[a] / DOP	TS	$M_n = 107000$ / 390	0	61.7	79 ± 2	1870 ± 60	5.63 ± 0.03	43
			5	52.3	71 ± 2	1720 ± 40	6.20 ± 0.12	
			7.5	40.9	65 ± 1	1660 ± 120	6.35 ± 0.02	
			10	41.6	60 ± 2	1480 ± 140	147 ± 5	
			12.5	40.4	44 ± 1	1330 ± 180	221 ± 11	
			15	41.3	46 ± 2	1300 ± 90	225 ± 9	
PLA / DOA	IM	$M_n = 63000$ / 371	0	58.2	47 ± 5	2000 ± 200	6 ± 2	68
			10	40.8	27 ± 4	1600 ± 100	259 ± 64	
			20	40.1	17 ± 1	1400 ± 100	295 ± 89	
PLA / TBC	IM	$M_n = 81000$ / 360	0	61	52 ± 2	1800 ± 150	6 ± 1	68
			20	20	20 ± 1	9 ± 1	320 ± 20	
PLA / DINCH	IM	$M_n = 142000$ / 425	0	59.4	69.5 ± 0.6	–	4 ± 1	69
			3	55.2	–	–	–	
			5	49.8	–	–	–	
			8	49.7	–	–	–	
			10	49.8	41.1 ± 0.2	–	129 ± 20	
			20	49.9	30.1 ± 0.1	–	200 ± 16	
TBC		360	10	38.5	57.1 ± 0.4	–	7 ± 1	
			20	24.0	37.2 ± 1.2	–	300 ± 15	
PLA[b] / TEC[c]	TS	$M_w = 204453$ / 360	0	58.01	41.7	3364	1.27	70
			12.44	41.12	38.3	2542	2.14	
			12.99	37.61	12.4	1248	> 100.2	
			15.58	32.1	6.4	53	> 100.16	
			22.52	21.74	9.9	55	> 102.77	
PLA[d] / TEC[e]		$M_w = 204453$ / 360	0	–	41.69	3364	1.27	
			12.44	39.45	41.95	2301	3.43	
			12.99	36.7	19.97	1337	> 99.98	
			15.58	30.37	7.67	313	> 100.03	
			22.52	33.95	11.95	159	100.08	
PLA	IM	–	0	60.7	–	–		71
[THTDP][DE]		655	5	49.5	–	–	brittle[f]	
			10	45.1	–	–		
[THTDP][BF$_4$]		571	5	55.9	–	–	ductile[f]	
			10	54.5	–	–		

Table 5.3 (*Continued*)

Plasticizer	Process	M (g mol⁻¹)	Content (%)	Tg (°C) DSC	σ (MPa)	E (MPa)	ε (%)	Ref.
PLA	IM	M_n = 90500	0	58	47 ± 7	1291 ± 60	8 ± 5	72
ATBC		402	2.5	52	40 ± 8	1300 ± 69	4 ± 1	
			5	47.3	41 ± 9	1306 ± 64	8 ± 3	
			9	41.5	30 ± 5	1039 ± 55	16 ± 10	
			13	35.1	22 ± 4	615 ± 41	300 ± 179	
			17	28.6	–	69 ± 18	503 ± 45	
PLA[g]	IM	M_n = 90500	0	58	46 ± 8	1216 ± 106	5 ± 2	55
ATBC		402	17	25.1	20 ± 2	347 ± 21	59 ± 24	
PLA	TS	M_w = 121400	0	58.6	69.8 ± 3.2	1777 ± 42	5.7 ± 0.3	73
GMS		358	5	52.7	44.8 ± 1.3	1570 ± 44	4.5 ± 0.5	
			10	52.5	41.9 ± 4.6	1200 ± 12	7.6 ± 2.4	
			15	52.3	39.7 ± 1.0	1270 ± 36	11.0 ± 5.0	
			20	51.9	35.1 ± 2.1	1210 ± 17	9.5 ± 6.5	
			25	52.1	32.4 ± 1.8	1190 ± 24	11.0 ± 3.1	
			30	51.5	29.9 ± 2.6	695 ± 38	45 ± 16	
PLA	IM	M_n = 96000	0	58	63 ± 4	2400 ± 100	49 ± 9	74
ATBC		402	20	33	50 ± 5	2400 ± 200	158 ± 27	
PLA/D43B[h] (3%)		–	0	58	59 ± 3	2800 ± 100	3 ± 0.2	
D43B[h] + ATBC		–	20	34	37 ± 5	1900 ± 300	167 ± 55	
PLA	TS	M_n = 100000	0	60	49 ± 4	3100 ± 150	3.4 ± 0.9	
ATBC		402	20	38	16 ± 2	1000 ± 140	143 ± 14	
PLA/D43B[h] (3%)		M_n = 59000	0	57	56 ± 5	3400 ± 200	2.3 ± 0.3	
D43B[h] + ATBC		–	20	31	34 ± 7	2100 ± 300	163 ± 35	

TS twin screw extrusion, IM internal mixer, E single screw extrusion, I injection molding, C casting, MM manual mixing.
[a]crosslinked PLA.
[b]aged 10 days.
[c]aged 20 days.
[d]aged 24 days.
[e]aged 32 days.
[f]behavior assessed by 3-point bending test.
[g]semicrystalline samples (degree de crystallinity 30%).
[h]Montmorillonite Dellite 43B.

Table 5.4 Glass transition and mechanical properties of PLA plasticized with polymeric external plasticizers.

Plasticizer	Process	M_w/M_n	Content (%)	T_g (°C) DSC	σ (MPa)	E (MPa)	ε (%)	Ref.
PLA	TS	$M_n = 74000$	0	54	57	3750	5	51
PEG 1500		$M_w = 1500$	2.5	–	50	3200	5	
			5	–	44	2500	7	
			10	28	38	1300	40	
PLA	IM + E	$M_w = 49000$	0	58	–	2050 ± 44	9 ± 2	59
OLA		–	10	37	–	1256 ± 38	34 ± 4	
			20	18	–	744 ± 22	200 ± 24	
M-PEG		$M_w = 400$	10	34	–	1571 ± 51	18 ± 2	
			20	21	–	1124 ± 33	142 ± 19	
PEG 400		$M_w = 400$	10	30	–	1488 ± 39	26 ± 5	
			20	12	–	976 ± 31	160 ± 12	
PEG 1500		$M_w = 1500$	10	41	–	–	–	
			20	30	–	–	–	
PLA	TS	$M_w = 100000$	0	54	–	–	6	41
TBC-3 oligomer		980	10	38	–	–	–	
			15	33	–	–	–	
			20	30	–	–	–	
TBC-7 oligomer		2240	15	43	–	–	–	
PLA	TS	$M_w = 100000$	0	52	–	–	–	45
DBM		220	15	29	–	–	–	
DBM-A-8 oligomer		$M_n = 4200$	15	42	–	–	–	
DBM-A-18 oligomer		$M_n = 5300$	15	39	–	–	–	
DBM-S-4 oligomer		$M_n = 1300$	15	36	–	–	–	
DBM-S-7 oligomer		$M_n = 2100$	15	40	–	–	–	
PLA	TS	$M_w = 160000$	0	58	53 ± 2	2200 ± 50	14 ± 1	77
PEG 8000		$M_w = 8000$	10	36	23 ± 1	950 ± 30	200 ± 10	
			15	30	16 ± 1	630 ± 20	260 ± 10	
			20	21	5 ± 1	180 ± 20	300 ± 20	
			30	9	–	5 ± 1	500 ± 20	

Table 5.4 (*Continued*)

Plasticizer	Process	M_w/M_n	Content (%)	T_g (°C) DSC	σ (MPa)	E (MPa)	ε (%)	Ref.
PLA	TS	M_w = 190000	0	60	68±2	2500±200	3±0,5	78
PEG 8000		M_w = 8000	10	39	26±1	900±50	180±10	
			20	21	4±0.5	150±20	260±20	
			30	12	–	20±2	300±30	
PLA	TS	M_w = 84000	0	60 (onset)	66	3300	2	60
PEG 400		M_w = 400	5	40	42	2500	2	
			10	23	33	1200	140	
			12.5	22	19	500	115	
			15	20	19	600	88	
			20	19	16	500	71	
PEG 1500		M_w = 1500	5	46	52	2900	4	
			10	42	47	2800	5	
			12.5	30	19	700	194	
			15	27	24	800	216	
			20	20	22	600	235	
PEG 10000		M_w = 10000	5	49	54	2800	2	
			10	42	49	2800	3	
			15	36	42	2500	4	
			20	34	22	700	130	
PLA	TS	M_n = 81800	0	55	–	–	–	79
PEG 1000		M_w = 1000	20	15	–	–	–	
PLA+CNa$^+$-3%/PEG		–	20	14	–	–	–	
PLA+C30B-3%/PEG		–	20	16	–	–	–	
PLA+C25A-3%/PEG		–	20	16	–	–	–	
PLA+C20A-3%/PEG		–	20	16	–	–	–	
PLA	IM	M_w = 160000	0	56	–	–	–	80
PEG 1500		M_w = 1500	10	45	–	–	–	
PLA+C25A-3%/PEG		–	10	44	–	–	–	
PLA	IM	M_n = 85000	0	58	–	–	–	81
PLA+C30B-3%	IM	–	0	58	–	–	–	
PLA+C30B-3%	MB+IM	–	0	61	–	–	–	

Sample	Method	M	Conc (wt%)						Ref
PLA + C30B-3%/PEG	MB + IM	$M_w = 1000$	20	26	–	–	–	–	82
PLA	IM	$M_n = 81800$	0	49.7	–	–	–	–	
PEG 1000		$M_w = 1000$	20	23.4	–	–	–	–	
PLA + C20A-3%/PEG		–	20	28.2	–	–	–	–	
PLA + C25A-3%/PEG		–	20	25.8	–	–	–	–	
PLA + C30B-3%/PEG		–	20	27.2	–	–	–	–	
PLA	IM	$M_w = 166000$	0	57	47	–	18	–	83
PEG 400		$M_w = 400$	5	–	–	–	–	–	
			10	47	39	–	300	–	
PEG 600		$M_w = 600$	5	36	18	–	25	–	
			10	43	35	–	550	–	
PEG-CH$_3$		$M_w = 550$	5	32	20	–	22	–	
			10	–	35	–	550	–	
PEG-CH$_3$		$M_w = 750$	5	–	18	–	>30	–	
			10	–	–	–	550	–	
PLA	IM	$M_w = 108000$	0	55.7	41.4 ± 1.5	–	64 ± 42	–	84
PPG 425		$M_w = 425$	5	44.5	31.3 ± 1.2	–	19 ± 8	–	
			7.5	38.5	29.0 ± 2.1	–	107 ± 50	–	
			10	33.1	17.4 ± 2.0	–	524 ± 66	–	
			12.5	26.8	6.0 ± 0.4	–	702 ± 31	–	
PPG 1000		$M_w = 1000$	5	44.8	32.3 ± 1.8	–	44 ± 3	–	
			7.5	39.0	28.4 ± 1.3	–	329 ± 20	–	
			10	34.0	$23.1 \pm .09$	–	473 ± 111	–	
			12.5	32.0	16.1 ± 1.6	–	496 ± 70	–	
PEG 600		$M_w = 600$	5	42.8	30.3 ± 1.8	–	67 ± 33	–	
			7.5	37.8	25.7 ± 0.2	–	360 ± 25	–	
			10	31.3	17.5 ± 1.7	–	427 ± 42	–	
			12.5	28.0	5.1 ± 0.8	–	622 ± 75	–	
PLA^{sc}	IM	$M_w = 108000$	0	56	55	–	10	–	85
PPG 425		$M_w = 425$	5	38	30	–	40	–	
			7.5	35	25	–	40	–	
			10	–	19	–	65	–	
			12.5	–	19	–	65	–	

Table 5.4 (*Continued*)

Plasticizer	Process	M_w/M_n	Content (%)	T_g (°C) DSC	σ (MPa)	E (MPa)	ε (%)	Ref.
PPG 1000		$M_w = 1000$	5	41	28	–	35	
			7.5	39.5	19	–	45	
			10	–	19	–	90	
			12.5	–	19	–	105	
PEG 600		$M_w = 600$	5	38	32	–	20	
			7.5	35	20	–	25	
			10	–	19	–	25	
			12.5	–	15	–	15	
PLA	IM	$M_w = 180000$	0	61.8	65	2600	5	86
PLA-MMT-Na-3%		–	0	66.2	56	2450	5	
PEG		$M_w = 1000$	10	41.3	38	1500	220	
PLA-MMT-Na-3%		–	10	40.2	30	1400	180	
Rikemal PL-710		–	10	42.4	45	2400	5	
PLA-MMT-Na-3%		–	10	43.5	39	1900	7	
PLA	IM	$M_w = 74000$	0	59.2	64 ± 1.5^b	2840 ± 50	3 ± 0.3	44
PBOH		$M_w = 2100$	10	47,6	$56,3 \pm 1.9^b$	2350 ± 50	3 ± 0.3	
			20	–48.5/30.1	30.2 ± 1.1^b	350 ± 20	302.5 ± 32	
			30	29.4/–45.0	25.2 ± 1.8^b	300 ± 50	390 ± 35	
PEG 200		$M_w = 200$	10	35.8	30 ± 4.1^b	1700 ± 100	2 ± 0.6	
PEG 400		$M_w = 400$	10	37.1	39 ± 0.3^b	1920 ± 53	2.4 ± 0.3	
			20	–50.2/18.6	16 ± 0.3^b	630 ± 20	21.2 ± 2.3	
PEG 1000		$M_w = 1000$	10	40.2	39.6 ± 5^b	1970 ± 120	2.7 ± 0.3	
			20	–62.7/22.4	21.6 ± 0.4^b	290 ± 50	200 ± 13	
			30	29.9/ – 68.9	4.7 ± 0.2^b	420 ± 40	1.5 ± 0.2	
PLA	TS + I	–	0	–	69.2 ± 0.41	3680 ± 130	6.0 ± 0.29	65
PEG		$M_w = 3350$	2	–	60.5 ± 0.76	3700 ± 140	10.5 ± 4.6	
			5	–	54.5 ± 0.96	3560 ± 100	9.1 ± 7.2	
			10	–	44.8 ± 0.97	3020 ± 70	39.9 ± 37.2	
PLA	IM	$M_n = 97000$	0	64	67	840	1	42
PEA		$M_n = 2000$	20	36	24.8	375	815	
PBA		$M_n = 2000$	20	47	42.6	675	428	

Sample	Method	Molecular weight	Plasticizer content (wt%)	T_g (°C)	Tensile strength (MPa)	E (MPa)	Elongation (%)	Ref.
PHA		$M_n = 2000$	20	55	36.8	710	19	87
PDEA		$M_n = 2000$	20	34	17.7	257	705	
PLA	IM	$M_n = 63000$	0	58 (DMA)	–	2000	6	
PA G206/2		$M_n = 1532$	10	44	–	1600	5	
			20	27	–	250	480	
PA G206/7		$M_n = 2565$	10	44	–	1700	8	
			20	32	–	400	480	
PLA	IM	$M_n = 63000$	0	58.2	47 ± 5	2000 ± 200	6 ± 2	43
PA G206/2		$M_n = 1532$	10	39.5	34 ± 2	1600 ± 100	5 ± 1	
			20	25.4	25 ± 4	200 ± 100	485 ± 65	
PA G206/7		$M_n = 2565$	10	42.1	36 ± 2	1700 ± 0.2	7 ± 5	
			20	30.6	28 ± 2	500 ± 100	491 ± 34	
PLA	IM	$M_n = 63000$	0	58	–	–	–	88
PA G206/7		$M_n = 2565$	15	37.6	–	660	250	
			20	31.1	–	300	310	
PLA-C30B-3%/G206/7		–	15	37.7	–	720	150	
			20	30.5	–	400	200	
PLA	IM	$M_n = 63000$	0	58	–	–	–	89
PA G206/2		$M_n = 1532$	15	33	–	430	320	
C30B-2.1%/G206/2		–	15	33	–	470	250	
PA G206/5		$M_n = 2209$	15	34	–	520	320	
C30B-2.1%/G206/5		–	15	35	–	550	230	
PA G206/7		$M_n = 2565$	15	36	–	620	320	
C30B-2.1%/G206/7		–	15	35	–	630	230	
PLA	IM	$M_n = 90500$	0	58	47.1 ± 6.9	1291 ± 60	8 ± 5	72
PEG 300		$M_n = 300$	2.5	48	41.1 ± 8.8	1335 ± 143	5 ± 1	
			5	40.5	38.1 ± 7.2	1338 ± 84	5 ± 2	
			9	39.5	29.6 ± 2.8	908 ± 36	7 ± 2	
			13	39	18.1 ± 1.5	547 ± 34	99 ± 43	
			17	38	15.8 ± 1.2	323 ± 44	137 ± 34	
PLA	IM	$M_w = 104000$	0	64 (DMA)	61	–	6	90
PPG		$M_w = 1000$	10	50	51	–	7	
			15	50	54	–	7	

Table 5.4 (Continued)

Plasticizer	Process	M_w/M_n	Content (%)	T_g (°C) DSC	σ (MPa)	E (MPa)	ε (%)	Ref.
EPE11		$M_w = 1180$	10	49	56	–	7	
			15	48	32	–	510	
EPE19		$M_w = 1940$	10	46	21	–	460	
			15	43	22	–	510	
PLA	IM	–	0	61	74	1700	5	91
PEG 1000		$M_w = 1000$	15	–	42	1100	120	
PEGCA		–	15	52	40	1300	230	
PLA	TS	–	0	59.98	–	–	–	92
C30B-3%		–	0	61.1	–	–	–	
PEG		–	20	29.4	–	–	–	
C30B-3%/PEG		–	20	33.5	–	–	–	
PLA	–		0	62	49	4300	1.8	93
PEG		$M_w = 1000$	5	46	53	3000	2.9	
			10	38	51	3300	3.1	
			15	33	29	2100	4.2	
			20	33	22	1900	15.4	
PLA	IM	$M_n = 98000$	0	59.2	–	2500	8	94
OLA		$M_n = 957$	15	38.9	–	1600	180	
			20	33.7	–	750	300	
			25	25.8	–	250	310	

chemical families have been used, where by far the most prominent family is those of citrates. Adipates have also attracted some interest. Both chemical families present good compatibility with PLA, but, according to their solubility parameters (Table 5.2), they are not highly miscible in PLA. They will swell PLA rather than dissolve it and a risk of bleeding out is present. Molecules with closer solubility parameters, such as DBS[44] (sebacate) and DEHP[67] (phthalate) have been used with success in increasing tensile properties. The molecule with *a priori* the highest miscibility, LA is evidently a very good plasticizer for PLA, increasing the most the elongation at break, but it is characterized by low permanence due to its relatively high volatility.[51]

Miscibility is one of the key factors for the successful increase of tensile properties and decrease of T_g. Most literature studies used a maximum plasticizer concentration of 20 wt%, which allowed for the fabrication of *a priori* fully miscible blends. Correlation between yield stress and plasticizer concentration is a sign of miscible blends, as yield stress can serve as a measurement of macromolecular mobility.[75] Miscibility limits in PLA observed by some authors were 5 wt% DOA measured by DSC,[66] 10 wt% DINCH (by DSC and FESEM),[69] 20 wt% for DBS (DSC)[44] and GMS 15 wt% (DSC).[73] Furthermore, below 10 wt% DOA, Muriaru *et al.*[66] observed a decrease in the impact strength, explained by the antiplasticization effect.

A detailed investigation of miscibility can be undertaken for ATBC, where most data are available, using T_g as a descriptor and using DSC as analysis technique. The main research interest in ATBC stems from the fact that it proves most efficient in T_g depression and increase in elongation at break. Its branched structure allows apparently to bring sufficiently free volume to the polymer. Maximum ε values reach almost the one obtained by LA, but permanence of ATBC is supposed to be higher. Figure 5.1 plots together the results of Baiardo *et al.*[60] with those from other authors. The continuous line represents the Fox equation, which is often used for the description of the T_g decrease of miscible blends:

$$\frac{1}{Tg} = \frac{w_1}{Tg_1} + \frac{w_2}{Tg_2}, \tag{8}$$

where w_1 and w_2 are the weight fractions of plasticizer and polymer, and T_{g1} and T_{g2} are the glass transition temperature of plasticizer and polymer, respectively.

One of the advantages of the Fox equation is that it does not contain adjustable parameters. An agreement of experimental data with the prediction supports the conclusion of the high miscibility. Data from different authors are consistent and show that up to 45 wt% of ATBC content in PLA, the Fox equation is applicable. At concentrations equal to or higher than 60 wt%, ATBC phase separation is evidenced in a miscible PLA/ATBC blend of stable composition and a pure ATBC phase. Consequently two T_g values can be measured, where the high temperature transition becomes independent

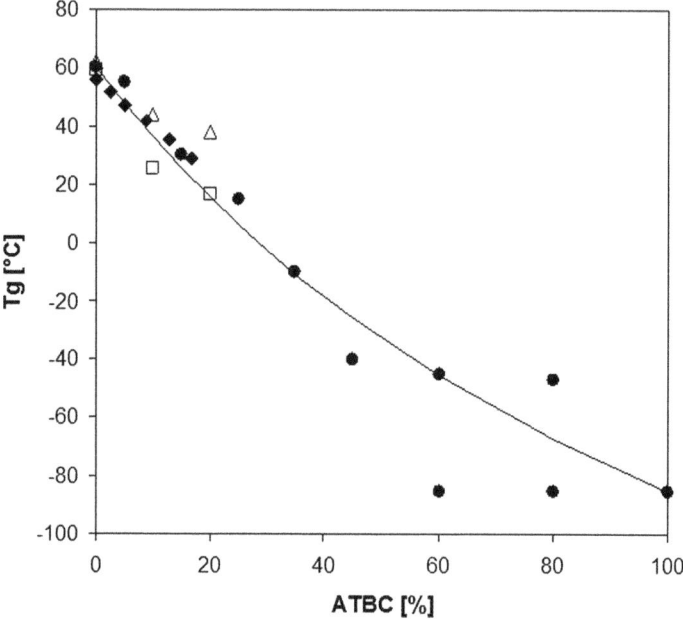

Figure 5.1   Glass transition temperature depression of PLA/ATBC blends as a function of ATBC content. Symbols: ● Data Ref. 60, ◆ Data Ref. 72, ☐ Data
Ref. 58, △ Data Ref. 66. The straight line represents the Fox equation,
modelled with the parameters given by Ref. 60.

of the ATBC content and the low temperature transition corresponds to the
T_g of pure ATBC.[60] These observations prove the phase separation in the
blend at ATBC concentrations higher than 45 wt%, which leads to the formation of a pure ATBC phase.

DSC is not a very sensitive technique to determine phase separation, as
micro-domains of a certain size (>20 nm)[44] need to be formed before being
significant in the heat flow signal. Courgneau *et al.*[76] probed phase separation in PLA/ATBC blends with the help of Electron Spin Resonance (ESR).
This non-invasive technique detects the local mobility of spin probes, generally stable nitroxyl radicals. ESR spectra allow reorientation dynamics of
the spin probes in the material to be measured and therefore to conclude on
domains of different mobility. By exposition of ATBC-plasticized PLA samples to vapours of two different spin probes at high temperature (70 °C for
29 days), the authors were able to assess mobility differences of the probes
with penetration depth into the polymer. PLA/ATBC samples were found to
be homogenous up to a concentration of 9 wt%. The sample at 17 wt%
showed mobility differences between sample surface and bulk, indicating a
micro-phase separation at the sample surface.

Another interesting feature of PLA plasticizing can be observed in
Figure 5.2, which plots ε values against T_g. In the case of the poorly compatible plasticizer GMS, even though some decrease in T_g is reached, not

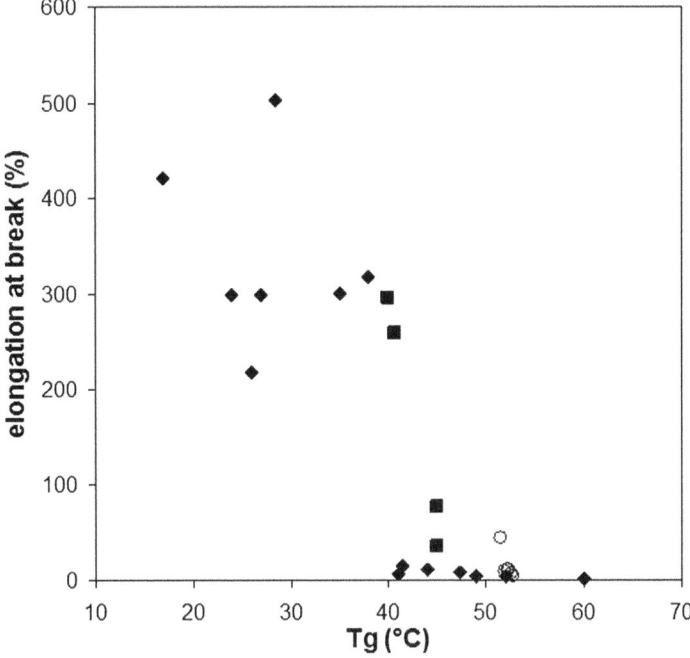

Figure 5.2 Strain at beak values plotted against T_g. Symbols: ◆ ATBC,[58,60,66,72] ■ DOA,[43,66] ○ GMS.[73]

much gain in ε can be obtained. Comparing di(octyl adipate), DOA, with ATBC shows that in the high T_g range, DOA outperforms ATBC. Apparently the more linear molecule DOA is able to bring more mobility to the macromolecular chains, which probably shows the importance of the molecular volume of the molecule. Indeed, DOA is bulkier than ATBC, although having lower molecular weight. Once the ATBC content reaches a critical limit, this leads to some increase in ε in the low T_g range. ATBC outperforms DOA, which cannot reach low T_g due to the miscibility limit. Moreover, properties change close to $T_g = 40$ °C, which means that the measuring temperature of the tensile testing (23 °C according to ASTM-D683) becomes to be close to the glass transition range of the material. However, ATBC, although it is the monomeric plasticizer yielding the highest increase in ε (Table 5.3), barely yields doubled impact strength.[66] However, DOA, even if not miscible with PLA produced an almost ten-fold gain in impact strength.[66]

5.4.1.2 Oligomeric Plasticizers

The main drawback of monomeric plasticizers is their tendency to migrate from the polymer or to be leached out by contacting media. In anticipation of permanence problems caused by monomeric plasticizers, oligomeric plasticizers have attracted the main research interest for toughening PLA. Furthermore, polymeric plasticizers often bring additional increase in impact strength. Drawbacks in the use of polymeric plasticizers have already been discussed and are, to sum up, the lower miscibility with the polymer, in consequence the higher propensity to phase separation, and lower efficiency in T_g depression. Table 5.4 displays a literature overview of the work carried out on oligomeric plasticizers. It shows the different chemical families studied, which are mainly oligomers of glycols, citrates, adipates and malonates. These molecules (except glycol) proved already to be interesting monomeric plasticizers.

The influence of the molecular weight on plasticizing efficiency and miscibility is an important point for oligomeric plasticizers. Solubility parameters have been used as a tool for prediction and explication of these properties. They are generally efficient to discriminate different chemical structures of oligomers, but not able to assess the influence of the increase in chain length for a given oligomer. For example, the solubility parameters were calculated for a series of TBC oligomers with increasing chain length. They were constant upon increase of monomer units up to $n = 7$, but the efficiency in T_g and melting point depression decreased.[41] Similar results were found in the case of DBM oligoesters and DBM oligoesteramides.[45,63] Polar amide groups introduced to the oligomeric chain of oligoesteramides increased plasticizing efficiency, which shows the importance of polar interactions and hydrogen bonding. These polar groups permitted the solubility coefficient to approach PLA[63] and, in consequence, blends had higher stability compared to oligoesters.[45,63] The use of citric acid branched with PEG, both individually good plasticizers, as macromolecular plasticizer was also evaluated.[91] Miscibility was rather poor between PLA and PEG-*co*-citric acid, and only a small decrease in T_g was observed. However, the highly branched structure brought significantly higher impact strength to PLA.

The miscibility of polyadipates with PLA was also investigated in the function of M_n. It decreased with increasing chain length. The largest molecule certainly has the least efficiency in T_g depression, but all oligomers produced large gains in PLA ductility.[64,87,88] A trade-off in oligomer chain length needs to be found between plasticizing efficiency and miscibility. Excellent properties were also obtained for different polyester adipates (PEA, PBA and PHA, see Table 5.4).[42] The best result was shown by a PLA/PEA blend having elongation at break of 800% and relatively high tensile strength (24.8 MPa).[42] The solubility parameter δ_s calculated at 180 °C explained the miscibility differences between PEA and PHA, but not between PBA and PDEA. Although δ_s values were equal, PBA was found to be only partially miscible, while the PDEA was fully miscible. The authors proposed that the higher configurational entropy of PDEA compared to PBA might be the reason for enhanced miscibility, by inducing higher entropy gain upon mixing.

The most investigated family of polymeric plasticizers is polyethylene glycols (PEGs). These linear polymers can have favourable solubility parameters (Table 5.2), and they are commercially available in a large range of chain lengths. Depending on the molecular weight average, PEGs can be a crystalline solid ($M_w > 1500$ Da), waxy ($M_w \approx 1000$ Da) or liquid ($M_w < 1000$ Da) at room temperature.[60] PEGs are capable of greatly enhancing the toughness of PLA. Table 5.4 shows that PEG of small chain length (M_w 300 Da[72] or 400 Da[60], for example) raises steeply from the brittle behaviour of PLA to a maximum value, while PEG of longer chain length (M_w 8000 Da,[77] *e.g.*) yields a more gradual property change. Insight into the influence of chain length on PEG miscibility with PLA can be gained from the T_g data of Pillin *et al.*[44] and Baiardo *et al.*[60] in comparison to values of different literature sources (Figure 5.3).

The miscibility limit of PEG in a M_n range of 200 Da to $M_n = 3400$ Da was established at 20 wt% with the help of DSC measurements.[44,60] PEG with $M_n = 10\ 000$ Da demixed already at 15 wt%.[60] Hu *et al.*[77,78] showed miscibility of PEG 8000 with PLA up to 30 wt% with the help of DSC and DMA measurements. Courgneau *et al.*[72] found exudation of PEG 300 already after 9 wt% with the help of DMA measurements. In their follow-up paper Courgneau *et al.*[76] used ESR to observe exudation of PEG 300 from the film samples already at 9 wt% PEG 300. However, their samples needed to be stored at a temperature higher than T_g for several weeks in order to allow

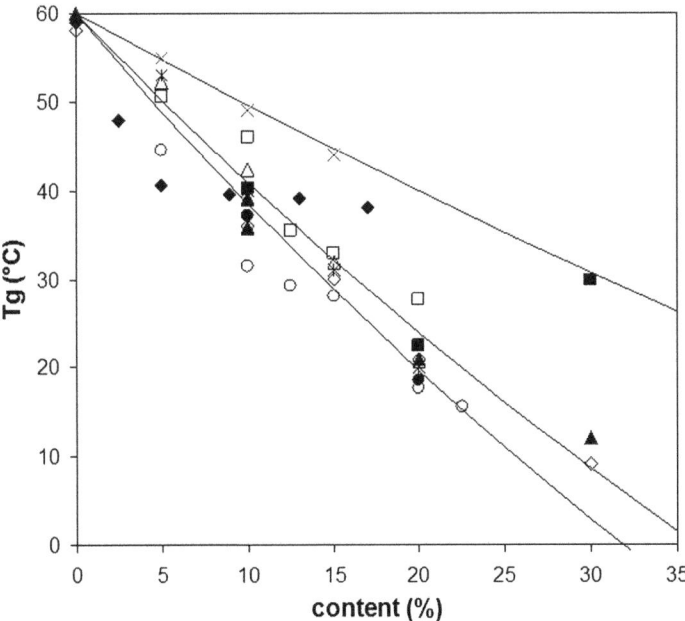

Figure 5.3 Glass transition temperature depression as a function of content and average molecular weight of PEG. Symbols: ▲ PEG 200,[44] ◆ PEG 300,[72] ● PEG 400,[44] ○ PEG 400,[60] ■ PEG 1000,[44] □ PEG 1500,[60] * PEG 2000,[60] △ PEG 3400,[60] ◇ PEG 8000,[77] × PEG 10000.[60] The straight lines correspond to the Fox equation for PEG having M_w of 10000, 8000 and 400 (upper to lower).

sorption of the ESR spin probes. Therefore, the differences between observations of different research teams might be linked to ageing. Instabilities in the PLA/PEG system will be discussed in more detail in Section 5.5 "long-term stability".

PPGs and their copolymers were studied in comparison to PEGs and showed good ability to increase PLA toughness.[68,84,85,90,93] PPGs have the advantage that their T_g is lower and even at $M_w = 1000$ Da they are viscous liquids at room temperature. Therefore T_g depression after the addition of PPGs is more efficient compared to PEG[85] (Table 5.4). Miscibility of PPG 400 with PLA was found for all the studied concentrations up to 12.5 wt%, while phase separation was found at 12.5 wt% in the PLA/PPG 1000 blend.[84,85] Using the PLA/PEG and two PLA/PPG systems, Kulinski *at al.*[84] and Piorkowska *et al.*[85] investigated the combined effect of the plasticizer content and crystallization on PLA. Figure 5.4, reprinted from Ref. 84, shows the stress/strain curves of neat and plasticized amorphous PLA. Plasticizing improved drawability and decreased yield stress due to enhanced macromolecular mobility. In the case of drawing PLA in the glassy state (neat or 5 wt% plasticizer) the plastic deformation was caused by crazing. Beginning with 7.5 wt% of plasticizer, T_g approached drawing temperature and the blends crystallized resulting in strain hardening (Figure 5.4). PPG 1000 separated from the amorphous PLA at 12.5 wt% and formed small droplets in the polymer, a phenomenon that was accelerated by the PLA crystallization during drawing. These liquid inclusions did, most interestingly, not downgrade drawability, but seemed to be beneficial by the plasticizer accumulation in the amorphous phase due to expulsion of their molecules from the originating crystallites.[84]

The effect of crystallization on PLA drawability is generally negative,[55,85] but it enhances other important properties, such as modulus, heat stability and barrier properties. The plasticizer separated in the amorphous phase during PLA crystallization accumulates at the crystalline phase boundary,[83] broadening the glass transition region. In the case of the PLA/PPG and PLA/PEG systems, different consequences were observed. In the case of PLA/PPG 400, the T_g value of the amorphous phase decreased due to the increase of the amount of plasticizer in the amorphous phase. In the case of PLA/PPG 1000 and PLA/PEG 600, the T_g value increased after PLA crystallization. Phase separation of the plasticizer was observed for every composition, but separate liquid inclusions inside the amorphous PLA matrix were revealed uniquely for PPG 1000.[85] Figure 5.5 plots the stress/strain curves of the corresponding samples. Both PPGs are far more efficient in preserving high elongation at break even for semicrystalline PLA. Yielding in PLA/PPG 400 and PLA/PEG 600 disappeared at a plasticizer content of 7.5 wt%. At high contents, flow stress decreased without further ε increase. Cracks propagated preferentially through interspherulitic crazes, as the interspherulitic boundaries were the weakest element. PEG 600 accumulation in front of a growing spherulite was higher compared to PPG 400, which explained the earlier fracture of PLA/PEG 600. In the case of PPG 1000, extensive yielding

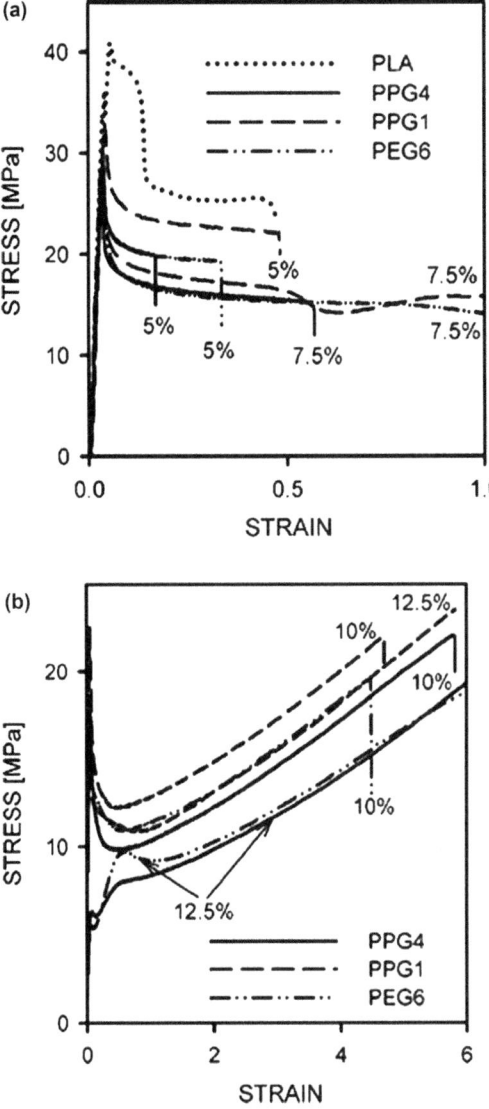

Figure 5.4 Stress-strain plots of initially amorphous neat PLA and amorphous PLA
plasticized with 5 wt% and 7.5 wt% of PPG 400, PPG 1000 and PEG 600
(a) and for PLA plasticized with 10 wt% and 12.5 wt% of PPG 400, PPG
1000 and PEG 600 (b). Reprinted with permission from Kulinski *et al.*,
Biomacromolecules, 2006, 7, 2128.[84]
Copyright 2013 American Chemical Society.

was observed. The phase separation of PPG 1000 in droplets prevented the
accumulation of the plasticizer in the amorphous phase and resulted in
comparatively higher T_g of PLA. This more uniform distribution of PPG 1000
apparently enhanced drawability.[84,85]

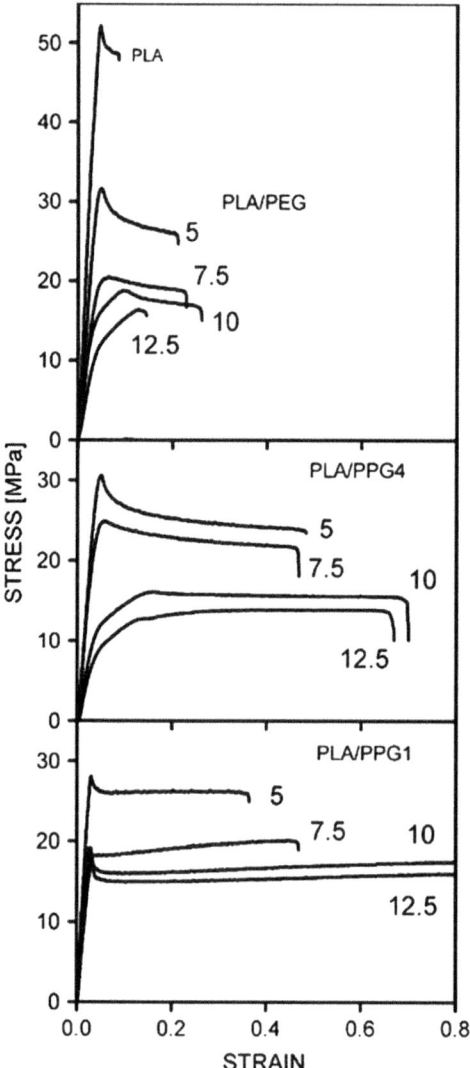

Figure 5.5 Stress-strain plots of initially semicrystalline neat and plasticized PLA, plasticized at 5 wt%, 7.5 wt%, 10 wt% and 12.5 wt% of PEG 600 (top), PPG 400 (middle) and PPG 1000 (bottom).
Reprinted from Piorkowska *et al.*, *Polymer*, 2006, **47**, 7178[85] with permission from Elsevier.

In summary, the accumulation of plasticizer can importantly weaken tensile properties in plasticized amorphous PLA. Solid inclusions and crystallizing plasticizers (such as PEG with high M_w) also deteriorate properties, while small inclusions of a liquid plasticizer phase can be beneficial for drawability.

5.4.1.3   Mixed Plasticizers

Knowledge from extensive studies of PVC plasticizing and results obtained from the different studies on plasticized PLA render evident the complementary character of monomeric and polymeric plasticizers. In fact, most PVC plasticizing techniques rely on mixtures of primary and secondary plasticizers. Primary plasticizers are soluble at high concentrations and secondary plasticizers show a limited compatibility, but they are blended with primary plasticizers to improve product properties or reduce the cost. The use of a plasticizer mixture in PLA formulations has not been extensively reported.[68,95] Mixtures of TBC and different PLA-*b*-PEG copolymers were studied to combine the positive effect of TBC on ε and the increase in σ given by the copolymer.[68] Among different copolymer structures, the linear molecule (PDLLA-*b*-PEG750) was one of the most successful. Furthermore, the use of the copolymers increased the impact strength in samples that did not break during testing. The use of a mixed plasticizer of GTA and oligomeric poly(1,3-butylene glycol adipate) showed moderate T_g depression (Table 5.4) but substantial increase in elongation at break ($T_g = 52$ °C, $\varepsilon = 320\%$).[95] The topographic AFM analysis revealed a fine "network" of crystalline lamellae all over the sample, the strands of which grew thinner when the plasticizer quantity increased. This morphology contributed to the high elongation at break values, although samples were clearly crystalline. Unfortunately, the authors gave no comment on the miscibility of the plasticizer and PLA, so a comparison of this system with the phase separated PPG system (Refs 84,85) cannot be drawn. Both approaches produced systems with high elongation at break and tensile strength. The use of plasticizer mixtures seems therefore a promising way to tune PLA mechanical properties.

5.4.2   Plasticizers as Processing Aids

Different properties of plasticizers are useful in the aim of easing PLA processing. Melting point depression and reduction of T_g results in decreasing the processing temperature. T_g depression was extensively discussed in previous sections. In the limit of miscibility, melting point depression is also efficient to speed up crystallization rates. Furthermore, plasticizers decrease polymers melt viscosity and allow better flow in the different process equipment. Some data are available on the melt viscosity of plasticized PLA, not aimed at foaming or spinning applications. Examples show the effect of the most popular PLA plasticizers, *i.e.* citrate derivatives and PEG. In general, PLA exhibits shear thinning behaviour in the high shear range and plasticizers decrease the viscosity values at all shear rates. Liu *et al.*[96] tested apparent viscosity of PLA plasticized with ATBC using a capillary rheometer and observed shear thinning all over the used apparent shear rate range (100 to 1000 s^{-1}). Apparent melt viscosity at 165 °C decreased from 1789 Pa s^{-1} at 100 s^{-1} apparent shear rate to approximately 400 Pa s^{-1} at 1000 s^{-1} apparent shear rate. The addition of 3 wt% ATBC decreased the apparent melt viscosity at 100 s^{-1} apparent shear rate to 1350 Pa s^{-1}, shear thinning curves being parallel to neat PLA. The melt flow rate and melt flow volume of PLA/ATBC in a twin screw extruder were examined by Scatto *et al.*[74] The authors found a significant increase in melt flow rate from 3.8 to 10.7 g (10 min^{-1}) for neat and PLA containing 20 wt% ATBC, respectively. Melt flow rate was furthermore found independent from the twin screw speed. Impact of plasticizers on zero shear viscosity of PLA in the Newtonian plateau was investigated for ATEC,[65] PEG 1000[93] and PEG 3350.[65] The authors showed an important decrease of zero shear viscosity upon blending with PEG, from 5000 to approximately 800 Pa s^{-1} for 10 wt% PEG 3350[65] and from 870 to 310 Pa s^{-1} for 10 wt% PEG 1000.[93] In both cases the decrease in the curve levelled off at about 20 wt% of PEG, when the miscibility limit was reached. ATEC gave a higher zero shear viscosity decrease from 5000 to approximately 100 Pa s^{-1} for 10 wt% ATEC.[65]

The decreased viscosity of the plasticized melt also helps greatly in the incorporation of fillers. Plasticizers have been used as processing aids for the fabrication of PLA nanocomposites by melt inclusion techniques. PEG is the plasticizer mostly used as a processing aid for the preparation of PLA nanocomposites. PEG has the ability to preferentially intercalate between layers of the silicates and thereby help nanoclay exfoliation.[79–82,91,97] Due to this preferential intercalation, PEG permits the opening of the tactoid structure of non-organomodified clays, such as Cloisite® Na$^+$.[79] It was observed that at high nanoclay loads, T_g of PLA increased because of the depletion of PEG from the polymer phase by intercalation between nanoclay stacks.[79] The preferential intercalation of PEG depends on the type of modification of the clay, where the most apolar structures present the lowest efficiency.[82] In organomodified nanoclays, for example Cloisite® 20A and Cloisite® 30B, co-intercalation of PLA with PEG during melt compounding can be found.[82] PEG is clearly a processing aid for the successful production of PLA/clay nanocomposites.

Polyadipates were also shown to be good processing aids for exfoliation of layered aluminosilicates.[88,89] On the contrary to PEG oligomers, polyadipates show no preferential interaction with the nanoclays.[89] Polyadipates of different chain length are able to swell nanoclays, as observed for Cloisite® 30B.[89] The higher molecular weight polyadiapate ($M_n = 3460$ Da) was found to be slightly less efficient than the smaller oligomer ($M_n = 2400$ Da). However, the first one has the advantage of lower migration propensity due to increased chain length.[89] Scatto *et al.*[74] not only showed the interest of ATBC as an aid for exfoliation of layered nanoclays in PLA but worked on different processing techniques and concluded that the best method for the nanoclay introduction is by forming a previous mixture with the plasticizer. The comparison between mixing in an internal mixer and by twin-screw extrusion showed the superiority of this latter processing for nanoclay dispersion.

5.5   Long-term Stability of Plasticized PLA

The long-term stability of plasticized materials is a major concern for their successful application and it can be depreciated by chemical degradation of the polymer and/or plasticizer or by physical changes in the bulk properties of materials.

5.5.1   Physicochemical Studies of Long-term Stability

Physical ageing is commonly understood as a phenomenon occurring during storage of a polymer below its glass transition, where only restricted motion as rotation of small groups of atoms and vibration and rotation motion of chemical bonds along the polymer chain is permitted. Consequently, the polymer relaxes slowly to its equilibrium, causing changes in bulk properties, such as volume or enthalpy.[100] Above the glass transition temperature, the amorphous phase is thought to be in its equilibrium and no ageing caused by relaxation phenomena is thought to occur, although for semicrystalline polymers it has been shown that the constrained amorphous phase close to the crystalline lamellae may give rise to ageing effects.[100] However, in the case of semicrystalline polymers phenomena like cold crystallization can occur.

Most studies on ageing effects of PLA including physical phenomena were carried out by storage of samples in ambient conditions with minor control of environmental parameters, although most of the authors are careful in sealing the samples in polyethylene bags to protect from ambient humidity. This method is of main interest as it corresponds to real conditions plasticized PLA materials might experience during their storage and service time, noting that major PLA applications correspond to the use at ambient temperature. The distance between glass transition temperature and ageing temperature varied among studies and often even changed during the experiment time. Therefore all the above-mentioned physical phenomena can be expected. Table 5.5 gives an overview of different studies available and sums up very briefly the main physical changes of the samples observed for each experiment.

Within the family of monomeric plasticizers, the long-term stability of GTA and TBC was investigated.[62,69] Phase separation and cold crystallization were evidenced in all cases, where ageing was carried out very close to T_g or even at T_g (Table 5.5). Storage of PLA/GTA and PLA/TBC at ambient conditions resulted in substantial chain scission of PLA (M_w divided by 5 after 123 days).[62] The authors suggested that the hygroscopic character of the plasticizers permitted the water sorption in the film triggering PLA hydrolysis. The chain scission then caused some increase in crystallinity, decrease in T_g and embrittlement of the aged material. Furthermore, plasticizers exuded during storage causing the material surface to become sticky.[62] Immersion of PLA/TBC films in water showed the gradual loss of TBC and the increase of T_g. Upon material storage in dry environment, cold

Table 5.5 Overview of physical ageing behaviour of PLA/plasticizer compounds.

	Plasticizer	t_a [d]	T_a [°C]	T_g [°C]		Ref.
PLA	GTA	123	RT	29	M_w degradation, T_g decrease, cold crystallization, phase separation, migration	62
PLA	TBC	123	RT	29	Cold crystallization	45
	DBM	120	22	29	Phase separation, cold crystallization	
	DBM-A	120	22	39	No phase separation	
PLA	TBC-E	120	22	33	Phase separation	41
PLA	TBC	120	RT	24.0	T_g increase, cold crystallization, embrittlement	69
	DINCH	120	RT	49.9	No T_g change, embrittlement	
PLA, 13% D	PEG 8 000	75	23	9	PEG crystallization as long as $T_a > T_g$, thus gradual increase of T_g, embrittlement	77
PLA, 5% D	PEG 8 000	125	23	12	Phase separation, increase of T_g, low crystallization of PEG, embrittlement	78
PLA	PEG 10 000	60	20	50.5 / 32.0 / 25.0	No change in T_g / T_g increase, cold crystallization, embrittlement / T_g increase, cold crystallization, PEG crystallization, embrittlement	98
	PEPG 12 000	60	RT	50.9	No change in T_g	
PLA	PEG 1500	90	RT	23.4	T_g increase, cold crystallization, phase separation, embrittlement	80
PLA-C25A-3%	PEG 1500	90	RT	44.8	T_g increase, cold crystallization	
PLA	PEG 1000	365	RT	44.0	T_g increase, cold crystallization	92
PLA-C30B-3%	PEG 1000	365	RT	29.4	T_g increase, small enthalpic relaxation peak, phase separation	
	PEG 1000			33.5	T_g increase, enthalpic relaxation, phase separation, tactoid clay structure increase	
PLA	PEG 1000	1460	4	23.4	T_g increase, cold crystallization phase separation, crystalline PEG layer outside	99
PLA-C20A-10%	PEG 1000			27.6	T_g increase, smaller cold crystallization, phase separation delay by clay presence	
PLA-C25A-10% PLA-C30B-10%				28.5	T_g increase, cold crystallization reduced by clay presence, phase separation delayed, all samples show strong embrittlement	
PLA	PA G206/2	150	28	39.5 / 25.4	No change in T_g, no phase separation / Cold crystallization, embrittlement, increase in barrier properties	87
PLA	PA G206/7	60	RT	37.6	No T_g change	88
PLA-C30B-3%				37.7	No T_g change, cold crystallization, embrittlement, increase in barrier properties	
PLA	OLA	90	25	38.9	No change in T_g, embrittlement, increase in barrier properties	94

t_a ageing time, T_a ageing temperature.

crystallization of PLA was still observed and the T_g increased with time.[69] DINCH, although it was much less efficient than TBC to depress T_g, had higher permanence in the PLA structure upon immersion in water and during ageing in dry conditions over 4 months almost no change in T_g was detected.[69] Due to the large difference in T_g of the initial blend, the storage of PLA/TBC was carried out inside the glass transition range, while the PLA/DINCH was glassy, where transport phenomena are strongly slowed down.

With respect to polymeric plasticizers, the PLA/PEG system received some interest in ageing studies. A common result of all of them is that PEG/PLA blends are not stable. In the case of PEG with small M_n and waxy at room temperature, phase separation by ageing was observed.[80,92,99] Pluta *et al.*[99] undertook an extensive study of ageing of PLA/PEG 1000 and PLA/nanoclay/PEG 1000 systems for 1 to 4 years below glass transition temperature. They investigated first the evolution of M_n, which showed that the ageing of 1 year decreased importantly the macromolecular chain length to only 18% of the initial value. A small quantity of organomodified nanoclay seemed meanwhile to have a positive effect on the chemical stability, as M_n decreased less in that case. Due to the substantial decrease of the macromolecular chain length, samples were highly crystalline and brittle after 4 years of storage. Interestingly, T_g after 4 years was higher than after one year, which was the effect of phase separation inside the sample and migration of PEG 1000 onto the sample surface. The authors evaluated the amount of PEG 1000 leached out by diffusion and reported that approximately 7.4% of PEG was lost from the bulk. It gathered on the sample surface forming a crystalline layer. The dispersion of nanoclays at low concentration (<10 wt%) inside PLA delayed the PEG 1000 diffusion, most probably due to the increase in the tortuosity of the molecules' pathway. Furthermore, the PEG 1000 diffusion from the PLA/Cloisite® 30B/PEG 1000 system was always low. A retention effect due to interactions of hydroxyl groups of PEG and ammonium cations contained in the organomodifier was proposed. Besides diffusion, the nanoclays hindered also the phase separation while preferential PEG/nanoclay interactions depleted the amorphous phase of the blends. The effect is that the PEG-rich phase in the nanocomposites always had higher T_g compared to the blank sample.[99] Another study of physical ageing in PLA/Cloisite® 30B nanocomposites plasticized with PEG 1000 for one year in the glassy state showed that the initially almost fully exfoliated structure of Cloisite® 30 B recovered a tactoid structure after ageing for one year.[92] Due to the phase segregation taking place the clay galleries got depleted again and recovered an aggregated structure.

In the case of PEG 8000, being a crystalline solid at room temperature, crystallization of the plasticizer inside the PLA matrix was observed as long as the storage temperature was higher than glass transition.[77,78,98] The crystallization depleted the amorphous phase from the plasticizer, which caused the T_g to rise. At the point at which the T_g became higher than the storage temperature, the crystallization of PEG 8000 ceased.[77] The material simultaneously experienced an increase of the glassy modulus by the reinforcing effect of PEG 8000 crystals. At high PEG 8000 concentrations (30 wt%), an influence of PLA stereo-regularity on the PLA/PEG 8000 mixing behaviour could be observed. While PLA with low stereo-regularity/PEG 8000 showed miscibility at room temperature, PLA with higher stereo-regularity readily phase separated.[78] In this case, phase separation, characterized by the appearance of two glass transition temperatures, was the primary ageing mechanism and completed after 48 h. No crystallization of PEG 8000 was observed.[78]

Faced with long-term instability of PLA/PEG blends, alternatives were studied and PPGs were considered good candidates. Jia *et al.*[98] tested the ageing of a PEG-PPG (PEPG-12000) copolymer against PEG-1000. PEG-1000 showed crystallization of PEG, as discussed earlier. The PEPG-12000 blend phase separated gradually over time. In the case of ageing in the glassy state, the authors reported the important increase of the enthalpic relaxation peak resulting in the drastic reduction of elongation at break.

The positive effect of increasing the chain length of oligomers on the long-term stability of plasticized PLA was shown in the case of oligomeric TBC,[41] malonates,[45] lactic acid[94] and adipates.[87,88] The ageing of PLA/TBC-E-3 (3 sub-units) near T_g showed phase separation and cold crystallization with time, but the material became non-sticky on the surface even after 4 months, showing the lower trend of the oligomer to migrate to the polymer surface.[41] The strategy of increasing M_n for long-term stability was successful in the case of malonate oligomers.[45] While DBM yielded phase separation, embrittlement, cold crystallization of PLA and migrated to the surface of the polymer, the oligomers $(n = 8)$ were very stable for an ageing period of 4 months without cold crystallization and changes in T_g. The ageing of PLA/OLA blends was done at temperatures at or slightly lower than T_g forcing minor decrease in ε.[94] The strongest embrittlement occurred upon storage at temperatures below T_g. In this case, ε decreased from almost 200% to nearly the value of neat PLA (5%) and enthalpic relaxation could be observed. A very interesting result was obtained by measuring the oxygen barrier properties of films during ageing. Plasticizing generally decreases barrier properties due to bringing free volume to the polymer, but ageing let the barrier properties increase again. In fact, PLA blends containing 20 wt% OLA $(T_g = 34\ °C, T_a = 25\ °C,$ Table 5.5) almost recovered the barrier properties of the neat PLA. The authors suggested that PLA experienced a gradual structure rearrangement causing densification, but without crystallization.[94] The same research group investigated the ageing of PLA plasticized with poly-adipates of different M_n at T_g or below.[87,88] Ageing at T_g let cold crystallization of PLA happen, causing the elastic modulus to rise and drawability to decrease. Phase separation in the material was solely observed for the addition of the lowest M_n polyadipate at 20 wt%. All samples showed exudation to the sample surface, which became sticky with time. In a follow-up study, the ageing of PLA/polyadipates was carried out for 2 months at temperatures slightly below T_g (Table 5.5). Interestingly, putting both studies together, the enthalpic relaxation peak was seen only for the 10 wt% blend with a difference of $T_g - T_a = 14\ ^{\circ}\text{C}$.[87] The 15 wt% blend with a difference of $T_g - T_a = 12\ ^{\circ}\text{C}$ and the 20 wt% blend having a difference of $T_g - T_a = 6\ ^{\circ}\text{C}$ showed no enthalpic relaxation.[88] Nevertheless, an increase in barrier properties was observed. The oxygen transmission rate thus seems a very fine probe for densification phenomena occurring during physical ageing. The absence of change in oxygen barrier properties over two months of storage of the PLA/Cloisite® 30B/polyadipate ternary nanocomposite proved that the delay of ageing brought by nanoclays applied also to polyadipate plasticizers.[88]

5.5.2 Chemical Stability and Degradation

Plasticizers are additives that could be present up to 30 wt% in PLA formulations. Their impact on the chemical ageing of PLA could be considered by either a negative or a positive role depending on their chemical architecture and hydrophilicity and also on their use in medical or commodity applications.

Lim *et al.*[101] showed that the melt mixing of PLA with additives often results in a decrease of the molecular weight, which can be due to the high sensitivity of PLA to the thermo-mechanical input, to the moisture and/or to trans-esterification reactions with additives. Furthermore, water should be considered and controlled as being a third active component during PLA plasticization process and acting in chemical ageing. Interchange reactions in polyesters are rapid in the melt during processing and in particular the mechanism of trans-esterification, which is an associative-type mechanism where bonds breaking and formation occur simultaneously. Jamshidi *et al.*[102] proved that low-molecular-weight compounds associated with the polymer seemed to play an important role in lowering its M_n at high temperatures. These molecules were identified as water, monomers, oligomers and polymerization catalysts. Afterwards Hyon *et al.*[103] and Cam and Marucci[104] showed the negative impact of residual monomers on the thermal stability of polylactides, but otherwise cyclic or aliphatic monomers are viewed as efficient plasticizers. Thus, removal of the non-polymeric contents and blocking the hydroxyl end-group enhanced the thermal stability of polymers. Obviously, the presence of plasticizers with ester-groups, OH- or COOH-derived groups could emphasize the degradation routes by trans-esterification.

Murariu *et al.*[66] studied the stability of a PLA grade (L/D isomer ratio 96/4, high molecular weight) when melt-mixed with selected plasticizers, namely DOA, GTA, ATBC up to 20 wt%. In particular, molecular weight parameters (number average molar mass, M_n, and polydispersity index, M_w/M_n) of neat PLA and plasticized PLA were determined. The authors concluded that no major decrease in the M_n of PLA was observed whatever the plasticizer. However, they were careful to use moderate and controlled mixing conditions by using an internal mixer and they specified that plasticizers were carefully dried under vacuum before blending with dry PLA. Careful drying is necessary due to the risk of hydrolysis reactions of ester linkages depending on residual water content and/or plasticizer leading to reduction in molecular weight. Minimizing moisture content by intensively drying all components represents a first step to reduce losses by hydrolysis and to preserve the polyester molecular weight as high as possible.

In the same manner, Courgneau *et al.*[72] showed that when PLA is extensively dried, only a slight decrease in molecular weight is observed after melt-blending at 160 °C. The addition of plasticizer brings about a notable decrease (40%) in M_w in the case of 9 wt% PEG. The addition of PEG results in the molecular weight decrease, which may be due to the degradation of PLA chains coupled to main chain scission and trans-esterification reactions between PLA and PEG. ATBC, on the contrary, does not induce a decrease in M_n at low contents and the addition of this plasticizer up to 17 wt% PLA results only in a slight decrease in M_n.[103]

Positive outcomes from the addition of a plasticizer into PLA are the increase in the environmental degradability at the end-of-life treatments. In fact, the slow degradation rate of neat PLA is often considered to be a major drawback for biomedical applications,[105] leading to long *in vivo* life-time, which could be up to years in some cases. Solutions to increase the abiotic degradation rate in biomedical applications could be an inspiration to optimize the degradation of PLA in other applications, such as food packaging.

However, choosing a hydrophobic plasticizer like ATBC for PLA may have an impact on the degradation rate of PLA. Hoglund *et al.*[106] studied the hydrolysis of PLLA at 37 and 60 °C for up to 364 days to compare the effect of hydrophobic and hydrophilic bulk properties. PLLA plasticized with ATBC had a slower degradation rate compared with pure PLA because of the increased hydrophobicity of the material. This result is in agreement with Labrecque *et al.*,[58] but it was proved by ESI-MS experiments that after just 1 day of immersion of plasticized PLA in water at 37 and 60 °C ATBC, water-soluble lactic acid oligomers and acetic acid were present. This proved the fast hydrolysis of ATBC in the aqueous environment. ATBC hydrolysis was demonstrated by analyzing the pure plasticizer in a parallel control experiment.[106] These results clearly showed that the combination of several factors, such as plasticizer solubility, T_g of all the involved materials and change in crystallization during degradation, play a role in the degradation rates of PLA.

5.6   PLA Plasticizers Derived from Biomass

One interesting way to modify the mechanical properties and preserve the environmentally friendly aspect of polylactides is by introducing bio-based compounds within the PLA matrix. Furthermore, the use of non-edible biomass could constitute a wide source of additives, and could provide an economically viable way to get added-value from residues.[107] Numerous studies to transform biomass by-products of the agriculture and food industries have been conducted recently[108–111] and some eco-friendly products have been successfully used to plasticize vitreous polymers, such as PVC.[112]

Nevertheless, due to the large diversity of molecules constituting a product issued from the biomass, isolating one of them is often costly and renders the process economically uncompetitive. Two main approaches using bio-based products to enhance PLA mechanical properties are currently investigated and examples of each one are given below.

The first approach can be summed up by the use of specific molecules extracted from the biomass. Thus, the chosen molecules, according to the classical plasticization concepts, have to be highly miscible within the polymer network to dissolve and/or swell it. Among bio-based molecules, Jacobsen et al.[51] investigated the plasticizing efficiency of glucose-monoester and partial fatty acids esters. However, no significant improvement in plasticization was observed. An eventual explanation of this result could be the poor miscibility of those molecules in PLA inducing a non-homogeneous material. Martin et al.[59] studied the plasticization ability of glycerol and OLA. The first one exhibited a poor compatibility with PLA since T_g was only decreased by 5 °C for a PLA/glycerol blend 80/20 wt%. On the contrary, adding 20 wt% of OLA strongly diminished the T_g to 18 °C, improved the elongation at break to 200% and greatly reduced the elastic modulus from 2050 to 744 MPa. More recently, Azwar et al.[113] proposed the use of liquefied wood flour and liquefied rice bran (which were obtained by acid-catalyzed reaction) to plasticize polylactide. They showed that liquefied wood flour, mainly constituted by low-molecular-weight polyols and mono/disaccharides was able to penetrate the matrix. Indeed, the T_g was decreased below room temperature (*i.e.* to 27 °C and 12 °C for 90/10 wt% and 70/30 wt% of PLA/additive blends). Furthermore, by adding 30 wt% of this product the elongation at break rose to 300%. Liquefied rice bran did not show any of these plasticizing abilities. According to size exclusion chromatography study, they assumed that the liquefied rice bran was composed of too high molecular weight oligosaccharides to permit an efficient plasticizing effect. Limonene, which is a common by-product of citrus juice processing,[114] has also been used as plasticizer for PLA. Indeed, Arrieta et al.[115] showed that the T_g of PLA was lowered by up to 30 °C (from 60.3 °C to 30.2 °C) when adding 15 wt% of limonene, and the elongation at break improved up to 150%. Blends containing 20 wt% of limonene were also prepared, exhibiting a T_g value slightly higher than the previous blend (33.8 °C) and elongation at break of 165%, very close to the previous result. These last observations could possibly point out the miscibility limit of limonene in PLA. Moreover, Sawalha et al.[116] studied the effect of the incorporation of volatile alkanes on the PLA morphology. They also observed the plasticizing efficiency of limonene. Although the amount of limonene remaining in the PLA matrix was unknown, they noticed that the T_g was lowered to 30 °C, while the elongation at break was enhanced up to 200%. The compatibility of limonene and PLA can be explained by the Hansen solubility parameters (HSP). Indeed, the limonene parameters are $(\delta_D = 17.2; \delta_P = 1.8; \delta_H = 4.3)$[37] and those for PLA are $(\delta_D = 18.6; \delta_P = 9.9; \delta_H = 6.0;$ radius 10.7).[38] This confers limonene compatibility with PLA similar to ATBC or DOA. Recent researches to study positive effects, such as antioxidant and plasticizing properties, have been conducted among bio-based additives. Lopez-Rubio et al.[117] investigated β-carotene (which is a well-known natural antioxidant[118] and UV-absorber[119]) by adding 0.4 wt% into PLA. They reported a nucleating effect and a decrease in cold crystallization temperature from 113.5 °C to 80.5 °C. They also observed a significant improvement in elongation at break from 13% to 250%. Furthermore, the tensile modulus was lowered from 2060 MPa to 570 MPa, meaning that β-carotene could exhibit good plasticizing efficiency. Nevertheless, another hypothetical explanation of these effects could come from the eventual residues of solvent remaining in the PLA matrix after casting. Hwang et al.[120] tested α-tocopherol and reservatrol, which are natural antioxidant molecules, present in soybean, cotton-seed, sunflower and grapes. They added both additives to PLA in several proportions representing a constant total quantity of 5 wt%. Nevertheless, no plasticizing abilities were observed, since the T_g values and ductility showed no significant changes.

The second strategy consists of improving the PLA ductility by using products that are directly issued from low-cost biomass derivatives but have only partial miscibility with PLA. In this case, the mechanical properties improvement is partly based on the modification of the polymer breakage behaviour. Indeed, when products with low compatibility are incorporated within PLA, some micro-voids can be formed during the tensile deformation. They appear at the interface between PLA and non-dissolved additive droplets, resulting in a stress whitening phenomenon.[83,85] In this case, a ductility improvement and a decrease in tensile strength while T_g remains relatively high can be observed. Piorkowska et al.[85] investigated blends of PLA and PPG and rationalized this behaviour by a local plasticization of PLA due to the presence of micro-droplets of the additive in the polymer structure. Moreover, the presence of these droplets in the polymer matrix might enable dissipation of the fracture energy through the interfaces between matrix and droplets, raising the maximum absorbable energy of the material.[85] Additives that involve such mechanisms to increase the polymer ductility are sometimes called toughening agents.

Epoxidized vegetable oils have received much recent interest. It has been shown that functional groups, such as ester or epoxy, could be degraded by micro-organisms,[107] making epoxidized vegetable oils interesting additives to maintain biodegradability of blends with PLA. In particular, epoxidized soybean oil (ESO), which has also been used as an effective plasticizer for PVC,[121] was studied. Ali et al.[122] reported that blending ESO with PLA reduced T_g by about 4 °C, 8 °C and 9 °C for 95/5 wt%, 80/20 wt% and 70/30 wt% blends, respectively. Partial miscibility of epoxy resins with thermoplastic polyesters with carboxyl end groups could be explained by hydrogen bonding between ester groups or the terminal hydroxyls of PLA with the epoxy group of the additive.[123–125] Regarding ductility, a maximum enhancement was obtained for the 80/20 wt% blend exhibiting nearly 40% elongation at break, while 90/10 wt% and 70/30 wt% blends showed values around 20%.

The 95/5 wt% PLA/ESO blend did not point to any ductility improvement to values for neat PLA (5% elongation at break), but stress at yield was lowered from 60 MPa to 45 MPa. Al-Mulla et al.[126] showed that modified montmorillonites with fatty nitrogen compounds obtained from vegetable oils could enhance the ductility of PLA/ESO blends by increasing their compatibility. Xiong et al.[127] studied a 90/10 wt% PLA/ESO blend but did not report a ductility increase. However, they noticed some decrease in T_g (no values mentioned) and a slight depression in elastic moduli and tensile strength from 3020 MPa and 69 MPa to 2730 MPa and 62 MPa, respectively. Xu et al.[124] introduced ESO at five concentrations into PLA (3, 6, 9, 12 and 15 wt%) and did not notice ductility improvements, but there was some decrease in tensile modulus and tensile strength.

Based on the same principles, epoxidized palm oil (EPO) was studied, partly because of the great economic importance of palm oil production.[128] Silverjah et al.[129] mixed EPO with PLA and observed an important increase in ductility, while T_g was reduced only by a few degrees and micro-voids were noticed. Indeed, the PLA containing 5 wt% of EPO exhibited a T_g value of 56.8 °C and elongation at break up to 100%, while neat PLA was measured at 63.5 °C and 6.3%, respectively. In another study,[130] the same authors examined the effect of the EPO epoxy content and observed that the higher it was, the more important was the improvement in mechanical properties. Al-Mulla et al.[131] also reported elongation at break of 210% for blends of PLA/EPO (80/20 wt%), which decreased with higher contents of EPO. In general terms, an ideal concentration of 20 wt% of epoxidized oils was found to maximize the ductility enhancement. Furthermore, Al-Mulla et al.[132] also found that a higher thermal stability was obtained when adding EPO to PLA, probably due to reduced heat sensitivity.[133]

5.7 Conclusion

Plasticization of PLA has proved to be very effective in improving mechanical and other important properties. The type and quantity of the external plasticizer depends on the target application of the compounded PLA-based material. Very schematically, if the service-life of the material is intended to be short, or even if the first service of the polymer is degradation, as in many biomedical applications, the use of a monomeric plasticizer bearing no toxicological risk is preferable. Monomeric plasticizers have generally higher efficiency in T_g depression and drawability increase. The use of solubility parameters is a good guide for choosing the adequate molecule, but other factors, such as molar volume, should be taken into account. At equal molar volume, the molecule with smaller molecular weight will be more efficient by the higher number of terminal groups at an equal mass concentration. Their drawback is the small permanence, so if the service-life of the material should be long (typically several years), such as in most commodity applications, polymeric plasticizers are very likely preferable. In this case, miscibility prediction by solubility parameters often fails, as the underlying models cannot take into account the effect of the macromolecular chain length. Partial miscibility between the polymeric plasticizers and PLA is not negative for properties in any case, though. It has been shown that inclusions of small droplets of liquid plasticizers within a certain proportion (typically below 20 wt%) can have a positive impact on drawability. The molecules function in this case as toughening agents. Many novel attempts to toughen PLA with molecules derived from biomass and not necessarily miscible with the polymer rely on this mechanism. More research needs to be carried out on the topic.

PVC plasticizing technology frequently uses mixtures of monomeric and polymeric plasticizers. This approach has the potential to increase PLA toughness in combining efficiencies of both types of plasticizer. Monomeric plasticizers help to increase miscibility of polymeric molecules with the polymer matrix. This strategy has been attempted very rarely for PLA, but the pioneering studies show very promising results. Another approach to solve permanence and miscibility problems is the use of internal plasticizers, which are grafted on the polymer chain or copolymerized into it.

Symbols and Abbreviations

ε	Elongation at break
σ	Tensile strength
E	Elastic modulus
HDPE	High density poly(ethylene)
HSP	Hansen solubility parameters
LDPE	Low density poly(ethylene)
PET	Poly(ethylene terephthalate)
PLA	Poly(lactide) or poly(lactic acid)
PP	Poly(propylene)
PS	Poly(styrene)
T_g	Glass transition temperature

Monomeric Plasticizers

AGM	Acetyl glycerol monolaurate
ATBC	Acetyl tributyl citrate
ATEC	Acetyl triethyl citrate
DBM	Diethyl bishydroxymethyl malonate
DBS	Dibutyl sebacate
Dehydat VPA 1726	Glucose monoester (Henkel)
DINCH	Di(isononyl) cyclohexane-1,2-dicarboxylate
DOA	Dioctyl adipate
DOP	Dioctyl phthalate
GMS	Glycerol monostearate
GTA	Glycerine triacetate or triacetine
LA	Lactic acid

Loxiol GMS 95	Fatty acid ester (Henkel)
TBC	Tributyl citrate
TEC	Triethyl citrate
[THTDP][DE]	Trihexyl tetradecyl phosphonium decanoate
[THTDP][BF4]	Trihexyl tetradecyl phosphonium tetrafluoroborate

Oligomeric Plasticizers

DBM-A-8 (or 18)	Diethyl bishydroxymethyl malonate-adipoyl dichloride (m = 8 or 18)
DBM-S-4 (or 7)	Diethyl bishydroxymethyl malonate-succinyl dichloride (m = 4 or 7)
EPE11	Triblock polymer PEG-*b*-PPG-*b*-PEG (Pluronic L31) with PEG 1100 g mol^{-1} and PPG 990 g mol^{-1}
EPE19	Triblock polymer PEG-*b*-PPG-*b*-PEG (Pluronic L35) with PEG 1900 g mol^{-1} and PPG 950 g mol^{-1}
M-PEG	PEG monolaurate
OLA	Oligomeric lactic acid
PA G206/2 (or 5 or 7)	Polyadipate Glyplast® 206/2 (5 or 7)
PBA	Poly(buthylene adipate)
PBOH	Poly(1,3-butanediol) (PBOH)
PDEA	Poly(diethylene adipate)
PDLLA-*b*-PEG750	Poly(D,L-lactide-*b*-ethylene glycol) with M_w of PEG = 750 g mol^{-1}
PEA	Poly(ethylene adipate)
PEGCA	Poly(ethyleneglycol-*co*-citric acid)
PEG-CH$_3$	PEG monomethylether
PHA	Poly(hexamethylene adipate)
PPG	Polypropylene glycolRikemal PL-710 Diglycerine tetraacetate
TBC-E	Tributylcitrate diethylene glycol oligomers

References

1. R. A. Auras, B. Harte, S. Selke and R. Hernandez, *J. Plast. Film Sheeting*, 2003, **19**, 123.
2. H. Z. Liu and J. W. Zhang, *J. Polym. Sci. Polym. Phys.*, 2011, **49**, 1051.
3. J. Mark (ed.), *Polymer Data Handbook*, Oxford University Press Inc., New York, USA, 1999.
4. S. Saeidlou, M. A. Huneault, H. Li and C. B. Park, *Progr. Polym. Sci.*, 2012, **37**, 1657.
5. G. Wypych, *Handbook of Plasticizers*, ChemTech Publishing, Toronto-Scarborough, 2004.
6. A. Kirkpatrick, *J. Appl. Phys.*, 1940, **11**, 255.
7. F. W. Clark, *Chem. Ind.*, 1941, **60**, 225.

8. R. Houwink, in *Proc. XI Int. Cong. Pure Appl. Chem.*, London, 1947, pp. 575–583.

9. D. J. Mead, R. L. Tichenor and R. M. Fuoss, *J. Am. Chem. Soc.*, 1942, **64**, 283.

10. W. Aiken, T. Alfrey, A. Janssen and H. Mark, *J. Polym. Sci.*, 1947, **2**, 178.

11. A. K. Doolittle, *Ind. Eng. Chem.*, 1944, **36**, 239.

12. A. K. Doolittle, *Ind. Eng. Chem.*, 1946, **38**, 535.

13. A. K. Doolittle, in *The Technology of Solvents and Plasticizers*, John Wiley & Sons, New York, 1954.

14. A. K. Doolittle, in *Plasticizer Technology*, ed. P. F. Bruins, Reinhold, New York, 1965.

15. P. B. Stickney and L. E. Cheyney, *J. Polym. Sci.*, 1948, **3**, 231.

16. T. Alfrey Jr., N. Wiederhorn, R. Stein and A. Tobolsky, *J. Colloid Sci.*, 1949, **4**, 211.

17. T. C. Moorshead, in *Advances in PVC Compounding and Processing*, ed. M. Kaufman & Sons, London, 1962.

18. M. Cohen and D. Turnbull, *J. Chem. Phys.*, 1959, **31**, 1164.

19. H. Fujita, *Adv. Polym. Sci.*, 1961, **3**, 1.

20. P. J. Flory, *J. Am. Chem. Soc.*, 1940, **62**, 1057.

21. T. G. Fox and P. J. Flory, *J. Am. Chem. Soc.*, 1948, **70**, 2384.

22. T. G. Fox and P. J. Flory, *J. Appl. Phys.*, 1950, **21**, 581.

23. M. L. Williams, R. F. Landel and J. D. Ferry, *J. Am. Chem. Soc.*, 1955, 77, 3701.

24. K. Ueberreiter and G. Kanig, *J. Colloid Sci.*, 1952, 7, 569.

25. J. Kern Sears and J. R. Darby, in *Handbook of Plasticizers*, ed. G. Wypych John Wiley & Sons, New York, 1982.

26. R. Todeschini and V. Consonni, *Handbook of Molecular Descriptors*, Wiley-VCH, Weinheim, 2000.

27. T. Oishi and J. M. Prausnitz, *Ind. Eng. Chem. Process Des. Dev.*, 1978, **17**, 333.

28. P. J. Flory, *Principles of Polymer Chemistry*, Cornell University Press, New York, 1953.

29. M. L. Huggins, *J. Chem. Phys.*, 1941, **9**, 440.

30. P. J. Flory, *J. Chem. Phys.*, 1941, **9**, 660.

31. P. J. Flory, *J. Chem. Phys.*, 1942, **10**, 51.

32. J. L. M. Abboud and R. Notari, *Pure Appl. Chem.*, 1999, **71**, 645.

33. P. A. Small, *J. Appl. Chem.*, 1953, **3**, 71.

34. D. W. Van Krevelen and N. T. Klaas, *Properties of Polym.s: Their Correlation with Chemical Structure, their Numerical Estimation & Prediction from Additive Group Contributions*, Elsevier Publishing, Amsterdam, 1972.

35. R. F. Fedors, *Polym. Eng. Sci.*, 1974, **14**, 147.

36. C. M. Hansen, *J. Paint Technol.*, 1967, **39**, 104.

37. C. M. Hansen, *Hansen Solubility Parameters: A User's Handbook*, CRC Press, Boca Raton, Florida, 2007.

38. S. J. Abbott, in *Poly(lactic acid): Synthesis, Structures, Properties, Processing, and Applications*, Wiley, 2010, pp. 83–93.

39. www.chemspider.com. Accessed September 2013.

40. S. Brouillet and J.-L. Fugit, *Polym. Bull.*, 2009, **62**, 843.

41. N. Ljungberg and B. Wesslen, *Polymer*, 2003, **44**, 7679.

42. K. Okamoto, T. Ichikawa, T. Yokohara and M. Yamaguchi, *Eur. Polym. J.*, 2009, **45**, 2304.

43. V. P. Martino, A. Jiménez and R. A. Ruseckaite, *J. Appl. Polym. Sci.*, 2009, **112**, 2010.

44. I. Pillin, N. Montrelay and Y. Grohens, *Polymer*, 2006, **47**, 4676.

45. N. Ljungberg and B. Wesslen, *J. Appl. Polym. Sci.*, 2004, **94**, 2140.

46. A. R. Berens, *Pure Appl. Chem.*, 1981, **53**, 365.

47. P. Dole, A. E. Feigenbaum, C. De la Cruz, S. Pastorelli, P. Paseiro, T. Hankemeier, Y. Voulzatis, S. Aucejo, P. Saillard and C. Papaspyrides, *Food Addit. Contam.*, 2006, **23**, 202.

48. X. Fang, S. Domenek, V. Ducruet, M. Refregiers and O. Vitrac, *Macromolecules*, 2013, **46**, 874.

49. O. Piringer and A. Baner, *Plastic Packaging: Interactions with Food and Pharmaceuticals*, Wiley-VCH, Weinheim, 2008.

50. S. R. Andersson, M. Hakkarainen and A. C. Albertsson, *Biomacromolecules*, 2010, **11**, 3617.

51. S. Jacobsen and H. G. Fritz, *Polym. Eng. Sci.*, 1999, **39**, 1303.

52. Y. Hu, Y. S. Hu, V. Topolkaraev, A. Hiltner and E. Baer, *Polymer*, 2003, **44**, 5681.

53. V. Ducruet, O. Vitrac, P. Saillard, E. Guichard, A. Feigenbaum and N. Fournier, *Food Addit. Contam.*, 2007, **24**, 1.

54. G. Colomines, V. Ducruet, C. Courgneau, A. Guinault and S. Domenek, *Polym. Int.*, 2010, **59**, 818.

55. C. Courgneau, S. Domenek, R. Lebosse, A. Guinault, L. Averous and V. Ducruet, *Polym. Int.*, 2012, **61**, 180.

56. R. Salazar, S. Domenek, C. Courgneau and V. Ducruet, *Polym. Degrad. Stabil.*, 2012, **97**, 1871.

57. R. G. Sinclair, *J. Macromol. Sci. Pure Appl. Chem.*, 1996, **A33**, 585.

58. L. V. Labrecque, R. A. Kumar, V. Davé, R. A. Gross and S. P. McCarthy, *J. Appl. Polym. Sci.*, 1997, **66**, 1507.

59. O. Martin and L. Averous, *Polymer*, 2001, **42**, 6209.

60. M. Baiardo, G. Frisoni, M. Scandola, M. Rimelen, D. Lips, K. Ruffieux and E. Wintermantel, *J. Appl. Polym. Sci.*, 2003, **90**, 1731.

61. N. Ljungberg and B. Wesslén, *J. Appl. Polym. Sci.*, 2002, **86**, 1227.

62. N. Ljungberg, T. Andersson and B. Wesslén, *J. Appl. Polym. Sci.*, 2003, **88**, 3239.

63. N. Ljungberg and B. Wesslen, *Biomacromolecules*, 2005, **6**, 1789.

64. V. P. Martino, R. A. Ruseckaite and A. Jiménez, *J. Therm. Anal. Calorim.*, 2006, **86**, 707.

65. H. Li and M. A. Huneault, *Polymer*, 2007, **48**, 6855.

66. M. Murariu, A. D. S. Ferreira, M. Alexandre and P. Dubois, *Polym. Adv. Tech.*, 2008, **19**, 636.

67. S.-L. Yang, Z. H. Wu, B. Meng and W. Yang, *J. Polym. Sci. Polym. Phys.*, 2009, **47**, 1136.
68. Y. Lemmouchi, M. Murariu, A. M. Dos Santos, A. J. Amass, E. Schacht and P. Dubois, *Eur. Polym. J.*, 2009, **45**, 2839.
69. R. Wang, C. Wan, S. Wang and Y. Zhang, *Polym. Eng. Sci.*, 2009, **49**, 2414.
70. J. Sierra, M. Noriega, E. Cardona and S. Ospina, *ANTEC Proceedings, Society of Plastics Engineering*, 2010.
71. K. Park, J. U. Ha and M. Xanthos, *Polym. Eng. Sci.*, 2010, **50**, 1105.
72. C. Courgneau, S. Domenek, A. Guinault, L. Averous and V. Ducruet, *J. Polym. Environ.*, 2011, **19**, 362.
73. H. H. Ge, F. Yang, Y. P. Hao, G. F. Wu, H. L. Zhang and L. S. Dong, *J. Appl. Polym. Sci.*, 2013, **127**, 2832.
74. M. Scatto, E. Salmini, S. Castiello, M. B. Coltelli, L. Conzatti, P. Stagnaro, L. Andreotti and S. Bronco, *J. Appl. Polym. Sci.*, 2013, **127**, 4947.
75. A. Galeski, in *Mechanical Properties of Polym.s Based on Nanostructure and Morphology*, ed. G. H. Michler and F. J. Balta-Calleja, Taylor & Francis Group, London, New York, Singapore, 2005, pp. 175–229.
76. C. Courgneau, O. Vitrac, V. Ducruet and A. M. Riquet, *J. Magn. Reson.*, 2013, **233**, 37.
77. Y. Hu, M. Rogunova, V. Topolkaraev, A. Hiltner and E. Baer, *Polymer*, 2003, **44**, 5701.
78. Y. Hu, Y. S. Hu, V. Topolkaraev, A. Hiltner and E. Baer, *Polymer*, 2003, **44**, 5711.
79. M. A. Paul, M. Alexandre, P. Degee, C. Henrist, A. Rulmont and P. Dubois, *Polymer*, 2003, **44**, 443.
80. M. Pluta, *Polymer*, 2004, **45**, 8239.
81. M. A. Paul, C. Delcourt, M. Alexandre, P. Degee, F. Monteverde, A. Rulmont and P. Dubois, *Macromol. Chem. Phys.*, 2005, **206**, 484.
82. M. Pluta, M. A. Paul, M. Alexandre and P. Dubois, *J. Polym. Sci. Polym. Phys.*, 2006, **44**, 299.
83. Z. Kulinski and E. Piorkowska, *Polymer*, 2005, **46**, 10290.
84. Z. Kulinski, E. Piorkowska, K. Gadzinowska and M. Stasiak, *Biomacromolecules*, 2006, **7**, 2128.
85. E. Piorkowska, Z. Kulinski, A. Galeski and R. Masirek, *Polymer*, 2006, **47**, 7178.
86. M. Shibata, Y. Someya, M. Orihara and M. Miyoshi, *J. Appl. Polym. Sci.*, 2006, **99**, 2594.
87. V. P. Martino, R. A. Ruseckaite and A. Jiménez, *Polym. Int*, 2009, **58**, 437.
88. V. P. Martino, R. A. Ruseckaite, A. Jimenez and L. Averous, *Macromol. Mater. Eng.*, 2010, **295**, 551.
89. V. P. Martino, A. Jimenez, R. A. Ruseckaite and L. Averous, *Polym. Adv. Technol.*, 2011, **22**, 2206.
90. M. Kowalczyk, M. Pluta, E. Piorkowska and N. Krasnikova, *J. Appl. Polym. Sci.*, 2012, **125**, 4292.

91. Z. Y. Gui, Y. Y. Xu, Y. Gao, C. Lu and S. J. Cheng, *Material Lett.*, 2012, **71**, 63.
92. S. Gumus, G. Ozkoc and A. Aytac, *J. Appl. Polym. Sci.*, 2012, **123**, 2837.
93. K. Sungsanit, N. Kao and S. N. Bhattacharya, *Polym. Eng. Sci.*, 2012, **52**, 108.
94. N. Burgos, V. P. Martino and A. Jiménez, *Polym. Degrad. Stabil.*, 2013, **98**, 651.
95. Z. J. Ren, L. S. Dong and Y. M. Yang, *J. Appl. Polym. Sci.*, 2006, **101**, 1583.
96. B. Liu, L. Jiang and J. W. Zhang, *J. Polym. Environ.*, 2011, **19**, 239.
97. S. Rodriguez-Llamazares, B. L. Rivas, M. Perez and F. Perrin-Sarazin, *High Perform. Polym.*, 2012, **24**, 254.
98. Z. Jia, J. Tan, C. Han, Y. Yang and L. Dong, *J. Appl. Polym. Sci.*, 2009, **114**, 1105.
99. M. Pluta, M. A. Paul, M. Alexandre and P. Dubois, *J. Polym. Sci. Polym. Phys.*, 2006, **44**, 312.
100. J. M. Hutchinson, *Progr. Polym. Sci.*, 1995, **20**, 703.
101. L. T. Lim, R. Auras and M. Rubino, *Progr. Polym. Sci.*, 2008, **33**, 820.
102. K. Jamshidi, S. H. Hyon and Y. Ikada, *Polymer*, 1988, **29**, 2229.
103. S. H. Hyon, K. Jamshidi and Y. Ikada, *Polym. Int.*, 1998, **46**, 196.
104. D. Cam and M. Marucci, *Polymer*, 1997, **38**, 1879.
105. R. M. Rasal, A. V. Janorkar and D. E. Hirt, *Progr. Polym. Sci.*, 2010, **35**, 338.
106. A. Hoglund, M. Hakkarainen and A. C. Albertsson, *Biomacromolecules*, 2010, **11**, 277.
107. M. Rahman and C. S. Brazel, *Progr. Polym. Sci.*, 2004, **29**, 1223.
108. R. Briones, L. Serrano and J. Labidi, *J. Chem. Tech. Biotechnol.*, 2012, **87**, 244.
109. Y. Kurimoto, A. Koizumi, S. Doi, Y. Tamura and H. Ono, *Biomass Bioenergy*, 2001, **21**, 381.
110. S. H. Lee, M. Yoshioka and N. Shiraishi, *J. Appl. Polym. Sci.*, 2000, **78**, 311.
111. S.-H. Lee, Y. Teramoto and N. Shiraishi, *J. Appl. Polym. Sci.*, 2002, **83**, 1473.
112. B. Yin and M. Hakkarainen, *J. Appl. Polym. Sci.*, 2010, **119**, 2400.
113. E. Azwar, B. Yin and M. Hakkarainen, *J. Chem. Tech. Biotechnol.*, 2012, **88**, 897.
114. A. G. Perez, P. Luaces, J. Oliva, J. J. Rios and C. Sanz, *Food Chem.*, 2005, **91**, 19.
115. M. P. Arrieta, J. Lopez, S. Ferrandiz and M. A. Peltzer, *Polym. Test*, 2013, **32**, 760.
116. H. Sawalha, K. Schroën and R. Boom, *J. Appl. Polym. Sci.*, 2008, **107**, 82.
117. A. Lopez-Rubio and J. M. Lagaron, *Polym. Degrad. Stabil.*, 2010, **95**, 2162.
118. A. Telfer, S. Dhami, S. M. Bishop, D. Phillips and J. Barber, *Biochemistry*, 1994, **33**, 14469.

119. V. B. Ivanov and V. Y. Shlyapintokh, *Polym. Degrad. Stabil.*, 1990, **28**, 249.
120. S. W. Hwang, J. K. Shim, S. E. M. Selke, H. Soto-Valdez, L. Matuana, M. Rubino and R. Auras, *Polym. Int.*, 2012, **61**, 418.
121. U. S. Ishiaku, A. Shaharum, H. Ismail and Z. A. M. Ishak, *Polym. Int.*, 1998, **45**, 83.
122. F. Ali, Y.-W. Chang, S. C. Kang and J. Y. Yoon, *Polym. Bull.*, 2009, **62**, 91.
123. S. Kim, W. Jo, J. Kim, S. Lim and C. Choe, *J. Mater. Sci.*, 1999, **34**, 161.
124. Y. Q. Xu and J. P. Qu, *J. Appl. Polym. Sci.*, 2009, **112**, 3185.
125. G. H. Yew, A. M. Mohd Yusof, Z. A. Mohd Ishak and U. S. Ishiaku, *Polym. Degrad. Stabil.*, 2005, **90**, 488.
126. E. A. J. Al-Mulla, A. H. Suhail and S. A. Aowda, *Ind. Crop. Prod.*, 2010, **33**, 23.
127. Z. Xiong, Y. Yang, J. Feng, X. Zhang, C. Zhang, Z. Tang and J. Zhu, *Carbohydr. Polym.*, 2012, **92**, 810.
128. W. D. Wan Rosli, R. N. Kumar, S. Mek Zah and M. M. Hilmi, *Eur. Polym. J.*, 2003, **39**, 593.
129. V. S. G. Silverajah, N. A. Ibrahim, N. Zainuddin, W. Yunus and H. Abu Hassan, *Molecules*, 2012, **17**, 11729.
130. V. S. G. Silverajah, N. A. Ibrahim, W. Yunus, H. Abu Hassan and C. B. Woei, *Int. J. Mol. Sci.*, 2012, **13**, 5878.
131. E. A. J. Al-Mulla, W. Yunus, N. A. B. Ibrahim and M. Z. Ab Rahman, *J. Mater. Sci.*, 2010, **45**, 1942–1946.
132. S. Al-Malaika, C. Goodwin, S. Issenhuth and D. Burdick, *Polym. Degrad. Stabil.*, 1999, **64**, 145.
133. S. N. Lee, M. Y. Lee and W. H. Park, *J. Appl. Polym. Sci.*, 2002, **83**, 2945.

CHAPTER 6

Electrospinning of PLA

LAURA PEPONI,[*a] ALICIA MÚJICA-GARCÍA[b] AND
JOSÉ M. KENNY[a,b]

[a] Instituto de Ciencia y Tecnología de Polímeros, ICTP-CSIC, c. Juan de la
Cierva, 3, 28006 Madrid, Spain; [b] Materials Engineering Center, UDR
INSTM, University of Perugia, Str. di Pentima 4, 05100 Terni, Italy
*Email: lpeponi@ictp.csic.es

6.1 Production of PLA fibres by Electrospinning

The *electrospinning process* is a useful technique to produce polymeric fibres
with diameters from micro- to nanometre scales invented in 1934.[1,2] In
particular, in this chapter we review and analyze the main electrospinning
processes for the production of electrospun-PLA fibres. This research topic is
a current trending topic in PLA research as confirmed by the increasing
number of scientific publications as reported by ISI Web of Knowledge. The
first publication in a scientific journal regarding PLA electrospinning appeared in 2001, and the exponential trend of the scientific publications from
its beginning up to 2012 is reported in Figure 6.1.

The main difference between the conventional mechanical spinning process and the electrospinning process is related to the driving force. In fact, in
the first case, a mechanical force is applied by a rotating mandrel to the
fluid, which is transformed in hydraulic pressure that pushes the forming
jet, while in the electrospinning process volumetric electrical forces are
applied to the charged jet. Moreover, the electrospinning process is very
versatile and can be used to obtain electrospun fibres with diameters
ranging from 3 nanometres to several microns of over 200 synthetic and
natural polymers.[3]

RSC Polymer Chemistry Series No. 12
Poly(lactic acid) Science and Technology: Processing, Properties, Additives and Applications
Edited by Alfonso Jiménez, Mercedes Peltzer and Roxana Ruseckaite
© The Royal Society of Chemistry 2015
Published by the Royal Society of Chemistry, www.rsc.org

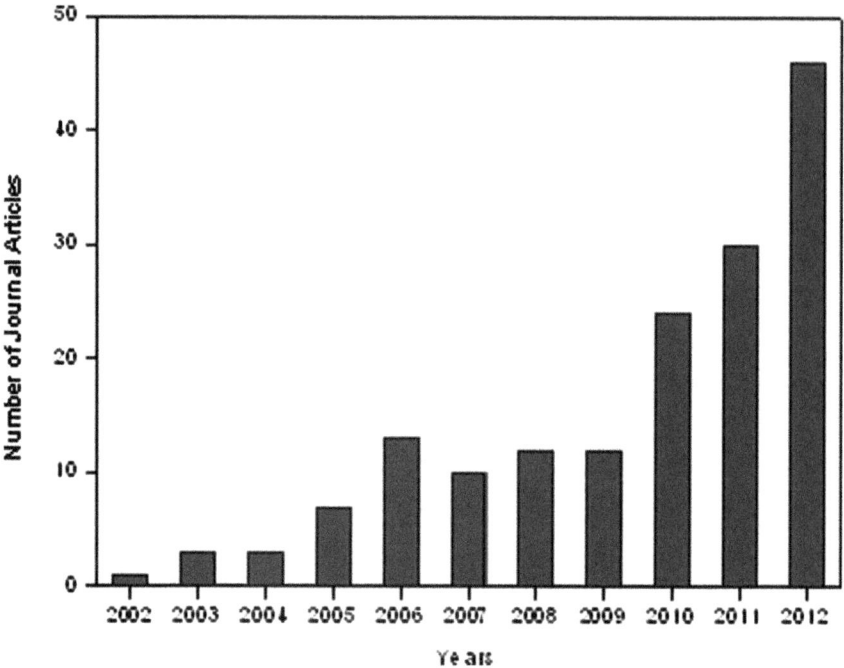

Figure 6.1   Number of scientific papers published on PLA electrospinning (data from ISI Web of Knowledge).

Electrospinning is normally applied to a polymer solution but, in a few cases, it has also been applied to polymer melts. The basic components of an electrospinner can be summarized as follows: a high-voltage DC supply, a spinneret, a pump to deliver the polymer solution (or the polymer melt) to the spinneret and a collector, where fibres are obtained and stored (Figure 6.2). An electrostatic voltage is applied between the spinneret, which contains the polymer solution (or the melt polymer), and the collector. Consequently, the polymer solution is charged and the induced electrostatic repulsion is in contrast with the surface tension of the solution (or the melt) producing an elongation of the drop in the polymer solution until a conical shape is assumed in the capillary tip called a Taylor cone. Once the Taylor cone is stable, the charged polymer solution is attracted towards the collector and the jet gets narrower because of the elongation as well as the solvent evaporation. The electrospun fibres are obtained and they can be collected in the form of single fibres as well as random or oriented mats. Typical collectors used for oriented fibres could be, among others, a rotating drum, a disc or parallel electrodes.[4,5]

Finally, depending on the needle used, that is simple or concentric, we refer to conventional or coaxial electrospinning, respectively.[6] In general, two groups of parameters have to be taken into account in order to optimize the electrospinning process: (i) those related to the solution properties, such

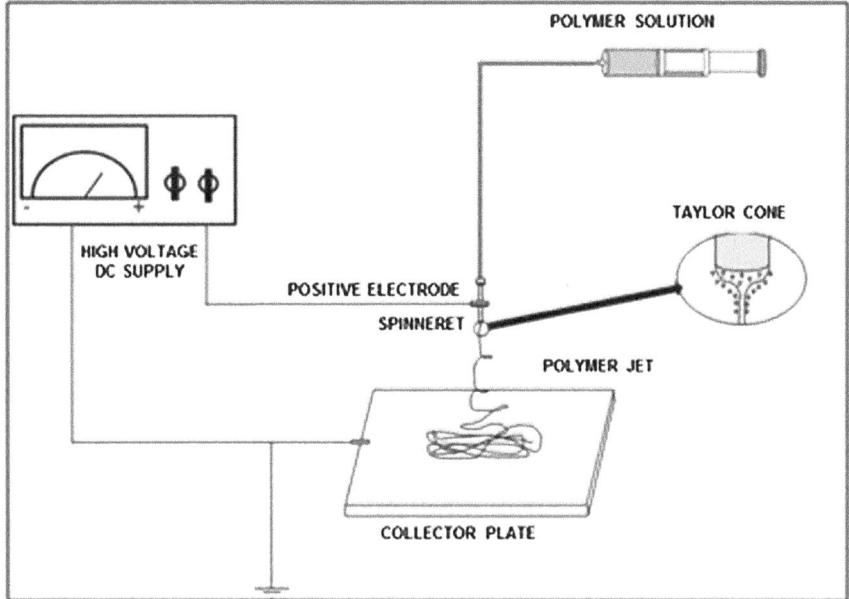

Figure 6.2 Main components of a typical electrospinning equipment.

as viscosity, conductivity and surface tension; and (ii) those related to the processing variables, such as flow-rate, applied voltage and working distance between the tip and the collector.[1] It is important to underline that experimental and atmospheric conditions can affect the morphology and final properties of the electrospun fibres. In solution electrospinning, the fibres' size, structure and morphology is characterized by rapid evaporation of the solvent, while in melt electrospinning, the rapid solidification during drawing determines the final properties of fibres.[5,6]

Solvent-based electrospinning is less expensive and simpler than melt-based electrospinning, but the latter does not have the environmental issues related to solvent handling and evaporation. The fibres obtained from melting usually have larger diameters, caused by the low conductivity and the high viscosity, while solution electrospinning produces thinner fibres taking advantage of the control of the solution properties. Numerous polymer fibres with submicron diameters have been obtained by solution electrospinning, the viscosity being the most important property related to the final fibre diameter. In general, the lower is the solution viscosity the thinner are the fibres. In addition, composites and hollow nanofibres have also been produced by coaxial electrospinning.[1,3,7] One of the common problems related to the solvent-based electrospinning is the formation of defects in the uniformity of the fibre diameter, called beads.

Melt electrospun PLA sub-micron fibre mats with no residual solvent can be employed as filter media and tissue scaffolds preferred to those prepared by solution electrospinning processes, due to the elimination of possible toxicity and environmental impact caused by solvent use. Moreover, no mass losses due to solvent evaporation are observed. Finally, the melt-based electrospinning process can produce polymer fibres with high resistance at ambient conditions. Despite all these advantages, not many studies on fibres obtained by melt-electrospinning have been reported, due to the high working temperature (180–255 °C), the high viscosity, the low conductivity of polymer melts and, above all, the thickness of the obtained fibres, higher than those prepared by solution electrospinning. Zhou *et al.*[8] produced sub-micron scale PLA fibres by melt-electrospinning showing that temperatures at the spinneret and in the spinning region are critical to produce sub-micron size fibres. Finally, the melt-based electrospinning process did not show any change in the polymeric chemical composition and PLA fibres produced by melt-electrospinning are mostly amorphous as a consequence of quenching during the rapid solidification.[8] Moreover, the prolonged exposure to heat in melt-electrospinning could produce the fibres' thermal degradation, but a CO_2-laser beam to produce local melting of the polymer can minimize this degradation problem.[9] This is useful for polymers having relatively high melting points, such as poly(lactide), poly(ethylene terephthalate) and thermotropic liquid-crystalline polymers. This technique is able to produce fibres with diameters smaller than 1 mm as reported by Ogata *et al.*[7] In fact, they obtained PLA fibres by a melt-electrospinning system equipped with a CO_2-laser, with an average diameter smaller than 1 mm. When laser output power increased, the fibre diameter decreased as well as its melting point and molecular weight. Finally, related to the crystalline nature of the fibres, they observed that fibres presented an amorphous state and the annealed fibres displayed an isotropic crystal orientation.[4]

Coaxial electrospinning (co-electrospinning) is a technique developed to obtain core-shell continuous electrospun fibres. This technique is normally used to add inclusions of therapeutic drugs into the polymer nanofibres. In the co-electrospinning process two different solutions flow through a coaxial needle, thus forming, after the solvent evaporation, core-shell micro and nanofibres. It is important to point out that the mixture of both the core and the shell polymers is avoided during the co-electrospinning process taking advantage of the very fast process preventing them from mixing. Moreover, polymers showing difficulties in electrospinning can be processed with the aid of a second polymer to produce core-shell electrospun fibres.[8] Core-shell fibres can be used in the field of microelectronics, optics and medicine, even if some applications are associated to the improvement of the biocompatibility of encapsulated materials and to the increment of the mechanical properties of certain nanofibre materials.[8] Moreover, these fibres could be used to the isolation and/or encapsulation of sensitive components for tissue engineering and drug release applications, where the polymer shell provides temporal protection for drug molecules and offers their controlled release.[5,10,11] Kriel *et al.*[11] designed proper processing conditions to prepare core-shell fibres of semicrystalline PLLA (core) and amorphous PDLLA (shell) through the coaxial electrospinning of two miscible solutions.

6.1.1 PLA Electrospun Nanofibres: Diameter, Morphology and Orientation

One of the most important properties targeted when working with the electrospinning process is the fibre diameter. In general, in the solvent-based electrospinning process many factors are responsible for the final behaviour of the electrospun fibres, such as the solvent quality, diffusion coefficient, flow rate and those parameters affecting the solvent evaporation as well as temperature and environmental conditions. All these factors strongly influence the fibres' diameter and their morphology. In Figure 6.3 an example of PLA electrospun fibres obtained by us is reported.

In particular, the most important parameters are the polymer molecular weight and its concentration in the solution. In fact, by varying both parameters it is possible to obtain electrospun fibres with diameters in the range of 50–100 nm for most polymeric systems. These parameters directly affect the polymer solution viscosity and spinnability influencing jet thinning. Furthermore, a threshold concentration depending on the polymer molecular weight exists. In fact, when the molecular weight is high a low concentration is required to obtain electrospun fibres.[3] In general, electrospun nanofibres are typically obtained when intermediate concentration and molecular weight values are employed. In fact, the use of low polymer concentrations induces the formation of microsize droplets known as beads, along the fibre length due to the varicose jet instability.[3] Surface tension, dielectric and electrical properties also affect the jet flow and, as a consequence, they influence the fibres' diameter and their morphology. Chew *et al.*[3] added surfactants or salts to the polymer solution to obtain nanofibres without beading. The influence of the variation of these parameters on the electrospun fibres' behaviour was studied by Tan *et al.*[12] These authors reported a processing map by studying the variation of the polymer

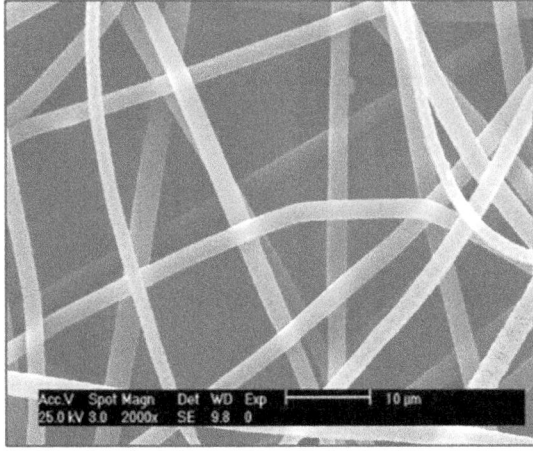

Figure 6.3 PLA electrospun fibres.

concentration, the molecular weight and the solution's electrical conductivity to control the morphology of the electrospun polymer fibres. They prepared polymer nanofibres as small as 9 nm in diameter. In addition, for some biomedical applications it is important to take into account the nanofibres' surface texture as well as their porosity. Indeed, nanoscale surface texture is strongly related to cell attachment and proliferation, while for drug delivery it is necessary to control the nanofibre porosity, which affects the specific fibre surface area. The solvent choice is very important, considering that nanofibres with designed porosity can be obtained by employing volatile solvents.[3] Qi *et al.*[13] produced highly porous PLLA fibres by electrospinning for drug delivery and tissue engineering by using a ternary system of non-solvent/solvent/PLLA. The non-solvent was butyl alcohol and the solvent was dichloromethane. They reported that porous fibres can be easily achieved if the applied voltage is low enough and the solvent is more volatile than the non-solvent used in the spinning solution. Wright *et al.*[14] added NaCl crystals to nanofibre mats during the electrospinning process to artificially create large voids or pores in the electrospun PLLA mats for three-dimensional tissue engineering applications to improve the proliferation and the infiltration of cells into the scaffold.

It is also important to note that the fibres' alignment, required in many biomedical applications, can be induced by using different collectors, such as rotating drum, mandrel, rotating disc, rotating cylinder, *etc.*[3] When using a rotating drum the nanofibre mats are thicker and their alignment is partial, even at very high rotation rate (Figure 6.4a). A rotating disc induces better fibre alignment but it is difficult to produce high quality nanofibre mats or 3D assemblies desirable for tissue engineering. Electrical forces acting on the charged jet segment can also be used to control the nanofibres' alignment. In fact, the use of split collecting electrodes to obtain highly

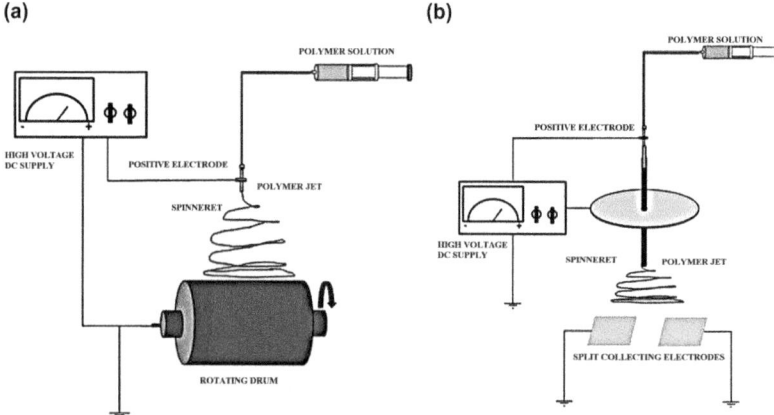

Figure 6.4   a) Conventional system with a rotating drum and b) split electrode (gap) for alignment nanofibre collection.

aligned 3D nanofibre mats is the more effective system to align electrospun fibres (Figure 6.4b).[3]

6.1.2 PLA Electrospun Nanofibres: Crystallinity

The control of the polymer fibre crystallinity is important, since it should be taken into account that it can strongly affect the mechanical properties, such as stiffness, yield stress, elastic modulus and tensile strength, as well as the degradation rate, solubility and the optical and electrical properties of the PLA-electrospun fibres. By optimizing the solution concentration and the processing parameters of PLA-based electrospun fibres, it is possible to tailor the degree of crystallinity, which decreases when using a more concentrated solution.[21] However, there is an optimum electrospinning voltage to obtain the maximum degree of crystallinity for each polymer solution concentration.[15] Liao *et al.*[16] used an additional centrifugal field in the preparation of PLA fibres by electrospinning with the aim to increase the nanofibres' crystallization. The combination of the strong stretching force parallel to the additional centrifugal field and an electrostatic field produced the alignment of the PLA chains, obtaining higher crystalline features, molecular orientation and conformation as well as good mechanical properties. Zong *et al.*[17] showed that the fibres morphology depends on the strength of the electric field as well as on the charge density of the solution. The formation of uniform nanofibres with no bead-like textures was favoured by using higher polymer concentrations and higher charge densities. Besides, higher electrospinning voltages as well as higher feeding rates produce an increase in the nanofibres' diameter. However, contrary effects of the electrospinning voltage have been reported by Wang *et al.*[1] The jet diameter decreases, obtaining thinner PDLLA microfibres when higher voltage is applied (from 11.7 μm at 13.5 kV to 7.9 μm at 16.5 kV).

6.1.3 PLA Electrospun Nanofibres: Mechanical Properties

The fibres' geometry and their preparation conditions strongly influence the mechanical properties of the PLA-electrospun fibres. Moreover, subsequent treatments of fibres can produce enhancements in their properties. Many studies relating the fibres diameter with the mechanical properties of the final electrospun mats are reported in the literature. Among others, You *et al.*[18] obtained PLA nanofibres by electrospinning, which were thermally treated at 180 °C for different periods. While the fibres' structure did not change, their diameters increased starting from 310 nm for the as-spun PLA nanofibres, due to the melting of the fibre surface. Mechanical properties, in terms of tensile strength and elongation at break, also increased with the thermal treatment as reported in Table 6.1.

Chen *et al.*[19] obtained fibres by using a rotating cylinder as collector at a line speed of 18 m min^{-1}, with flow-rate 0.1 mL h^{-1}, 10 kV of voltage and a distance between the needle and the cylinder about 100 mm. They obtained

Table 6.1 Mechanical properties of the neat and thermally treated PLA electrospun nanofibre mats.[18]

Sample	Tensile strength (MPa)	Breaking strain (%)
As-spun PLA	3.59	12.8
Thermally treated PLA (180 °C, 15 min.)	6.09	116.0
Thermally treated PLA (180 °C, 30 min.)	4.14	102.5
Thermally treated PLA (180 °C, 60 min.)	4.19	35.5

Table 6.2 Mechanical properties of electrospun-PLA mats at different PLA molecular weight and take-up rates.[20]

M_w (g mol^{-1})	Take-up velocity (m min^{-1})	Average fibre diameter (nm)	Young's modulus (GPa)	Tensile strength (MPa)	Breaking strain (%)
100 K	63	498 ± 67	4.1 ± 1.7	119 ± 50	121 ± 23
	630	382 ± 124	3.6 ± 1.2	113 ± 22	42 ± 24
	1890	346 ± 147	7.8 ± 2.0	343 ± 81	23 ± 2
300 K	63	401 ± 18	2.5 ± 0.8	60 ± 18	163 ± 10
	630	388 ± 19	3.1 ± 0.2	122 ± 2	148 ± 1
	1890	357 ± 31	5.2 ± 2.2	280 ± 37	34 ± 5
700 K	63	496 ± 93	1.5 ± 0.2	55 ± 13	135 ± 2
	630	486 ± 25	4.6 ± 2.0	143 ± 84	97 ± 5
	1890	399 ± 26	3.6 ± 0.3	150 ± 7	42 ± 1

electrospun fibres with an average diameter about 252 nm, with a tensile strength 12.7 MPa and a breaking strain 55.5%. Table 6.2 shows the values of the mechanical properties obtained by Zhang *et al.*[20] for oriented PLA-electrospun mats prepared by variation of the PLA molecular weight at three different take-up rates. A rotational disc was used as collector system, with flow-rate 0.1 mL h^{-1}, 10 kV and 150 mm between the needle and the cylinder.

It is not trivial to find a general correlation between the PLA molecular weight and the take-up rate to obtain the highest mechanical properties. The smallest diameter values are generally obtained when the highest take-up rate is used, but these diameters do not correspond to high mechanical properties. In fact, in the case of the highest molecular weight used by Zhang *et al*,[20] the best mechanical response was obtained for an intermediate take-up rate and diameter. Tomaszewski *et al.*[21] studied the mechanical properties of oriented PLA-electrospun mats after different annealing treatments, as reported in Table 6.3. The annealing of the raw PLA-electrospun micro-fibrous mats was carried out by positioning them between two hot plates at 90 °C, at atmospheric pressure, for 20 min. The mats positioned between two hot plates at 80 °C were compressed at about 20–30 G cm^{-2} for 10 min. In this case, the PLLA electrospun fibres were collected in a rotating collector, with a voltage 20 kV and a distance between the needle and the cylinder of

Table 6.3 Mechanical properties of electrospun PLA mats with different treatment methods.[21]

Treatment method	Average fibre diameter (µm)	Breaking strain (%)	Tenacity (MPa)	Ball piercing force (N)
PLA raw	1.23	27.5	1.47	4.90
PLA annealed	1.23	23.3	3.19	8.58

Table 6.4 Mechanical properties of oriented PLA electrospun mats obtained from different solvents.[16]

PLA nanofibres	Average fibre diameter (nm)	Young's modulus (GPa)	Avg. hardness/ modulus (GPa)
PLA in $CHCl_3$[a]	255	1.25	0.12/1.60
PLA in $CHCl_3$/Acetone $(1/1)$[b]	188	1.60	0.14/1.94
PLA in $CHCl_3$/THF $(1/1)$[b]	203	2.64	0.26/3.30
PLA in $CHCl_3$/DMF $(1/1)$[b]	137	0.68	0.04/0.63
PLA in $CHCl_3$/DMF $(2/1)$[b]	139	1.01	0.10/1.52

150 mm. The annealing treatment strongly influences the mechanical behaviour of the PLLA mats, even if it is not affected by the average diameter of the electrospun fibres.

Finally, the mechanical properties of PLA-electrospun mats obtained by using different solvents were studied by Liao *et al.*[16] In particular, fibres obtained from a solution of PLA and chloroform ($CHCl_3$) were prepared by using a collector disc with a centrifugal field of 1800 rpm. The experimental conditions used in this case, indicated with (*a*) in Table 6.4, are voltage 15 kV, flow-rate 0.25 mL h^{-1} and distance between the needle and the cylinder 200 mm. Other samples, indicated with (*b*) in Table 6.4, were obtained from a PLA solution with a solvents mixture and a collector disc with centrifugal field 1800 rpm, flow-rate 0.25 mL h^{-1}, 30 kV and distance between the needle and the cylinder 200 mm. From these results, it is worthwhile to note that the solvent and its evaporation rate strongly influence the PLA-electrospun fibres' behaviour. The best results were obtained when a mix of tetrahydrofuran and chloroform was used.

6.2 PLA Matrix Nanocomposite Electrospun Fibres

PLA nanocomposite fibres can be obtained by following different methods, such as melt-spinning, dry-spinning and electrospinning, depending on the final application. The electrospinning process is the most suitable for drug delivery and tissue engineering applications of PLA-based nanocomposites.[22] Some examples of PLA-electrospun nanocomposite fibres found in the scientific literature are reported below.

6.2.1 PLA Electrospun Nanocomposite Fibres with Montmorillonites

Montmorillonite (MMT), one of the most common layered silicates, belongs to the family of phyllosilicates and it is formed by two tetrahedral silica thin layers with a central octahedral sheet of magnesia.[23] Lengths of silicate sheets are typically in the order of 200–250 nm and sheet thickness is 1 nm.[24] MMT nanosized platelets were incorporated into PLLA solutions by Lee *et al.*[25] with the aim of increasing strength and structural integrity during the biodegradation process. But, in particular cases where transport of metabolic nutrients and wastes through the nanosized pores is necessary as well as for the cell implantation and blood vessel invasion through the micro-sized pores, a fabrication technique combining electrospinning and salt leaching/gas foaming processes by incorporating MMT nanoplatelets was used. The polymer solution was electrospun and mechanically entangled by a cold compression moulding process to obtain a robust 3D scaffold structure. The electrospun fibres showed an increase of mechanical and physical properties including higher strength, greater stiffness, higher heat resistance and higher UV resistance maintaining transparency when a small quantity of nanosized MMT was added.[25]

6.2.2 PLA Electrospun Nanocomposite Fibres with Halloysite Nanotubes

Halloysite nanotubes (HNT) are clays belonging to the kaolinite family with tubular form, diameter in the range of 100–300 nm and length 1–3 mm. Their structure is similar to carbon nanotubes and they are formed by rolling the layers of tetrahedral sheets of silica and octahedral sheets of alumina. Touny *et al.*[22] produced PLA nanocomposite fibres reinforced with halloysite nanotubes by electrospinning determining the optimum of the polymer solution concentration, HNT content and feed rates to produce uniform bead-free nanofibres. They found that the PLA crystallization was increased by the addition of HNT, while the fibres' average diameter was reduced from the micrometre range for pure PLA to the nanometre range for PLA/HNT nanocomposites, due to the increase in the solution electrical conductivity and viscosity by the addition of the inorganic clay. Some fibres were produced with beads and agglomerations when nanocomposite fibres contained 2.5 wt% HNT, due to the instability caused by the low viscosity of the solution leading to the formation of droplets instead of fibres. But, in nanocomposite fibres containing 15 wt% HNT, fewer beads were observed and aligned fibres with diameters in the nanoscale range were produced. These results are due to the increase in the solution electrical conductivity and viscosity by the addition of the inorganic clay. The charge density was increased on the solution surface during electrospinning by the presence of HNT. The accumulated charges overcome the cohesive forces and increase the repulsive forces leading to the formation of bead-free fibres. Besides, the increase in electrical conductivity leads to a significant decrease in the average diameter of the polymer nanocomposite fibres.[22]

6.2.3 PLA Electrospun Nanocomposite Fibres with Hydroxyapatite

PLLA is a polymer with good biocompatibility and it can be modified with inorganic materials, such as hydroxyapatite (HA), to improve its biological properties for tissue engineering purposes.[26] HA is an inorganic component of bones with bioactive, biodegradable and osteoconductive properties. HA is widely used in biomedical implant applications and bone regeneration. The mechanical properties of bones are improved due to the nanosize of the inorganic component.[27] Moreover, the use of PLA/HA membranes fabricated *via* electrospinning to improve the *in vitro* bone-cell responses and *in vivo* bone-forming ability has been reported. The preparation of ceramic/polymer composites for bone regeneration is carried out by different methods, such as *in situ* polymerization of ceramic/monomer mixtures or by dispersing the ceramic component in the polymer solutions followed by drying and hot-pressing. Another method, called biomimetic mineralization, mimics the biological mineral growth in natural bones from a biocompatible aqueous solution under mild reaction conditions. This method consists of the partially carbonated HA (the bone-like mineral) growing onto the polymer scaffold in a simulated body fluid. The carbonated HA has low crystallinity and nanoscale size, for the reabsorption and remodelling properties in bones. This method could also be used for surface modifications of 3D scaffolds in other tissue engineering applications. Chen *et al.*[28] used a highly porous PLLA electrospun nanofibre scaffold as matrix for the HA biomineralization. Seyedjafari *et al.*[29] also prepared PLLA/HA nanofibre scaffolds showing the ability of ectopic bone formation in the absence of exogenous cells.

Natural biomaterials as well as synthetic biocompatible polymers for biomedical applications can be prepared by electrospinning, obtaining electrospun scaffolds with a topology similar to the fibrillar structure of the extracellular matrix. The enhancement of the cell adhesion and their proliferation is due to the large specific surface area and the small pore size of the non-woven system formed by electrospun fibres. Sui *et al.*[27] prepared PLLA/HA hybrid membranes by electrospinning and reported that the addition of HA nanoparticles increased the specific surface area and mechanical properties while slowing down the degradation rate of the PLLA/HA membrane. Spadaccio *et al.*[26] prepared PLLA/HA hybrid membranes by electrospinning obtained from dispersed nanopowders of HA in PLLA solutions. This nanocomposite was used to promote and to facilitate osteoblast cell adhesion and growth and it is a potential candidate for bone tissue engineering. Prabhakaran *et al.*[30] prepared PLLA, PLLA/HA and PLLA/collagen/HA nanofibres by electrospinning, showing their possible application as structural scaffolds with suitable cell recognition sites, biocompatibility, osteoconductivity and sufficient mechanical strength for bone tissue engineering. Li *et al.*[31] prepared electrospun PLLA/HA fibre mats by laser-melt electrospinning to minimize the polymer degradation by using CO_2 laser heating. In fact, for *in vivo* tissue engineering applications it is important to remove the residual solvent. Melt-based electrospinning is a technique to obtain scaffolds without using solvents, but it is possible that the polymer could suffer some thermal degradation. D'Angelo *et al.*[32] prepared nanocomposite fibrous mats of PLLA loaded with calcium-deficient nanohydroxyapatite (d-HAp) to demonstrate that human stem cells respond to nanocomposite fibrous mats. Their results showed that the interaction of human stem cells with PLLA/d-HAp mats in the absence of soluble osteogenic factors permitted the osteogenic differentiation. Sonseca *et al.*[33] prepared aligned mats of PLA nanofibres containing nanosized HA filler *via* electrospinning onto a rotating mandrel and they studied their morphology and final properties. They reported that electrospun nanocomposite fibres showed some enhancement in thermal and mechanical properties due to the nucleating effect of HA nanoparticles. Higher thermal stability was obtained for aligned mats with high filler content. Furthermore, the reduction in fibre diameter and some increase of fibre orientation by the highly crystalline structure of the collected mats was observed due to the stretching produced during the process. Jeong *et al.*[34] prepared 3D nanofibrous PLA/HA composite scaffolds by electrospinning and they studied the improvement on the mechanical properties of the electrospun fibres due to the HA incorporation. Tensile strength increased with the HA amount in the nanocomposite fibres. In fact, the scaffold containing 20 wt% HA displayed the highest tensile strength $(0.262 \pm 0.007 \text{ MPa})$ compared to other samples $(0.157 \pm 0.005 \text{ MPa for 5 wt\% HA, and } 0.063 \pm 0.005 \text{ MPa for neat PLA})$. The same trend was reflected on the elongation at break and the elastic modulus, confirming that the incorporation of HA, even at large loadings (20 wt%), enhances mechanical properties. The nanofibre scaffold with 20 wt% HA showed the largest elongation at break $(36 \pm 1\%)$, compared to $27 \pm 1\%$ for PLA and $30 \pm 1\%$ for PLA with 5 wt% HA. They related the increment in the mechanical response with the ability of the nanocomposite scaffolds to absorb a considerable amount of energy before failure. This characteristic can be attributed to the decrease in the fibre diameters with the increased amount of HA nanoparticles.[34]

6.2.4   PLA Electrospun Nanocomposite Fibres with Carbon Nanotubes

The addition of carbon nanotubes (CNT) causes some reduction in the fibre diameter and narrow diameter distribution of the electrospun fibres, by the uniform charge density in the solution during the electrospinning process. Frank *et al.*[35] prepared PLA fibres with single-walled carbon nanotubes

(SWCNTs) dispersed in dimethylformamide (DMF), which is an efficient solvent for SWCNTs, by co-electrospinning. The SWCNTs were randomly distributed in the PLA fibres, since only about 10% PLA fibres were found to contain SWCNTs forming spherical agglomerates. This heterogeneous distribution of SWCNTs is due to the agglomerated microstructure, which is detrimental to the mechanical and electronic properties of the polymer fibres. On the contrary, multi-walled carbon nanotubes (MWCNTs) produced a clear improvement on the fibre's elastic modulus, toughness and tensile strength.[36] In general terms, the use of carbon nanotubes as reinforcement in nanocomposites provides some enhancement in the electrical conductivity. But, for biomedical applications, CNTs have a disadvantage in their cytotoxicity. It is possible to limit this drawback by the encapsulation of MWCNTs inside PLA nanofibres with diameters around 700 nm.[37] These authors reported some increase in the scaffold conductivity and reduction of the fibre diameter, providing a functional composite for tissue engineering. Shao *et al.*[38] prepared random oriented and aligned PLA/MWCNTs electrospun nanofibres, with potential applications in coupling bone regeneration and fracture healing with the use of appropriate electrical stimulations.

6.2.5 PLA Electrospun Nanocomposite Fibres with Graphene

Graphene is a flat monolayer of carbon atoms firmly packed into two-dimensional (2D) honeycomb lattices. It is the basis for different dimension graphitic materials, such as wrapped up into 0D fullerenes, rolled into 1D nanotubes or stacked into 3D graphite.[39] Ma *et al.*[40] prepared PLA, HA and graphene oxide (GO) nanofibres by electrospinning. GO has extraordinary mechanical properties, such as high elastic modulus and hardness, and excellent flexibility. HA and GO powder can be uniformly dispersed into PLA matrices. Electrospun PLA/HA/GO nanofibres present a porous fibrous 3D structure with relatively rough surface morphology, showing good biocompatibility and excellent mechanical properties. These results show that these fibres can be used in tissue engineering for cell attachment and proliferation in bone tissue regeneration.

6.3 Applications of Electrospun-PLA Fibres

Electrospun-PLA micro and nanofibres based on neat PLA as well as PLA blends and PLA nanocomposites are mainly used in biomedicine, textile and pharmaceutical fields. The specific applications will define the requirements in terms of aligned or random collected electrospun fibres, also influencing their morphology and final properties. Electrospinning offers a tailored production of very thin fibres with large surface areas, ease of functionalization for various purposes, superior mechanical properties and ease of process, showing a wide range of applications in the biomedical field with particular attention to tissue engineering, drug release, implants or biotransformations and wound healing.[41] The most significant and recent applications in scientific literature are reported below. In particular, attention is focused on the main applications, one related to the use of the PLA-electrospun fibres as matrix for tissue engineering, due to the high compatibility of the degradation products with the human body. Another relevant application is related to wound dressing and the use of electrospun-PLA fibres for drug delivery systems.

6.3.1 Tissue Engineering

Tissue engineering is an interdisciplinary field that combines knowledge of medicine, biology, engineering and materials science. Its main objective is the development of biological substitutes to restore, maintain or improve damaged tissue and organ functionality, through the use of scaffolds to provide support for cells and to regenerate a new extracellular matrix if destroyed by disease, injury or congenital defects without stimulating any immune response.[41,42] The use of polymer nano-biocomposites is suitable to mimic the surface properties of natural tissues and they are promising candidates to be used in the tissue engineering field. In fact, they present excellent behaviour compared with conventional materials, such as their high cytocompatibility and improved mechanical, electrical, optical, catalytic and magnetic properties.[42] For example, highly porous artificial extracellular matrices or scaffolds are required to accommodate mammalian cells and to guide their growth and the tissue regeneration in three dimensions. In this regard, electrospinning is generally considered a suitable technique to produce scaffolds for tissue engineering, allowing the production of functionalized materials with interconnected pores and high surface area in the same scale of the extracellular matrices.[36] In particular, scaffolds produced by electrospinning are 3D porous mats with high porosity and surface area, which can mimic extracellular matrix structures (ECM), formed by proteins and glycosaminoglycans. Non-woven electrospun mats can be used to biomimic physical dimensions of native natural ECM, able to separate different tissues forming a supportive meshwork around cells and providing anchorage to cells. Electrospun nanofibres can mimic the physical structure of the major constructive elements in the native ECM, synthesized and hierarchically organized into fibrous form with fibre dimensions down to the nanometre scale.[41,43] Moreover, electrospun PLA-based fibres can be used as carriers for bioactive agents, including antibiotics and anticancer drugs.[4] Furthermore, macroporous scaffolds consisting of randomly oriented or aligned PLA nanofibres can be easily produced by electrospinning, allowing the incorporation of drug delivery functionalities with optimal microenvironment for the seeded cells.[3]

In general terms, biodegradable polyesters such as PLA, and poly (ε-caprolactone) (PCL) are suitable choices for constructing nanofibrous scaffolds,[43] by their good processability and mechanical properties. In fact, electrospun fibres of these polymers could replicate the physical dimensions and morphology of the major components in native ECM. On the other hand, biopolymers used in these applications should present some requirements related to mechanical properties and degradation times, as well as cells proliferation and differentiation in the process of forming functional tissues.[41] In addition, electrospun nanofibres of many biodegradable and biocompatible polymers, such as PCL, PLA or poly(dl-lactic-*co*-glycolic acid) (PLGA), loaded with different drugs present large surface areas, have good mechanical properties and can be easily functionalized. Besides, a reduced drugs release has been observed in electrospun mats.[41] Nevertheless, electrospun-PLA is used for its biocompatibility and degradation properties.[36] Lu *et al.*[44] used electrospun-PLLA nanofibres combined with outgrowth endothelial cells (OECs) isolated from human peripheral blood. OECs are used as the seeding cells for engineering vascular crafts, as they can maintain stability in the different phenotypes. These authors observed that PLLA is biocompatible with OECs and that the aligned PLLA fibres were suitable as scaffolds and as cell growth promoters during vascular tissue engineering. Moreover, the cytocompatibility of PLA-electrospun fibre mats has been analyzed in terms of cell adhesion, proliferation and changes in cell morphology and cellular functions driven by topographical features provided by the nanofibrous scaffolds.[3] P(LLA-CL) copolymers have also been investigated for tissue engineering applications due to their high biocompatibility, high permeation rate for steroids and degradation rates controlled by adjusting their composition in the copolymer.

Mo *et al.*[45] prepared P(LLA-CL)-electrospun fibres where cells were well proliferated as monolayer cultures. Khatri *et al.*[46] produced PCL-PLLA tubular nanofibres by electrospinning using different PCL-PLLA ratios to obtain good mechanical stability combined with faster biodegradability. They concluded that these materials are viable in tissue engineering applications, due to their biocompatibility, and can be considered as good substrates for *in vivo* or *in vitro* research.

Furthermore, scaffolds based on synthetic biomaterials have been developed for the reconstruction of large bone defects, which usually require the use of graft materials.[47] Electrospun fibres of biodegradable polyesters can be used to prepare scaffolds for bone tissue engineering, since they could mimic the nanoscale features of the ECM presenting a suitable bone-implant interface for the host.[48] Besides, it is possible the incorporation of particles into the fibres to obtain a controlled drugs release, necessary in tissue engineering such as growth factors or other substances. Electrospun nanofibres have shown better cell adhesion and proliferation than electrospun microfibres, considering that the large surface area of the constructs offer properties similar to the natural ECM.

The 3D structure of the electrospun scaffolds should allow the cells to be fully differentiated and should permit them to migrate freely.[49] The influence of 3D PLLA nanofibre scaffolds on *in vivo* bone formation was studied by Schofer *et al.*,[50] besides the effect of the incorporation of bone morphogenetic protein 2 (BMP-2) to enhance efficiency. BMP-2 is involved in the development of bone and cartilage tissues and induces osteoblast differentiation in a variety of cell types. These authors reported that the cell migration was easily produced in PLLA nanofibre scaffolds. In fact, PLLA/BMP-2 nanofibre scaffolds combine a suitable matrix for cell migration with an osteoinductive stimulus, able to close calvarial defects within 8 weeks. Li *et al.*[51] produced PLLA/keratin non-woven fibrous membranes by electrospinning to improve the PLLA cell affinity. They used wool keratin as natural protein and reported higher cell viability and differentiation with more osteoblast cells on PLLA/keratin membranes after one week. Yan *et al.*[52] prepared nanofibres of PLLA-gelatin blends by electrospinning to improve the hydrophilicity and mechanical properties of the PLLA nanofibrous mats. They observed that the strength of PLLA-gelatin nanofibrous mats was enhanced when compared to pure gelatin joined to some increase on the growth of osteoblastic cells.

Tissue engineering is also applied to develop feasible substitutes to aid in the clinical treatment associated with the human nervous system. Axons are nerve fibres, acting as slender projectors of nerve cells or neurons, which typically conduct electrical impulses away from the neuron's cell body. Much care on new medical therapies for the treatment of nerve tissue repair arising from spinal cord injury is necessary. In this regard, PLLA has been investigated for nerve regeneration, due to its biocompatibility and biodegradability.[53] Moreover, nanoscale architecture scaffolds prepared by electrospinning are also useful for repairing nervous tissues. Therefore, polymeric scaffolds should be functionalized to improve neural tissue repair. Electrospinning is a useful and easy method to modify biomaterials in order to produce biomimetic scaffolds with topographical and biochemical cues compared to conventional covalent binding and physical adsorption. Koh *et al.*[54] carried out scaffolds functionalization by coupling laminin onto PLLA-electrospun nanofibres. Laminin is a neurite promoting ECM protein and its incorporation onto the nanofibres is an alternative to mimic the biochemical properties of nervous tissues to create biomimetic scaffolds. Nerve guidance channels containing longitudinally aligned fibres could also be used to treat nerve injuries. In fact, the neurite growth of immortalized neural stem cells can be directed by aligned electrospun nanofibres.

Corey *et al.*[55] studied neurites from dorsal root ganglia explants on electrospun-PLLA nanofibres of high, intermediate and random alignment. Neurites grew radially outwards from ganglia upon contact on aligned fibres. Besides, neurites never left the fibres to grow on the surrounding cover slip. Highly aligned PLA-electrospun fibre scaffolds of varying diameter were prepared by Wang *et al.*[56] to assess neurite and Schwann cells (SC) behaviour from dorsal root ganglia explants. They observed that large diameter fibres promoted direct neurite extension and SC migration, while small diameter fibres did not show any apparent effect. The selection of certain geometrical parameters, such as fibre alignment, diameter and the space between fibres can assist to develop translational analogues to most efficiently direct and promote nerve regeneration.

He *et al.*[57] prepared PLLA-electrospun fibres and studied the influence of the fibre diameter, mesh size and arrangement on the viability/proliferation and neurite outgrowth of neural stem cells (NSC). On the aligned fibres, longer neurite lengths were obtained and cells were induced to spread along the longitudinal axis with the formation of focal contacts. However, randomly oriented cells and polygonal morphologies were observed on fibres with short neurite lengths. They concluded that the cellular behaviour is strongly influenced by the topographical characteristics of fibrous scaffolds, as well as by fibre arrangement, diameter and mesh size. Thus, cell proliferation or differentiation functions in the design of fibrous scaffolds can be enhanced by manipulating the substrate topographical features.

On the other hand, Corey *et al.*[58] prepared PLLA nanofibre substrates that support serum-free growth of primary motor and sensory neurons at low plating densities. The application of nanofibre scaffolds to different nervous system regions requires prior *in vitro* testing of scaffold designs with specific neuronal and glial cell types, which can be obtained by using primary neurons in serum-free media. They prepared a substrate with PLLA nanofibres electrospun directly on cover slips pre-coated with PLGA, retaining fibre alignment even if the fibre bundle detaches from the cover slip and keeping cells in the same focal plane. The alignment of neurons grown on this substrate followed the nanofibre alignment. Schofer *et al.*[50] reported enhancing synthetic poly(L-lactic acid) nanofibres by blending them with collagen I (COLI). The objective was to improve their ability to promote the *in vitro* growth and osteogenic differentiation of stem cells. These authors observed a decrease in hydrophobicity compared to neat PLLA nanofibres and obtained PLLA fibres with COLI with higher thermal stability and similar growth of stem cells to collagen fibres. Finally, Yang *et al.*[53] prepared aligned PLLA electrospun nano-microfibrous scaffolds by studying the optimum conditions to apply in nerve tissue engineering. They observed elongated neural stem cells and their neurite outgrew along the fibre for the aligned scaffolds. The fibre diameter dimension strongly affected the NSC differentiation rate being higher for nanofibres than for microfibres.

6.3.2 Wound Dressing

Wound dressing is used to protect wounds from the external environment and to accelerate their healing. These dressings should have highly porous structure but at the same time they should act as a good barrier to water in order to keep the wound moist, to accelerate healing and to prevent bacterial invasion, which can be caused by the accumulation of fluids between the wound and the dressing. Non-woven mats obtained by electrospinning have pores small enough to prevent bacterial penetration showing high surface area, suitable for fluid adsorption and dermal drug delivery.[59] Wound dressings are currently formed by biopolymer fibres with active compounds that promote healing, such as antimicrobial, antibacterial and anti-inflammatory agents. In the biopolymer nanofibres produced by electrospinning, different ingredients able to control the healing process of a wound can be incorporated. A reduction in the frequency for changing the dressing is provided by electrospun nanofibres wound dressing, since it is possible to obtain different functionalities into one blended layer by electrospinning.[60] Some properties of electrospun polymer membranes useful for wound-dressing, such as the controlled release of drugs, cultivation in the cells' physiological environment and very high increase in the reproduction and growth of live cells, make them more efficient than other modern materials for bandages, such as hydrocolloids, hydrogels and alginates.[60] Oxygen permeation and wound protection from infection and dehydration are required conditions for use in wound-dressing materials. These conditions can be reached by homogeneous scaffolds produced by electrospinning.[41] In fact, electrospun nanofibrous wound dressings promote the haemostasis phase, a process causing bleeding to stop, due to their small holes and highly effective surface area. These wound dressings offer conformability providing a better coverage and protection of wounds from infection. Their high area to volume ratio provides a better absorption, since nanofibrous wound dressings made of hydrophilic polymers absorb wound exudates more efficiently than the typical film dressings. In general, a wound dressing material has to be placed in the physiological and biological environment of the wound and present good cell conductivity to facilitate wound healing and skin regeneration.

On the other hand, the mechanical properties of wound dressings, such as tensile strength, sustainability, flexibility, bending and elastic properties should also be taken into account.[60] Gu *et al.*[59] prepared gelatin/PLLA porous structured electrospun mats for wound dressings, pharmaceuticals and adhesives owing to its biocompatibility, biodegradability, hydrogel characteristics, formability and cost efficiency. This structure showed a controlled evaporative water loss, a fluid drainage ability and excellent biocompatibility, so these materials can be used for wound dressing.

6.3.3 Drug Delivery

Electrospun ultrafine polymer fibres are widely used in the drug delivery field, since they provide drug release with a rapid, immediate and/or delayed release due to their high surface area per unit mass and small pore size.[10] Fibres obtained by electrospinning have some advantages to be suitable carriers for drug delivery, such as ultrathin diameters, small size porosity and suitable surface morphology. High surface area is also beneficial for mass transfer and efficient drug release, while short diffusion passage length is provided by the nanofibres' small diameter.[10] Besides, fibres prepared by electrospinning show some improvement in the therapeutic effect allowing a reduction of drugs toxicity and/or lower administration frequency. However, one of the most important requirements for biomaterials used in medicine is a suitable degradation rate.[61]

Electrospun fibres can be also used for anticancer drugs delivery, especially in post-surgery operations and local chemotherapy, taking advantage of the improvement in therapeutic effect, reduced toxicity and handling convenience. For example, BCNU (bis-chloroethylnitrosourea) was well incorporated and uniformly dispersed in biodegradable PEG-PLLA electrospun fibres.[62] In particular, BCNU is used in the treatment of several types of brain cancer, multiple myeloma and lymphoma as well as part of the chemotherapeutic protocols to prepare for haematological stem cell transplantation. The authors observed that long-term delivery of BCNU with high doses was possible, allowing the development of new implantable polymeric devices for malignant glioma.

In general, two types of electrospinning processes can be used to prepare drug-loaded fibres: compounding and coaxial electrospinning. In the first case, after mixing bioactive agents and polymers, they are electrospun together, providing a matrix for drug delivery devices. Typically, drug particles tend to be placed on the fibre surface, by their high ionic strength in solution and the quick solvent evaporation during processing, causing a burst release in the initial stage. This is suitable for antibiotic release, since the elimination of intruding bacteria before they begin to proliferate is necessary. But in the case of delivery of growth factors or other therapeutic drugs, their high concentration probably leads to adverse side effects decreasing the curative effect. A sustained and smooth release of bioactive agents is preferable, whereby drug concentrations in the therapeutic range are maintained during the required time.[63] Coaxial electrospinning is the preferred technique to produce composite fibres with a core layer encapsulated inside a shell and drugs or their mixtures are enclosed within the polymer shell to form a reservoir-type drug delivery device.[63]

As reported in other chapters in this volume, PLA is one of the most used materials for biomedical applications due to biocompatibility and biodegradability, but hydrophobic properties require the encapsulation of water-soluble agents to avoid decomposition and failure of the polymer as a drug-carrier substrate.[56] One of the advantages of coaxial electrospinning is that solid or liquid drugs can be incorporated into electrospun nanofibres and it is possible to encapsulate drugs into a structural shell.

Controlled release *via* diffusion properties, porosity or biodegradability of the shell can be adjusted by the selection of parameters in the coaxial electrospinning.[3] He *et al.*[10,63] produced drug-loaded fibres of tetracycline hydrochloride (TCH) into PLLA by both electrospinning processes (compounding and coaxial) that were subsequently submitted to hot-stretching treatment. The comparison of properties of the different fibres showed that the molecular orientation and crystalline structure depended strongly on the used electrospinning process. Fibres obtained by coaxial electrospinning provided a sustained release avoiding bursting, while burst release was produced with fibres obtained by compounding electrospinning. They also observed that fibre diameters could be tailored by controlling the shell solution concentrations. Finally, Qi *et al.* obtained fibers for the encapsulation of drug reservois by emulsion electrospinning.[64]

The use of polymer drug delivery systems provides some improvement in the therapeutic effects and toxicity reduction. Nevertheless, research is still needed to solve some issues regarding the low efficiency of drug delivery or to burst release of drugs in short times. Tailoring diameter sizes and uniformity of PLLA fibres with the addition of different surfactants with typical drugs into the PLLA solution was studied by Zeng *et al.*,[65] who reported that fibre diameters were reduced with the addition of surfactants. They prepared electrospun-PLLA fibres containing various concentrations of Rifampin® and showed that the drug was released constantly without bursting.

On the other hand, Yang *et al.* studied the ways to avoid the burst release phenomenon.[66] They concluded that the drugs' solubility and compatibility in the drug/polymer/solvent system should be taken into consideration for the preparation of the PLA electrospun fibre formulations with constant drugs release. They prepared PLLA-electrospun fibre mats in chloroform and carried out the encapsulation of two different drugs, lipophilic paclitaxel and hydrophilic doxorubicin hydrochloride, both drugs related to cancer treatments. They studied the release kinetics, taking into account the precipitation of paclitaxel on aqueous solutions. Encapsulation of paclitaxel showed good compatibility with PLLA, but doxorubicin hydrochloride was located on or near the surfaces of PLLA fibres. They also observed that the release of paclitaxel was nearly zero-order kinetics due to the fibres' degradation. On the other hand, doxorubicin hydrochloride showed a burst release as the consequence of the very fast diffusion of the drug on or near the fibre surface.

Drug, solvent and polymer should be compatible but complete solubility of the drug in the polymer solution must be avoided to favour diffusion, migration and release. Moreover, drugs should be located far from the fibre surfaces to avoid burst release.[66] However, Xu *et al.*[67] prepared multi-drug delivery PLA electrospun nanofibre mats blended with poly(ethylene glycol) (PEG) with paclitaxel and doxorubicin hydrochloride. Doxorubicin hydrochloride was easy to diffuse out from the fibres with faster release rate than paclitaxel, due to the high hydrophilicity of doxorubicin hydrochloride. In addition, they observed that the dual drug combination showed higher inhibition apoptosis. Xu *et al.*[68] carried out another research on the incorporation of doxorubicin hydrochloride. They prepared PEG-PLA nanofibres with different doxorubicin loadings by emulsion-electrospinning. Doxorubicin was molecularly distributed in the fibres' centre and the rate of doxorubicin release decreased as its content increased. The release rate of doxorubicin was slowed down due to the reservoir-type structure of the fibres.

Finally, the coaxial electrospinning technique is suitable to prepare core-shell fibres with high-molecular-weight molecules, such as proteins, incorporated into them. In this case the formation of the high-molecular-weight fibres is aided by the electrospinning of homogenous polymer solutions.[69]

Emulsion electrospinning can be also used. Maretschek *et al.*[69] produced protein-loaded nanofibre non-wovens based on PLLA *via* emulsion electrospinning. As the hydrophobicity of the non-woven nanofibres affects the release of macromolecules, PLLA was blended with hydrophilic polymers such as poly(ethylene imine) and poly(L-lysine) to modify the release profiles, resulting in an increase in fibre diameter and a decrease in the specific surface area and highly hydrophobic surfaces.

6.4 Conclusions

This chapter collects information on PLA and PLA-nanocomposites electrospun fibres with special attention to their processing, properties and main applications in the biomedical field. Electrospinning is a versatile technique to obtain micro- and nanofibres by applying electrical voltages. PLA-electrospun fibres show diameters from micro- to nanometre scale. Nanoparticles, such as montmorillonite, halloysite nanotubes, hydroxyapatite, cellulose, single- and multi-walled carbon nanotubes and graphene, have been used to improve the PLA-electrospun fibre properties, finding application in tissue engineering, wound dressing and drug delivery. Macroporous scaffolds produced by randomly oriented or aligned PLA-electrospun nanofibres can be used for the regeneration of heart, neural, bone and blood vessel tissues. These electrospun fibres can be used as carriers for bioactive agents, including antibiotics and anticancer drugs by their controlled release. Active compounds, such as antimicrobial, antibacterial and anti-inflammatory agents, can be introduced into the scaffolds, producing an improvement in wound healing abilities. The results reported in different studies based on the use of PLA-electrospun fibres can open new perspectives for development of the next generation of wound dressing materials with desirable properties.

References

1. C. Wang, H. S. Chien, K. W. Yan, C. L. Hung, K. L. Hung, S. J. Tsai and H. J. Jhang, *Polymer*, 2009, **50**, 6100.
2. F. Mei, J. Zhong, X. Yang, X. Ouyang, S. Zhang, X. Hu, Q. Ma, J. Lu, S. Ryu and X. Deng, *Biomacromolecules*, 2007, **8**, 3729.
3. S. Y. Chew, Y. Wen, Y. Dzenis and K. W. Leong, *Int. J. Curr. Pharm. Res.*, 2006, **12**, 4751.
4. L. T. Lim, R. Auras and M. Rubino, *Progr. Polym. Sci.*, 2008, **33**, 820.
5. M. Bognitzki, T. Frese, M. Steinhart, A. Greiner, J. H. Wendorff, A. Schaper and M. Hellwig, *Polym. Eng. Sci.*, 2001, **41**, 982–989.
6. R. Dersch, M. Steinhart, U. Boudriot, A. Greiner and J. H. Wendorff, *Polym. Adv. Tech.*, 2005, **16**, 276.
7. N. Ogata, S. Yamaguchi, N. Shimada, G. Lu, T. Iwata, K. Nakane and T. Ogihara, *J. Appl. Polym. Sci.*, 2007, **104**, 1640.
8. H. Zhou, T. B. Green and Y. L. Joo, *Polymer*, 2006, **47**, 7497.

9. A. K. Moghe and B. S. Gupta, *Polym. Rev.*, 2008, **48**, 353.
10. C. L. He, Z. M. Huang, X. J. Han, L. Liu, H. S. Zhang and L. S. Chen, *J. Macromol. Sci. Phys.*, 2006, **45**, 515.
11. H. Kriel, R. D. Sanderson and E. Smit, *Fibres Text. East Eur.*, 2012, **20**, 28.
12. S. H. Tan, R. Inai, M. Kotaki and S. Ramakrishna, *Polymer*, 2005, **46**, 6128.
13. Z. Qi, H. Yu, Y. Chen and M. Zhu, *Mater. Lett.*, 2009, **63**, 415.
14. L. D. Wright, T. Andric and J. W. Freeman, *Mater. Sci. Eng. C*, 2011, **31**, 30.
15. O. Ero-Phillips, M. Jenkins and A. Stamboulis, *Polymers*, 2012, **4**, 1331.
16. C. C. Liao, C. C. Wang and C. Y. Chen, *Polymer*, 2011, **52**, 4303.
17. X. H. Zong, K. Kim, D. F. Fang, S. F. Ran, B. S. Hsiao and B. Chu, *Polymer*, 2002, **43**, 4403.
18. Y. You, S. Won Lee, S. Jin Lee and W. H. Park, *Mater. Lett.*, 2006, **60**, 1331.
19. H. C. Chen, C. H. Tsai and M. C. Yang, *J. Polym. Res.*, 2010, **18**, 319.
20. X. Zhang, R. Nakagawa, K. H. K. Chan and M. Kotaki, *Macromolecules*, 2012, **45**, 5494.
21. W. Tomaszewski, W. Swieszkowski, M. Szadkowski, M. Kudra and D. Ciechanska, *J. Appl. Polym. Sci.*, 2012, **125**, 4261.
22. A. H. Touny, J. G. Lawrence, A. D. Jones and S. B. Bhaduri, *J. Mater. Res.*, 2011, **25**, 857.
23. J. H. Lee, T. G. Park, H. S. Park, D. S. Lee, Y. K. Lee, S. C. Yoon and J. D. Nam, *Biomaterials*, 2003, **24**, 2773.
24. J. H. Chang, Y. U. An, D. Cho and E. P. Giannelis, *Polymer*, 2003, **44**, 3715.
25. Y. H. Lee, J. H. Lee, I. G. An, C. Kim, D. S. Lee, Y. K. Lee and J. D. Nam, *Biomaterials*, 2005, **26**, 3165.
26. C. Spadaccio, A. Rainer, M. Trombetta, G. Vadala, M. Chello, E. Covino, V. Denaro, Y. Toyoda and J. A. Genovese, *Ann. Biomed. Eng.*, 2009, **37**, 1376.
27. G. Sui, X. Yang, F. Mei, X. Hu, G. Chen, X. Deng and S. Ryu, *J. Biomed. Mater. Res.*, 2007, **82**, 445.
28. J. Chen, B. Chu and B. S. Hsiao, *J. Biomed. Mater. Res.*, 2006, **79**, 307.
29. E. Seyedjafari, M. Soleimani, N. Ghaemi and I. Shabani, *Biomacromolecules*, 2010, **11**, 3118.
30. M. P. Prabhakaran, J. Venugopal and S. Ramakrishna, *Acta Biomater.*, 2009, **5**, 2884.
31. X. Li, H. Liu, J. Wang and C. Li, *Mater. Lett.*, 2012, **73**, 103.
32. F. D'Angelo, I. Armentano, I. Cacciotti, R. Tiribuzi, M. Quattrocelli, C. Del Gaudio, E. Fortunati, E. Saino, A. Caraffa, G. G. Cerulli, L. Visai, J. M. Kenny, M. Sampaolesi, A. Bianco, S. Martino and A. Orlacchio, *Biomacromolecules*, 2012, **13**, 1350.
33. A. Sonseca, L. Peponi, O. Sahuquillo, J. M. Kenny and E. Giménez, *Polym. Degrad. Stabil.*, 2012, **97**, 2052.

34. S. I. Jeong, E. K. Ko, J. Yum, C. H. Jung, Y. M. Lee and H. Shin, *Macromol. Biosci.*, 2008, **8**, 328.
35. Y. G. Frank Ko, A. Ali, N. Naguib, H. Ye, G. Yang, C. Li and P. Willis, *Adv. Mater.*, 2003, **15**, 13.
36. S. D. McCullen, K. L. Stano, D. R. Stevens, W. A. Roberts, N. A. Monteiro-Riviere, L. I. Clarke and R. E. Gorga, *J. Appl. Polym. Sci.*, 2007, **105**, 1668.
37. S. D. McCullen, D. R. Stevens, W. A. Roberts, L. I. Clarke, S. H. Bernacki, R. E. Gorga and E. G. Loboa, *Int. J. Nanomed.*, 2007, **2**, 253.
38. S. Shao, S. Zhou, L. Li, J. Li, C. Luo, J. Wang, X. Li and J. Weng, *Biomaterials*, 2011, **32**, 2821.
39. A. K. Geim and K. S. Novoselov, *Nat. Mater.*, 2007, **6**, 183.
40. H. Ma, W. Su, Z. Tai, D. Sun, X. Yan, B. Liu and Q. Xue, *Chin. Sci. Bull.*, 2012, **57**, 3051.
41. S. Agarwal, J. H. Wendorff and A. Greiner, *Polymer*, 2008, **49**, 5603.
42. L. Zhang and T. J. Webster, *Nano Today*, 2009, **4**, 66.
43. Y. Z. Zhang, B. Su, J. Venugopal, S. Ramakrishna and C. T. Lim, *Int. J. Nanomed.*, 2007, **2**, 623.
44. H. Lu, Z. Feng, Z. Gu and C. Liu, *J. Mater. Sci. Mater. Med.*, 2009, **20**, 1937.
45. X. M. Mo, C. Y. Xu, M. Kotaki and S. Ramakrishna, *Biomaterials*, 2004, **25**, 1883.
46. Z. Khatri, R. Nakashima, G. Mayakrishnan, K. H. Lee, Y. H. Park, K. Wei and I. S. Kim, *J. Mater. Sci.*, 2013, **48**, 3659.
47. M. D. Schofer, U. Boudriot, I. Leifeld, R. I. Sutterlin, M. Rudisile, J. H. Wendorff, A. Greiner, J. R. Paletta and S. Fuchs-Winkelmann, *Sci. World J.*, 2009, **9**, 118.
48. X. Li, J. Xie, X. Yuan and Y. Xia, *Langmuir*, 2008, **24**, 14145.
49. A. Thorvaldsson, H. Stenhamre, P. Gatenholm and P. Walkenstrom, *Biomacromolecules*, 2008, **9**, 1044.
50. M. D. Schofer, P. P. Roessler, J. Schaefer, C. Theisen, S. Schlimme, J. T. Heverhagen, M. Voelker, R. Dersch, S. Agarwal, S. Fuchs-Winkelmann and J. R. Paletta, *PLoS One*, 2011, **6**, 25462.
51. J. Li, Y. Li, L. Li, A. F. T. Mak, F. Ko and L. Qin, *Polym. Degrad. Stabil.*, 2009, **94**, 1800.
52. S. Yan, L. Xiaoqiang, L. Shuiping, W. Hongsheng and H. Chuanglong, *J. Appl. Polym. Sci.*, 2010, **117**, 1509.
53. F. Yang, R. Murugan, S. Wang and S. Ramakrishna, *Biomaterials*, 2005, **26**, 2603.
54. H. S. Koh, T. Yong, C. K. Chan and S. Ramakrishna, *Biomaterials*, 2008, **29**, 3574.
55. J. M. Corey, D. Y. Lin, K. B. Mycek, Q. Chen, S. Samuel, E. L. Feldman and D. C. Martin, *J. Biomed. Mater. Res.*, 2007, **83**, 636.
56. H. B. Wang, M. E. Mullins, J. M. Cregg, C. W. McCarthy and R. J. Gilbert, *Acta Biomater.*, 2010, **6**, 2970.
57. L. He, S. Liao, D. Quan, K. Ma, C. Chan, S. Ramakrishna and J. Lu, *Acta Biomater.*, 2010, **6**, 2960.

58. J. M. Corey, C. C. Gertz, B. S. Wang, L. K. Birrell, S. L. Johnson, D. C. Martin and E. L. Feldman, *Acta Biomater.*, 2008, **4**, 863.
59. S. Y. Gu, Z. M. Wang, J. Ren and C. Y. Zhang, *Mater. Sci. Eng. C*, 2009, **29**, 1822.
60. P. Zahedi, I. Rezaeian, S.-O. Ranaei-Siadat, S.-H. Jafari and P. Supaphol, *Polym. Adv. Tech.*, 2009, **21**, 77.
61. J. Zeng, X. Chen, Q. Liang, X. Xu and X. Jing, *Macromolecular Biosciences*, 2004, **4**, 1118.
62. X. Xu, X. Chen, T. Lu, X. Wang, L. Yang and X. Jing, *J. Contr. Release*, 2006, **114**, 307.
63. C. L. He, Z. M. Huang and X. J. Han, *J. Biomed. Mater. Res.*, 2009, **89**, 80.
64. H. Qi, P. Hu, J. Xu and A. Wang, *Biomacromolecules*, 2006, 7, 2327.
65. J. Zeng, X. Xu, X. Chen, Q. Liang, X. Bian, L. Yang and X. Jing, *J. Contr. Release*, 2003, **92**, 227.
66. J. Zeng, L. Yang, Q. Liang, X. Zhang, H. Guan, X. Xu, X. Chen and X. Jing, *J. Contr. Release*, 2005, **105**, 43.
67. X. Xu, X. Chen, Z. Wang and X. Jing, *Eur. J. Pharm. Biopharm.*, 2009, **72**, 18.
68. X. Xu, X. Chen, P. Ma, X. Wang and X. Jing, *Eur. J. Pharm. Biopharm.*, 2008, **70**, 165.
69. S. Maretschek, A. Greiner and T. Kissel, *J. Contr. Release*, 2008, **127**, 180.

CHAPTER 7

Modification of PLA by Blending with Elastomers

N. BITINIS, R. VERDEJO AND M. A. LÓPEZ-MANCHADO*

Instituto de Ciencia y Tecnología de Polímeros, ICTP-CSIC, c. Juan de la Cierva, 3, 28006 Madrid, Spain
*Email: lmanchado@ictp.csic.es

7.1 Copolymerization

Copolymerization has been studied to tailor the PLA tensile and impact performance. Ring-opening polymerization (ROP) with other cyclic monomers has been the preferred synthesis method as it allows better control of polymer reaction. Poly(ε-caprolactone) (PCL), a very ductile biodegradable polyester, is the most used polymer to be copolymerized with PLA.[1–5] Gripjma et al.[1,2] observed that the elongation at break increased for the copolymers in relation to the simple blends. Hiljanen-Vaio et al.[3] showed that the copolymers containing more than 40 wt% of PCL presented an elongation at break higher than 100% in comparison with values lower than 10% for neat PLA.

Other copolymers with interesting mechanical properties have been reported in the literature. Haynes et al.[6] copolymerized PLA with perfluoropolyether (PFPE) and observed an increase of the elongation at break up to 300% with 5% of PFPE. Another alternative was described by Pitet et al.,[7] who synthesized ABA triblock copolymers from 1,5-cyclooctadiene (COD) and D,L-lactide. The triblocks were considerably tougher than PDLA, especially for poly-COD low midblock contents. An elongation at break of 180% was obtained for a triblock containing 0.76 volume fraction of PLA.

RSC Polymer Chemistry Series No. 12
Poly(lactic acid) Science and Technology: Processing, Properties, Additives and Applications
Edited by Alfonso Jiménez, Mercedes Peltzer and Roxana Ruseckaite
© The Royal Society of Chemistry 2015
Published by the Royal Society of Chemistry, www.rsc.org

Although interesting results are obtained by direct copolymerization of PLA, direct blending with other polymers appears to be a less expensive and more practical strategy to overcome PLA drawbacks.

7.2   Blending with other Polymers

Blending PLA with other polymers to broaden its applications is a method that has been the object of intense research. Among the chosen polymers for blending with PLA, biodegradable and biocompatible materials are the most attractive alternatives. Nevertheless, some studies have been focused on blending with petroleum-based non-biodegradable polymers. Since most polymer blends are immiscible, a very important aspect of PLA blends is related to their morphology, which strongly determines the ultimate mechanical properties obtained. The selected toughening polymers are classified in three categories: biodegradable polymers, petroleum-based non-biodegradable polymers and finally elastomers, which will be the main subject of this chapter.

7.2.1   Blending with Biodegradable Polymers

PLA blending with PCL has been the first choice for modification of mechanical properties due to its high flexibility and ductility with a glass transition temperature (T_g) of about –60 °C.[8–14] PLA and PCL were demonstrated to be immiscible, and PCL formed droplets in the PLA matrix at a concentration of 25 wt%.[8,14] However, only marginal improvements of mechanical properties were obtained for the addition of less than 40 wt% of PCL.[14] Therefore, compatibilization methods should be used. Wang *et al.*[8] proposed the addition of triphenyl phosphite, which could act as a coupling agent through trans-esterification reaction during reactive blending. An interesting increase of elongation at break from 28% to 127% was observed when adding 2 wt% of triphenyl phosphate to a PLA/PCL 80/20 blend. More recently, Semba *et al.*[11] added dicumyl peroxide (DCP) as a crosslinking agent. A reduction of the droplet size was observed with the addition of DCP for a PCL content up to 30 wt%. The best results of elongation at break by blending with PCL (130%) were obtained with the addition of 0.1 and 0.2 phr of DCP while the incorporation of 0.3 phr of DCP showed an Izod impact strength 2.5 times higher than pristine PLA. Other reactive compatibilizers have been used for PLA/PCL systems, such as isocyanates, which could react with hydroxyl or carboxyl groups in the PLA structure.[12]

Poly(butylene adipate-*co*-terephthalate) (PBAT), poly(butylene succinate) (PBS) or poly(ethylene succinate) (PES) are flexible biodegradable polyesters that have also been considered for blending with PLA.[15–17] PBAT and PLA are immiscible polymers where PBAT phase formed small droplets in the PLA matrix.[15] Drastic increase of elongation at break was observed (200% with only 5 wt% of PBAT) due to interfacial debonding mechanisms as the interfacial adhesion between the two polymers was low. The impact toughness was only slightly increased, suggesting the need of a better compatibilization agent. Meanwhile, Shibata *et al.*[16] reported good mechanical properties for PLA/PBS blends with an elongation at break increasing up to 110% for PBS contents of 10 wt%. PLA/PES blends were studied by Lu *et al.*,[17] reporting an improvement of the elongation at break up to 140% at high PES contents, 40 wt%.

PLA/thermoplastic starch (TPS) blends have also been subject to intense research.[18–20] When starch is heated above 80–90 °C in the presence of a plasticizer, such as water or glycerol, a gelatinization process occurs and leads to the disappearance of the starch crystalline structure. Once starch is gelatinized and plasticized, a thermoplastic starch able to be processed in extruder is obtained. The mechanical properties of TPS strongly depend on the plasticizer content. Huneault *et al.*[18] used an extrusion process to obtain water-free glycerol plasticized TPS, which was then blended with a PLA matrix. However, the compatibility between both polymers was very poor and the authors grafted maleic anhydride onto PLA to improve the blend's morphology. Improvements of mechanical properties were then obtained and elongation at break 200% was attained for 20 wt% of TPS containing 36 wt% of glycerol. The TPS phase appeared to be too rigid for lower contents of glycerol and no improvement of the ductility of the final PLA/TPS blend was observed.

PLA blending with other biodegradable polymers leads to interesting mechanical properties. Nevertheless, the use of non-biodegradable or biocompatible compatibilizers is often required in order to optimize these properties.

7.2.2 Blending with Non-biodegradable Petroleum-based Polymers

Blending PLA with non-biodegradable petroleum-based polymers does not seem to be the best option in order to preserve the biodegradable and biocompatible character of PLA. However, this approach could have the advantage of enabling the production of low-cost toughened PLA. Thus, the addition of petroleum-based polymers, such as low density linear polyethylene (LLDPE),[21] thermoplastic polyolefins (TPO)[22,23] or poly(ethylene-*co*-vinyl acetate) (EVA)[24] have been reported in the literature. Anderson *et al.*[21] reported the mechanical properties of PLDA/LLDPE and PLLA/LLDPE blends compatibilized with a PLLA-PE block copolymer. A super-tough material was obtained for PLA/LLDPE 80/20 blend with 5 wt% of the copolymer (Izod impact resistance 460 J m^{-1} compared to 12 J m^{-1} for pure PLA and 36 J m^{-1} for the simple blend). In addition, Ho *et al.*[22] synthesized TPO-PLA copolymers to compatibilize the PLA/TPO blend. An increase of the elongation at break from 15% for the simple blend to 182% by adding 2.5 wt% of a copolymer with long PLA segments in PLA/TPO 80/20 blends was observed. Moreover, those samples containing copolymers did not break during the impact tests.

Another interesting study was reported by Ma *et al.*[24] where PLA was toughened by using poly(vinyl acetate) (EVA). EVA copolymer properties can be tailored from thermoplastic to rubber materials, by increasing the vinyl acetate content. Moreover, the great advantage of EVA is that its compatibility with PLA could also be tailored as poly(vinyl acetate) has been demonstrated to be miscible with PLA. Thus, the toughness of PLA/EVA 80/20 blends increased with VA content up to 50%, which was then the optimal VA content for EVA. The addition of 5 wt% EVA50 into PLA displayed an elongation at break of 300% and super-tough materials were obtained for 15% and more of EVA50 content.

However, most of these blends are immiscible and they need the addition of compatibilizers to improve their compatibility. In addition, most of the added polymers have no biocompatibility, which clearly limits the biomedical applications of these systems.

7.2.3 Blending with Elastomers

Rubber matrices have commonly been used as a second phase to improve the toughness of brittle thermoplastic materials, such as polypropylene and polyethylene.[25] These systems, commonly referred to as polyolefin thermoplastic elastomers (TPOs), are a special class of thermoplastic elastomers that combine the processing characteristic of plastics at elevated temperatures with the physical properties of conventional elastomers at service temperature, playing an increasingly important role in the polymer material industry.[26] Polyolefin blends attract additional interest due to the possibility of recycling plastic wastes, avoiding the complex and expensive processes of separation of the different components.

Well-dispersed rubber particles behave as stress concentrators enhancing the fracture energy absorption of brittle polymers and ultimately result in materials with improved toughness. In order to provide toughness to polymers, the elastomer must meet certain criteria: i) it has to be distributed as small domains in the continuous polymer phase; ii) it must have a good interfacial adhesion with the polymer matrix; iii) its glass transition temperature must be at least 20 °C lower than the processing temperature; iv) its molecular weight must not be low; v) it should not be miscible with the polymer matrix; and vi) it must be thermally stable at the polymer processing temperatures.[27–29] The physical and mechanical properties of PLA/rubber blends have received little attention until now. However, in the last four years there has been growing interest in the development of these systems as a viable alternative to produce ductile PLA.[30–58]

Several elastomer matrices have been used to improve the mechanical properties of PLA, but special attention has been paid to polyurethane elastomer (PU)[30–35] and natural rubber (NR).[37–47] It is well known that thermoplastic polyurethane elastomers are widely used in a diverse range of implantable medical devices due to a unique combination of toughness, durability, flexibility, biocompatibility and biostability.[59–62] In addition, it has been demonstrated that PU is miscible with some polyesters and poly-ethers, such as PEO and PPG, since the soft segments of PU elastomers are mainly polyester or polyether. All these features make them in an ideal candidate for the PLA toughening.

Li *et al.*[30] reported the first work on these PLA/PU blends prepared by melt mixing in a twin screw at several PU concentrations. The FE-SEM micro-graphs of the PLA/PU blends showed a clear phase-separated morphology with spherical particles of PU dispersed homogeneously in the continuous PLA matrix (Figure 7.1). The authors observed that all blends showed two glass transition temperatures indicating that they were not thermo-dynamically miscible. However, the T_g values of both polymers were slightly shifted towards each other as a function of the blend concentration, sug-gesting that both phases, PU and PLA, are partially miscible and molecular interactions between both components take place (Figure 7.2).

In addition, the authors observed that the addition of PU significantly changed the tensile behaviour of PLA, as shown in Figure 7.3. PLA is a rigid and brittle polymer with a very low elongation at break (around 4–5%). The material breaks after yield without necking. However, the stress-strain curves of blends exhibited an elastic deformation stress plateau. This indicates that the addition of PU (10 wt%) transforms the brittle fracture of neat PLA to the ductile fracture with elongation at break 225% and higher impact strength. Moreover, the elongation at break continuously increased with increasing PU content in the blend. All the blends showed clear yielding behaviour

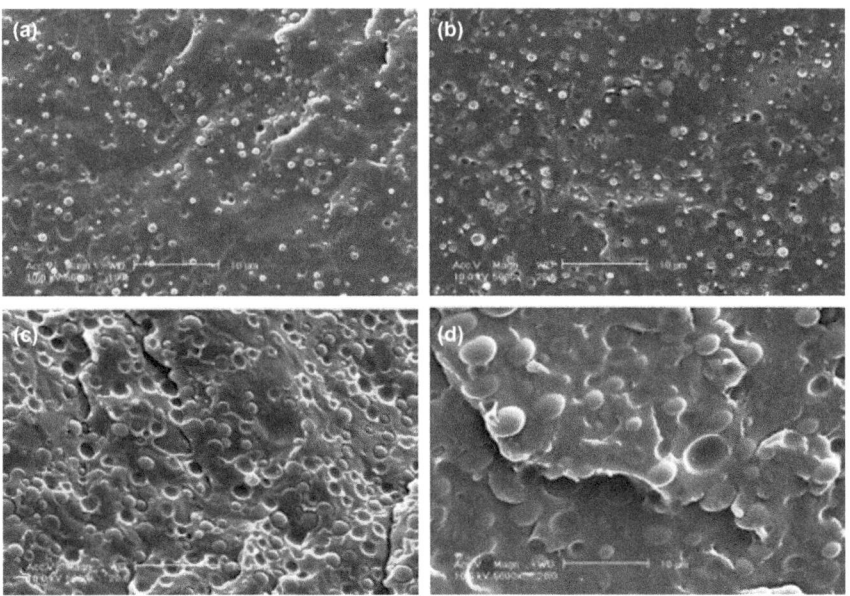

Figure 7.1 Phase morphologies of the PLA/PU blends with (a) 5, (b) 10, (c) 20 and (d) 30% PU.

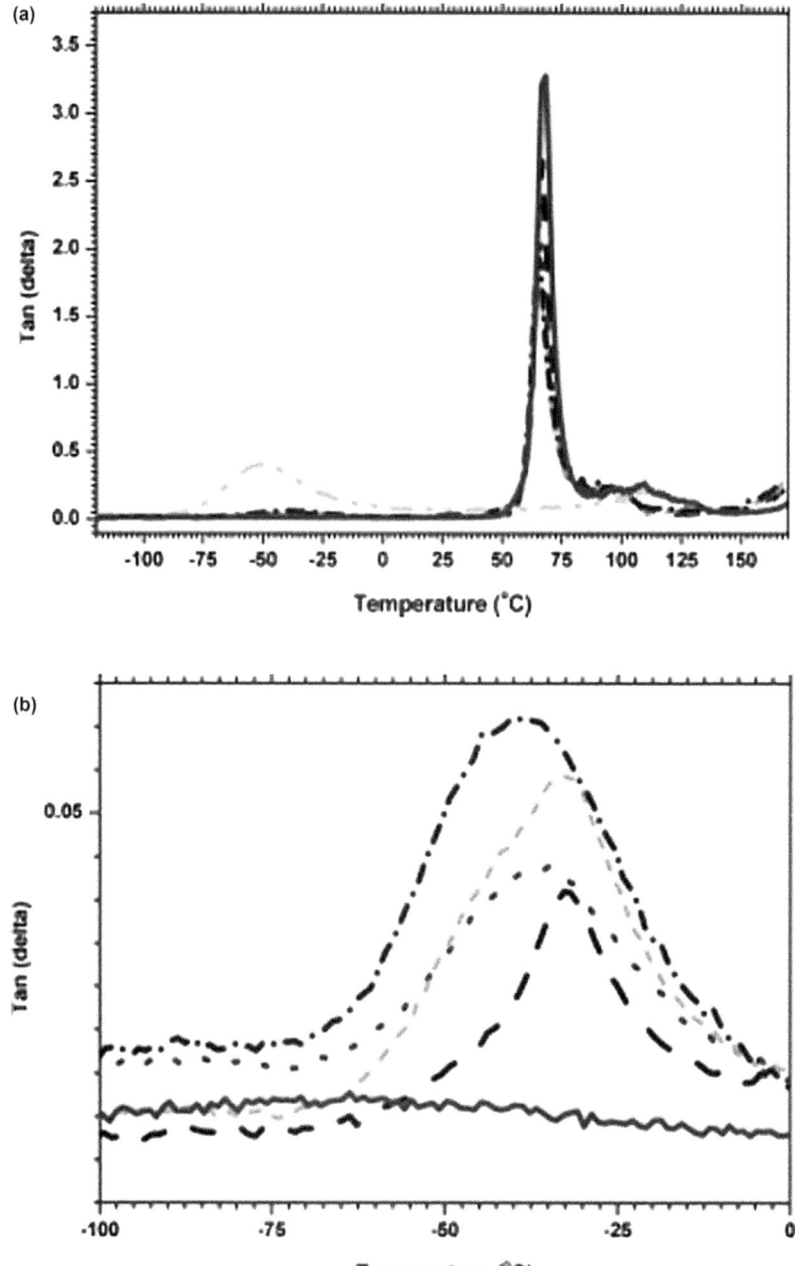

Figure 7.2 Dynamic viscoelastic curves for the PLA/PU blends: (a) tan δ *versus* temperature; (b) the enlarged part of the PU glass transition region.

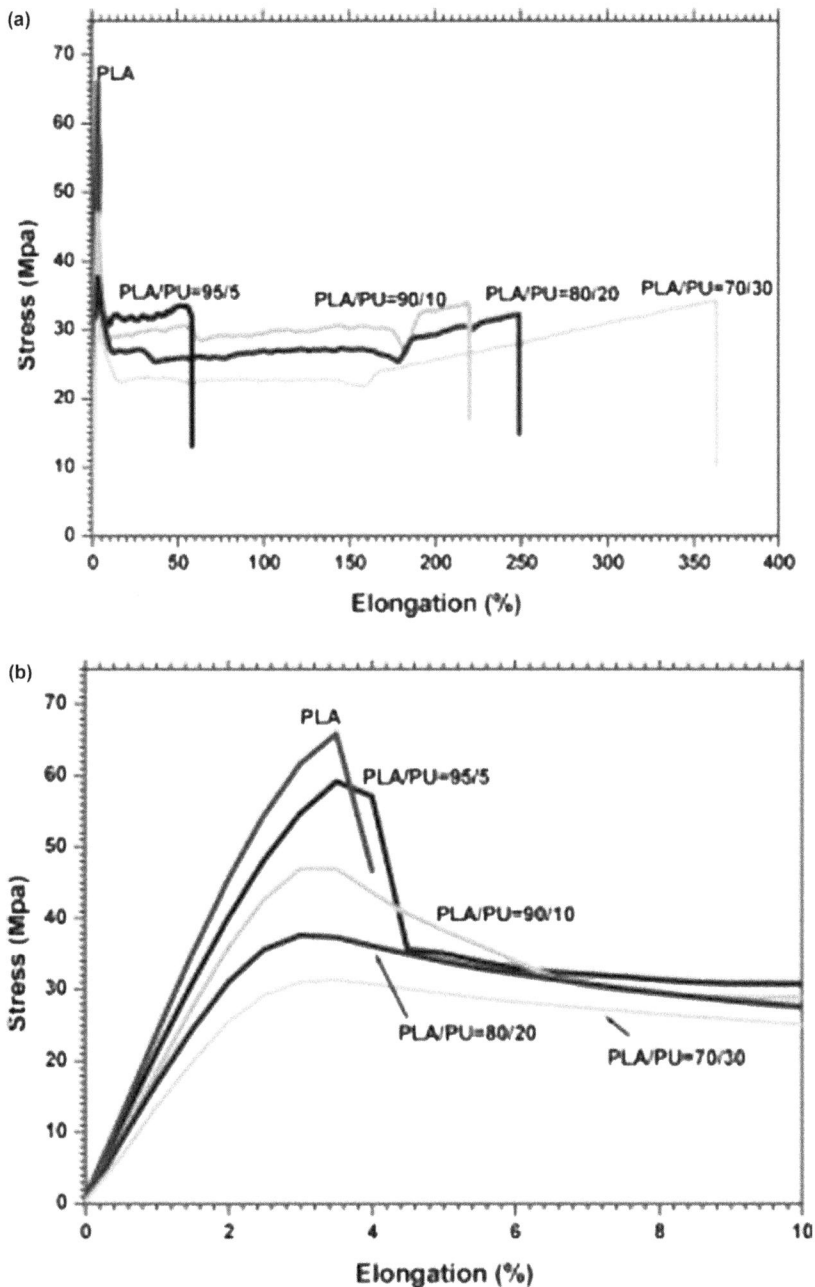

Figure 7.3 (a) Stress-strain curves of the blends with different PU contents, and (b) enlarged part of the curves near the yielding region.

upon stretching. After yielding occurred, the strain developed continuously, while the stress remained almost constant.

These authors also investigated the deformation mechanisms by analyzing the morphology of the different necking regions of the tensile tested specimens using scanning electron microscopy (SEM). They concluded that shear yielding of the PLA matrix induced by the presence of PU domains is the main toughening mechanism. The shear yielding may have been initiated by the stress concentrations associated with rubber particles; consequently, debonding at the matrix/particle interface releases the hydrostatic stresses and encourages shear yielding to proceed.

Similar results were observed by Han *et al.*[31] by analyzing PLA/PU blends prepared by melt blending. The incorporation of 30 wt% PU led to a significant increase of the elongation at break of the blend reaching more than 600%, and the samples could not be broken in the notched Izod impact tests at room temperature even with a larger pendulum bob. These results suggested that the fracture impact strength of the blend with 30 wt% PU would be larger than 40.7 kJ m^{-2}.

Feng *et al.*[32] successfully fabricated highly oriented self-reinforced 80/20 PLA/PU blends through solid hot stretching technology. These authors observed that the stretch ratio and the tensile strength increased while the elongation at break of the blend dramatically decreased. These samples showed superior mechanical properties with a notched Charpy impact strength of 150 kJ m^{-2} and a tensile strength of 197 MPa. With increasing the hot stretch ratio, the storage modulus increased, the glass transition temperatures of the PLA-rich phase and TPU-rich phase in the blends moved to higher temperatures and the melting temperature and crystallinity of the blend increased, indicating the stress-induced crystallization of the blend during drawing.

The longitudinal fracture surfaces morphology of PLA/PU blends with varying stretch ratios were analyzed. The isotropic sample shows that the spherical particles of PU were homogeneously dispersed in the biodegradable PLA matrix, and island structures were preferably formed. However, the fracture surfaces of stretched blends exhibited orderly arranged fibrillar bundle structures parallel to the stretch ratio. This fibrous structure was more evident as the stretch ratio was increased. The formation of these highly oriented microfibres contributed significantly to the higher strength and toughness of the blend.

Ishida *et al.*[36] reported melt blending of PLA with four types of common rubbers, ethylene-propylene copolymer (EPM), ethylene-acrylic rubber (EAM), acrylonitrile-butadiene rubber (NBR) and isoprene rubber (IR), to toughen PLA. All blends showed separated phase morphology where the elastomer phase was homogeneously distributed in the form of small droplets in the continuous PLA phase. Izod impact testing showed that toughening was achieved only when PLA was blended with NBR, which showed the smallest rubber particle size in the blends. In addition, the interfacial tension between both phases, PLA and NBR, was the lowest, suggesting that the high polarity rubber was more suitable for toughening PLA. The tensile behaviour demonstrated that the tensile modulus and strength decreased in a similar way for all rubbers. Nevertheless, a slight increase in elongation at break (around 20%) was observed for IR and NBR. These authors attributed this effect to the absence of crosslinks for these two rubbers compared to EPM and EAM, which were thermo-reversibly cross-linked. This result implies the high ability to induce plastic deformation before the break as well as high elongation properties. The authors concluded that the intrinsic mobility of rubber is crucial for the dissipation of the breaking energy.

7.2.4   Blending with Natural Rubber

Great interest has recently arisen in the use of natural rubber (NR) as elastomer phase for toughening PLA matrices.[37–47] NR was the first elastomer to be industrially exploited. It is a renewable resource derived from a milky colloidal suspension or latex found in the sap of some plants such as *Hevea Brasiliensis*. The latex is collected in cups and then subsequently co-agulated to obtain a dry rubber (Figure 7.4).

NR molecules consist mainly of *cis*-1,4-poly(isoprene) with practically no evidence for any *trans* structure in the natural product, in contrast to the synthetic poly(isoprene). It has a very high average molecular weight ranging from 3.4×10^6 g mol^{-1} to 10.2×10^6 g mol^{-1}. Commercial raw NR has an important number of non-rubber constituents, proteins, phospholipids, sugars and fatty acids which can influence the methods of coagulation, the vulcanization process and, hence, the physical properties of NR.[63] In addition, as a natural product, NR is subjected to biological mineralization cycles, and reports on its biodegradation have been published.[64,65] It is assumed that the degradation of the polymer backbone is initiated by an oxidative cleavage of the double bond. Moreover, rubber-degrading bacteria have been isolated but the rubber biodegradation, as well as the growth of bacteria using rubber as a sole carbon source, are low processes.[64,65] Therefore, natural rubber exhibits a unique combination of toughness, flexibility, biocompatibility and biodegradability that together with its low cost makes it an ideal candidate to improve the PLA brittleness.

Figure 7.4   Production steps of dry coagulated natural rubber.

Bitinis *et al.*[37] developed for the first time a novel and industrially scalable PLA-NR blend prepared by melt mixing blends at 5, 10 and 20 wt% of natural rubber to analyze the effect of the NR concentration on the blend morphology. Figure 7.5 shows SEM micrographs of the blends fracture surfaces where it is observed that the size of the rubber particles is similar for 5 and 10 wt% but increases for the blend at 20 wt% from 1.15 to 2.00 μm. In general, in an immiscible binary polymer blend, the size of the dispersed phase increases as a function of the concentration of the minor phase in the blend, due to coalescence phenomena.[66]

It is worthwhile to note that the transparency of PLA is not completely lost after blending (Figure 7.6). The addition of NR produces only a slight

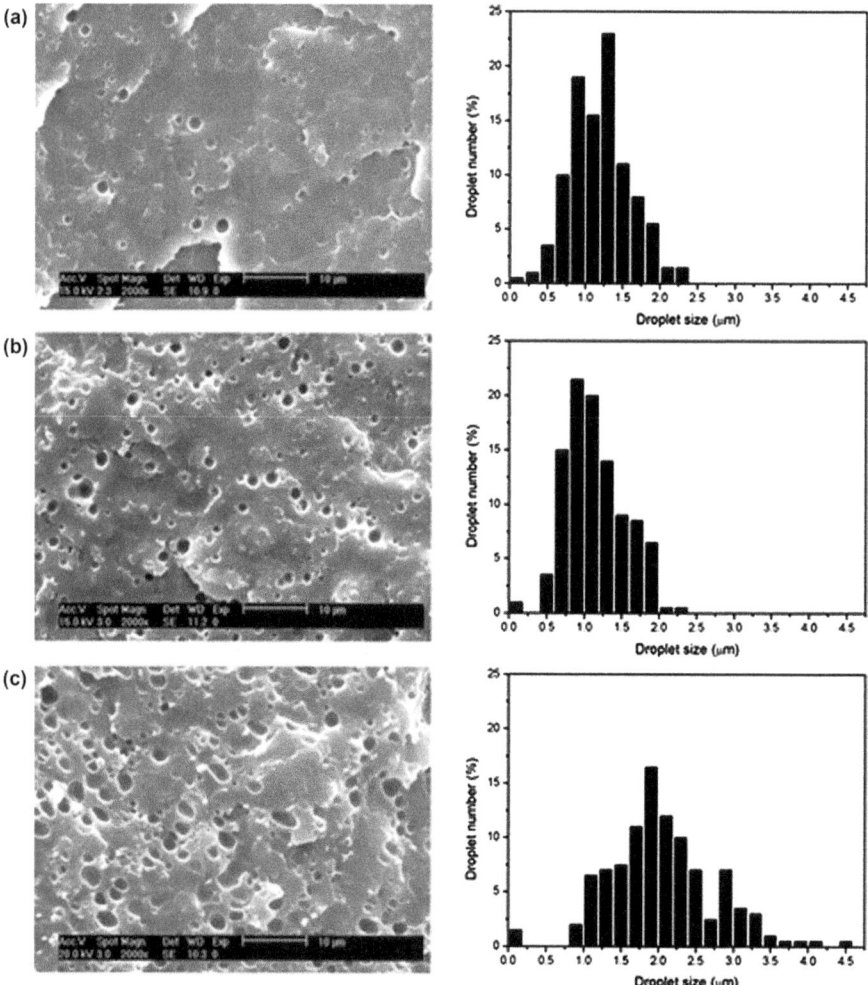

Figure 7.5 Morphology and distribution of rubber droplet size in PLA/NR blends at different NR concentrations (a) 5 wt%, (b) 10 wt% and (c) 20 wt%.

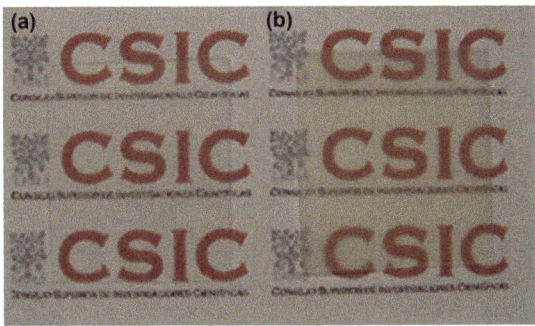

Figure 7.6 Photographs of films of about 150 μm for (a) PLA and (b) PLA/NR blend at 10 wt%.

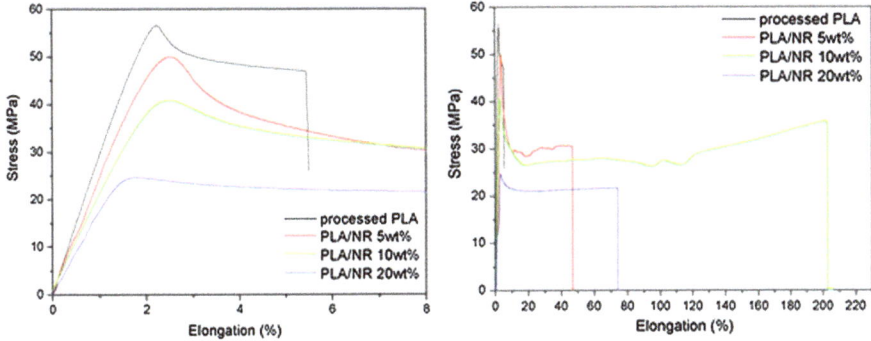

Figure 7.7 Stress-strain curves for PLA and PLA/NR blend at different rubber contents.

yellowish coloration of the blend and translucent materials are obtained with low thickness. Refractive indices of 1.4537 ± 0.0013, 1.5209 ± 0.0008 and 1.4579 ± 0.0013 were measured for PLA, NR and PLA/NR 10 wt% blend, respectively.

No changes in the glass transition temperature of PLA by the addition of NR were observed, confirming the immiscible behaviour of these blends. In addition, it was demonstrated that the NR acted as a nucleating agent enhancing the crystallization ability of PLA since a cold-crystallization exothermic peak was observed in the DSC heating curves.

The analysis of the tensile behaviour of these materials (Figure 7.7) resulted in the conclusion that the addition of NR to PLA matrices led to a drastic increase in the elongation at break from 5% for neat PLA to 200% for the blend. The optimal NR content to improve PLA brittleness was found to be 10 wt% since a further increase of the rubber content in the blend decreased the elongation at break to approximately 70% (Table 7.1). The addition of NR changed the common brittle fracture of neat PLA to ductile fracture with the formation and propagation of a neck while stretching and the yield fracture was suppressed. NR behaves as a stress concentrator

Table 7.1   Mechanical properties of PLA and PLA/NR blend.

	Elastic modulus, MPa	Tensile strength, MPa	Elongation at break, %
Neat PLA	2874 ± 108	63.1 ± 1.1	3.3 ± 0.4
Processed PLA	3136 ± 38	58.0 ± 1.5	5.3 ± 0.7
PLA/NR 5 wt%	2480 ± 61	50.4 ± 1.6	48 ± 22
PLA/NR 10 wt%	2036 ± 47	40.1 ± 1.5	200 ± 14
PLA/NR 20 wt%	1837 ± 78	24.9 ± 0.9	73 ± 45

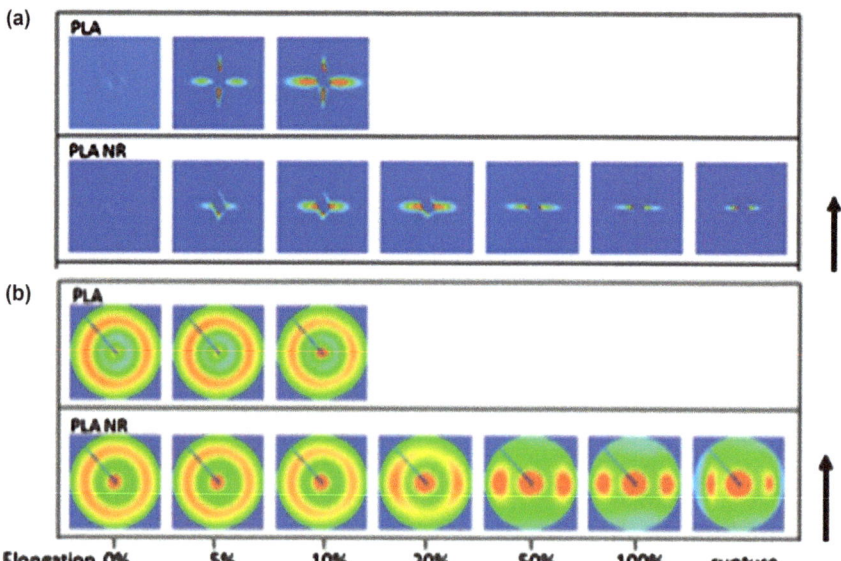

Figure 7.8   *In situ* simultaneous SAXS patterns (a) and WAXS patterns (b) at selected elongation values. The arrows indicate the stretching direction.

forming a yield point at the maximum strength value followed by a stable plastic deformation up to fracture, reaching 200% elongation. As expected, the elastic modulus and tensile strength of these blends decrease with the addition of NR.

The same authors[67] analyzed the deformation mechanisms by combining scattering techniques using synchrotron light with stress-strain experiments in a simultaneous fashion to understand the results of the mechanical properties. Figure 7.8 shows the simultaneous SAXS and WAXS patterns at selected elongation values of the neat PLA and its PLA/NR 90/10 blend. As deduced from SAXS patterns (Figure 7.8), PLA is a brittle material with failure resulting from the formation of crazes, which are defined as micro-craks bridged by small fibrils.[68] Crazing is a common process in polymer deformation and it is both a cause of failure and a mechanism of energy absorption. SAXS is an efficient technique to identify crazes by their

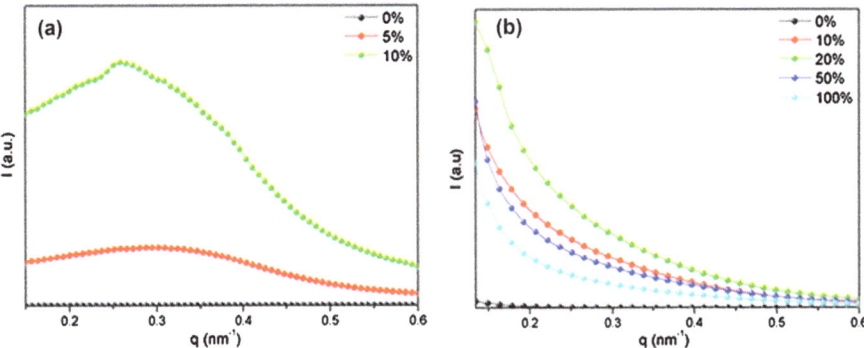

Figure 7.9 1D-SAXS intensity (linear scale) integrated in the equatorial region at different elongations for a) PLA and b) PLA/NR blend.

cross-like SAXS signature, resulting from the scattering of the craze fibrils in the equatorial direction and of the crack planes in the meridional direction. The maximum at $q = 0.26$ nm^{-1} for 10% strain suggests the development of crazes during stretching (Figure 7.9) corresponding to the average correlation length between consecutive fibrils formed across the craze plane.

SAXS patterns of stretched PLA/NR blend are also shown in Figure 7.8 (second row). Prior to deformation, the PLA/NR blend presents a slight isotropic scattered intensity. At the yield point, the blend exhibits a sharp meridional intensity and equatorial strike close to the beam stop. This indicates the formation of crazes in the PLA matrix prior to the yield point and the appearance of the first microvoids at the PLA/NR interface due to a debonding process. As stretching progresses (around 20% strain), the SAXS intensity increases in the equatorial direction, while no variations are observed in the meridional axis. According to these results, it was concluded that these microvoids are oriented parallel to the direction of the applied stress. At the macroscopic scale, this corresponds to the whitening of the blend due to the formation of microvoids and to the formation of a neck due to the microscopic localization of the stress at the yield point. Since the material is further stretched, the propagation of the neck through the sample hardly changes the SAXS patterns. Neither a peak nor shoulders are observed in the 1D-scattered intensity in the equatorial direction (Figure 7.9). However, some increase of the intensity at low q-values in the equatorial direction was observed. The absence of a maximum in the scattered equatorial intensity reflects changes in the deformation mechanism in comparison with neat PLA. The authors suggested that the crazing is not the main deformation process for the blend as happened for neat PLA. In this case, cavitation and debonding are the main deformation mechanisms proposed for the blend. Both processes have been proposed as the traditional deformation mechanisms for thermoplastic/elastomer blends, depending on the phase morphology and adhesion.[69] However, due to the poor

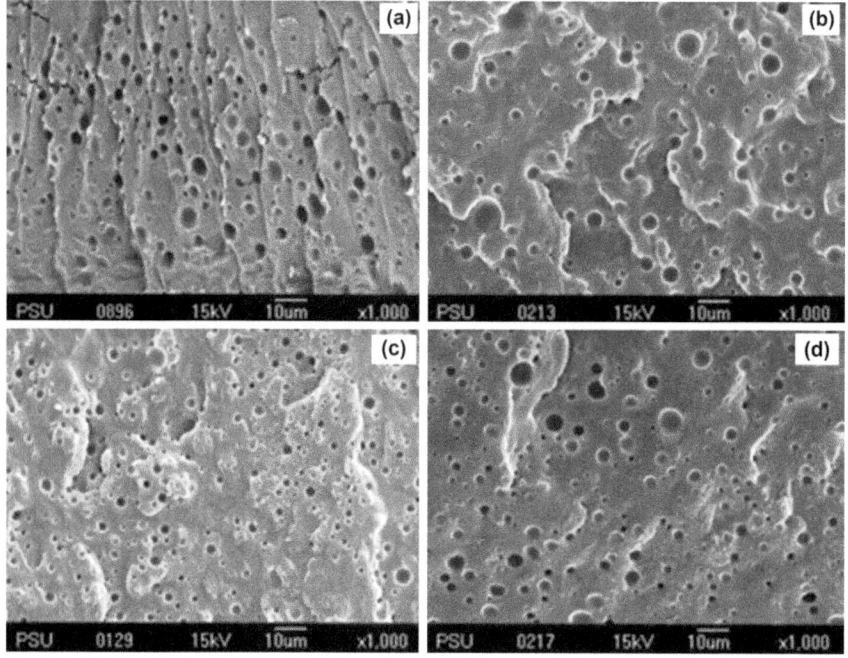

Figure 7.10   SEM micrographs of freeze-fractured surface of PLA/NR (a) and PLA/NR-*g*-PVAC blends at different PVAc contents: 1% (b), 5% (c) and 12% (d).

interfacial adhesion between PLA and NR, the debonding prevails as the main mechanism for PLA/NR blend.

Other strategies have been studied to improve the miscibility between both polymer phases and then the final properties of the PLA/NR blend. Chumeka *et al.*[38] synthetized a natural rubber grafted with poly(vinyl acetate) copolymer (NR-*g*-PVAc) by emulsion polymerization at different PVAc contents. These blends contained 10–20 wt% and they were prepared by melt-blending in a twin-screw extruder. The authors observed that the PVAc acted as an emulsifier increasing the miscibility of PLA and NR by decreasing the temperature of the maximum *tan δ* value of PLA in the blends. In consequence, the average rubber particle size in the blends decreased in the presence of PVAc as shown in Figure 7.10. The higher miscibility and smaller rubber particle size in the PLA/NR-*g*-PVAc blends led to a higher impact strength and elongation at break. The authors concluded that the NR-*g*-PVAc could be used as a toughening agent of PLA and as a compatibilizer of the PLA/NR blend.

7.3   Conclusions

Blending of PLA with different thermoplastics and elastomers is an excellent alternative to copolymerization to improve PLA mechanical properties, in particular those related with the stress-strain behaviour, *i.e.* elastic modulus and elongation at break. The addition of elastomers, and particularly natural rubber, permits to reach important improvements in tensile properties leading to open the applications window in PLA technologies.

References

1. D. W. Grijpma and A. J. Pennings, *Macromol. Chem. Phys.*, 1994, **195**, 1649.
2. D. W. Grijpma, R. D. A. Vanhofslot, H. Super, A. J. Nijenhuis and A. J. Pennings, *Polym. Eng. Sci.*, 1994, **34**, 1674.
3. M. Hiljanen-Vainio, T. Karjalainen and J. Seppala, *J. Appl. Polym. Sci.*, 1996, **59**, 1281.
4. J. H. Zhang, J. Xu, H. Y. Wang, W. Q. Jin and J. F. Li, *Mater. Sci. Eng. C*, 2009, **29**, 889.
5. T. Yu, J. Ren, S. Y. Gu and M. Yang, *Polym. Adv. Tech.*, 2010, **21**, 183.
6. D. Haynes, A. K. Naskar, A. Singh, C. C. Yang, K. J. Burg, M. Drews, G. Harrison and D. W. Smith, *Macromolecules*, 2007, **40**, 9354.
7. L. M. Pitet and M. A. Hillmyer, *Macromolecules*, 2009, **42**, 3674.
8. L. Wang, W. Ma, R. A. Gross and S. P. McCarthy, *Polym. Degrad. Stabil.*, 1998, **59**, 161.
9. M. E. Broz, D. L. Van der Hart and N. R. Washburn, *Biomaterials*, 2003, **24**, 4181.
10. N. Lopez-Rodriguez, A. Lopez-Arraiza, E. Meaurio and J. R. Sarasua, *Polym. Eng. Sci.*, 2006, **46**, 1299.
11. T. Semba, K. Kitagawa, U. S. Ishiaku and H. Hamada, *J. Appl. Polym. Sci.*, 2006, **101**, 1816.
12. T. Takayama and M. Todo, *J. Mater. Sci.*, 2006, **41**, 4989.
13. M. Harada, K. Iida, K. Okamoto, H. Hayashi and K. Hirano, *Polym. Eng. Sci.*, 2008, **48**, 1359.
14. C. L. Simoes, J. C. Viana and A. M. Cunha, *J. Appl. Polym. Sci.*, 2009, **112**, 345.
15. L. Jiang, M. P. Wolcott and J. W. Zhang, *Biomacromolecules*, 2006, 7, 199.
16. M. Shibata, Y. Inoue and M. Miyoshi, *Polymer*, 2006, **47**, 3557.
17. J. M. Lu, Z. B. Qiu and W. T. Yang, *Polymer*, 2007, **48**, 4196.
18. M. A. Huneault and H. B. Li, *Polymer*, 2007, **48**, 270.
19. P. Sarazin, G. Li, W. J. Orts and B. D. Favis, *Polymer*, 2008, **49**, 599.
20. J. Ren, H. Fu, T. Ren and W. Yuan, *Carbohydr. Polym.*, 2009, 77, 576.
21. K. S. Anderson, S. H. Lim and M. A. Hillmyer, *J. Appl. Polym. Sci.*, 2003, **89**, 3757.
22. C. H. Ho, C. H. Wang, C. I. Lin and Y. D. Lee, *Polymer*, 2008, **49**, 3902.
23. K. S. Anderson and M. A. Hillmyer, *Polymer*, 2004, **45**, 8809.
24. P. Ma, D. G. Hristova-Bogaerds, J. G. P. Goossens, A. B. Spoelstra, Y. Zhang and P. J. Lemstra, *Eur. Polym. J.*, 2012, **48**, 146.
25. L. A. Utracki, *Polymer Blends Handbook*, Kluwer Academic Publishers, Norwell, MA, USA, 2002.

26. D. J. Synrott, D. F. Sheidan and E. G. Kontos, in *Thermoplastic Elastomers from Rubber-Plastic Blends*, ed. S. K. De and A. K. Bhowmick, Ellis Horwood, New York, 1990.
27. S. H. Wu, *Polymer*, 1985, **26**, 1855.
28. G. M. Kim and G. H. Michler, *Polymer*, 1998, **39**, 5689.
29. G. M. Kim and G. H. Michler, *Polymer*, 1998, **39**, 5699.
30. Y. Li and H. Shimizu, *Macromolecular Bioscience*, 2007, 7, 921.
31. J. J. Han and H. X. Huang, *J. Appl. Polym. Sci.*, 2011, **120**, 3217.
32. F. Feng, X. Zhao and L. Ye, *J. Macromol. Sci. Phys.*, 2011, **50**, 1500.
33. F. Feng and L. Ye, *J. Appl. Polym. Sci.*, 2011, **119**, 2778.
34. H. Hong, J. Wei, Y. Yuan, F. P. Chen, J. Wang, X. Qu and C. S. Liu, *J. Appl. Polym. Sci.*, 2011, **121**, 855.
35. X. Zhao, L. Ye, P. Coates, F. Caton-Rose and M. Martyn, *Polym. Adv. Tech*, 2013, **24**, 853.
36. S. Ishida, R. Nagasaki, K. Chino, T. Dong and Y. Inoue, *J. Appl. Polym. Sci*, 2009, **113**, 558.
37. N. Bitinis, R. Verdejo, P. Cassagnau and M. A. Lopez-Manchado, *Mater. Chem. Phys.*, 2011, **129**, 823.
38. W. Chumeka, V. Tanrattanakul, J. F. Pilard and P. Pasetto, *J. Polym. Environ.*, 2013, **21**, 450.
39. R. Jaratrotkamjorn, C. Khaokong and V. Tanrattanakul, *J. Appl. Polym. Sci.*, 2012, **124**, 5027.
40. P. Juntuek, C. Ruksakulpiwat, P. Chumsamrong and Y. Ruksakulpiwat, *J. Appl. Polym. Sci.*, 2012, **125**, 745.
41. M. Kowalczyk and E. Piorkowska, *J. Appl. Polym. Sci.*, 2012, **124**, 4579.
42. B. Suksut and C. Deeprasertkul, *J. Polym. Environ.*, 2011, **19**, 288.
43. C. Zhang, C. Man, Y. Pan, W. Wang, L. Jiang and Y. Dan, *Polym. Int*, 2011, **60**, 1548.
44. Y. Huang, C. Zhang, Y. Pan, W. Wang, L. Jiang and Y. Dan, *J. Polym. Environ.*, 2013, **21**, 375.
45. K. Pongtanayut, C. Thongpin and O. Santawitee, *Energy Procedia*, 2013, **34**, 888.
46. C. Zhang, Y. Huang, C. Luo, L. Jiang and Y. Dan, *J. Polym. Res*, 2013, **20**, 121.
47. C. Zhang, W. Wang, Y. Huang, Y. Pan, L. Jiang, Y. Dan, Y. Luo and Z. Peng, *Mater. Des*, 2013, **45**, 198.
48. W. Zhang, L. Chen and Y. Zhang, *Polymer*, 2009, **50**, 1311.
49. H. Liu, F. Chen, B. Liu, G. Estep and J. Zhang, *Macromolecules*, 2010, **43**, 6058.
50. F. C. Pai, H. H. Chu and S. M. Lai, *J. Polym. Eng.*, 2011, **31**, 463.
51. H. U. Zaman, J. C. Song, L. S. Park, I. K. Kang, S. Y. Park, G. Kwak, B. Park and K. B. Yoon, *Polym. Bull.*, 2011, **67**, 187.
52. M. Kaavessinaa, I. Alic and S. M. Al-Zahrani, *Procedia Chem.*, 2012, **4**, 164.
53. H. Liu, L. Guo, X. Guo and J. Zhang, *Polymer*, 2012, **53**, 272.

54. H. Zhang, N. Liu, X. Ran, C. Han, L. Han, Y. Zhuang and L. Dong, *J. Appl. Polym. Sci.*, 2012, **125**, E550.
55. Y. Feng, Y. Hu, J. Yin, G. Zhao and W. Jiang, *Polym. Eng. Sci.*, 2013, **53**, 389.
56. J. Jiang, L. Su, K. Zhang and G. Wu, *J. Appl. Polym. Sci.*, 2013, **128**, 3993.
57. S. Ye, T. T. Lin, W. W. Tjiu, P. K. Wong and C. He, *J. Appl. Polym. Sci.*, 2013, **128**, 2541.
58. Q. Zhao, Y. Ding, B. Yang, N. Ning and Q. Fu, *Polym. Test*, 2013, **32**, 299.
59. N. M. K. Lamba, K. A. Woodhouse and S. L. Cooper, *Polyurethanes in Biomedical Applications*, CRC Press, Boca Raton, FL, USA, 1998.
60. R. A. Gunatillake, G. F. Meijs, S. J. McCarthy and R. Adhikari, *J. Appl. Polym. Sci.*, 2000, **76**, 2026.
61. A. Simmons, J. Hyvarinen and L. Poole-Warren, *Biomaterials*, 2006, **27**, 4484.
62. S. Gogolewski, K. Gorna and A. S. Turner, *J. Biomed. Mater. Res. A*, 2006, 77, 802.
63. A. K. Bhowmick and H. L. Stephens, *Handbook of Elastomers*, Marcel Dekker, New York, USA, 2nd edn, 2001.
64. H. B. Bode, K. Kerkhoff and D. Jendrossek, *Biomacromolecules*, 2001, **2**, 295.
65. K. Rose and A. Steinbuchel, *Appl. Environ. Microbiol.*, 2005, **71**, 2803.
66. U. Sundararaj and C. W. Macosko, *Macromolecules*, 1995, **28**, 2647.
67. N. Bitinis, A. Sanz, A. Nogales, R. Verdejo, M. A. Lopez-Manchado and T. A. Ezquerra, *Soft Matter*, 2012, **8**, 8990.
68. A. C. Renouf-Glauser, J. Rose, D. F. Farrar and R. E. Cameron, *Biomaterials*, 2005, **26**, 5771.
69. G. M. Kim and G. H. Michler, *Polymer*, 1998, **39**, 5689.

PLA-based Nano-biocomposites

CHAPTER 8

Polylactide (PLA)/Clay Nano-biocomposites

JOSE M. LAGARÓN*[a] AND LUIS CABEDO[b]

[a] Novel Materials and Nanotechnology Group, Institute of Agrochemistry and Food Technology (IATA), C.S.I.C., Apdo. Correos 73, 46100 Burjassot, Spain; [b] Polymers and Advanced Materials Group (PIMA), University Jaume I, 12071 Castelló, Spain
*Email: lagaron@iata.csic.es

8.1 Introduction

Polylactides (PLA) may have many potential applications for an important set of products but some of their properties should be improved to obtain similar performance to petroleum-based commodities. One of the most important current applications of PLA is food packaging, in particular for short-shelf-life products with common applications such as rigid containers, drinking cups, over-wrap and lamination films. PLA production and consumption are expected to increase; therefore research into the variation of PLA mechanical and barrier properties is currently very active. The control of barrier properties in PLA films is possible by modification of the polymer network through the formation of intramolecular and intermolecular covalent crosslinking, for example by applying thermal treatment, or by modifying the chemical composition. Another method is to incorporate fillers, in particular layered nanoclays, and this will be the subject of this chapter.

RSC Polymer Chemistry Series No. 12
Poly(lactic acid) Science and Technology: Processing, Properties, Additives and Applications
Edited by Alfonso Jiménez, Mercedes Peltzer and Roxana Ruseckaite
© The Royal Society of Chemistry 2015
Published by the Royal Society of Chemistry, www.rsc.org

8.2   Nanoclays for PLA Nanocomposites

Some of the most widely studied and technologically relevant polymer nanocomposites used in PLA technologies are those making use of nanoclays as reinforcing fillers. In order to achieve a full understanding of the nanocomposite behaviour and, therefore, of the system behaviour and performance, it is necessary to understand the structure and characteristics of nanoclays. With this objective, a small section of this chapter is devoted to analyzing the basic concepts of the nanoclays, with special interest in the most interesting/widely used clays for these applications.

Nanoclay is the common name used for the clay materials used as nanoadditives to obtain nanocomposites or nanofluids. However, the term nanoclay or even the term clay can be a fairly vague concept, thus a definition is necessary. The Clay Minerals Society (CMS) together with the Association Internationale pour l'Etude des Argiles (AIPEA) define a clay as follows: "The term 'clay' refers to a naturally occurring material composed primarily of fine-grained minerals, which is generally plastic at appropriate water contents and will harden when dried or fired. Although clay usually contains phyllosilicates, it may contain other materials that impart plasticity and harden when dried or fired. Associated phases in clay may include materials that do not impart plasticity and organic matter."[1]

From a mineralogical point of view, the term clay encompass a group of minerals (clay minerals) mainly composed of phyllosilicates, whose physicochemical properties depend on their structure and particle size, which is very small (in the nanometre range in at least one of their dimensions). For the particular application of clay materials as nanofillers for the synthesis of nanocomposites, the only interesting clay minerals present in clays are the phyllosilicates, any other clay mineral being an impurity for this purpose. Therefore, from now on, although clay and phyllosilicate are not strictly synonyms, both terms will be equally used.

Phyllosilicates are a family of silicates with a structure based on the stacking of interconnected $Si_2O_5^{-2}$ tetrahedral layers (T) with octahedral layers (O) of either Brucite $[Mg(OH)_3]$ or Gibbsite $[Al(OH)_3]$ type (Figure 8.1).

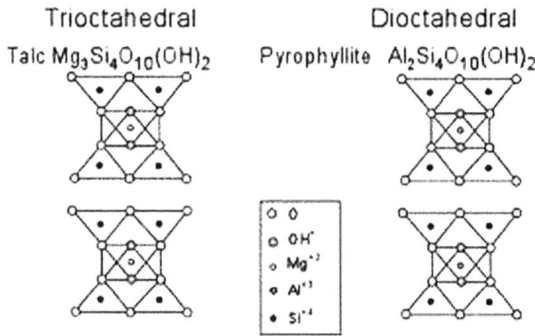

Figure 8.1   Typical structure of phyllosilicates.

In the tetrahedral layer, three vertices of the $Si_2O_5^{-2}$ tetrahedron are shared to make possible the formation of hexagonal rings, thus allowing distinguishing between basal oxygens (which are shared) and apical oxygen (not shared). These tetrahedral layers are attached to other octahedral gibbsite or brucite type (formed by Al^{3+} and Mg^{2+}, respectively). Considering the octahedral layer, there are two possibilities: trioctahedral (in this case all three octahedrons are occupied by a divalent cation, such as Mg^{2+}) or dioctahedral (in this case only two octahedrons are occupied by a trivalent cation, such as Al^{3+}). The octahedral layer can be bonded to another tetrahedral silicate layer on the opposite surface of the octahedral layer. Thus, phyllosilicates may be formed by two layers, tetrahedral plus octahedral (and they are called bilaminar, 1:1 or T:O), or by three layers, one octahedral and two tetrahedral (called trilaminar, 2:1 or T:O:T). The unit formed by the union of one octahedral layer and one or two tetrahedrons is called a sheet. The actual dimensions of tetrahedral and octahedral sheets can differ and, thus, a variety of structural adjustments are needed to permit the different sheets to fit together. Moreover, the substitution of cations by others with lower electrical charge is common: for instance aluminium for silicon in tetrahedral sheets, and magnesium or iron for aluminium in octahedral sheets. These substitutions can lead to a net charge in the sheet surface. In the case of the 1:1 clays these substitutions are usually fully compensated by others, to obtain no net layer charge. However, in the 2:1 clays, layers can carry a net negative charge on the surface. This charge may be compensated by cations, hydrated cations or hydroxides, which are located between the layers, *i.e.* in the interlayer or gallery.

Phyllosilicates can be classified according to several criteria (see Table 8.1): i) the type of sheet (1:1 or 2:1); ii) the nature of the interlayer, either empty, hydrated, interchangeable hydrated cations, non-hydrated cations or hydroxides layer; iii) the magnitude of any net negative layer charge resulting from atomic substitutions; iv) the subgroup, which is determined by the dioctahedral or trioctahedral character of the octahedral sheets. The distinction between species of the same subgroup is obtained in response to the chemical composition and the geometry of sheet stacking and interlayers.

Among all the phyllosilicates the most widely used as nanofillers in PLA nanocomposites are, by far, those of the montmorillonite subgroup. Montmorillonite belongs to the smectite group, and it has effective negative charge at the sheets surface, which is compensated for by the presence of hydrated cations in the gallery. This gives montmorillonite the property of swelling in the presence of water and the possibility of exchange of the cations present in the gallery by others. This property of easily exchanging the cations in the interlayer space allows the montmorillonite surface to be chemically modified in order to increase compatibility between the clay surface and the organophyllic nature of PLA. The modifiers are generally surfactants, having a cation to be linked to the clay surface, and an organic tail, which compatibilizes the surface of the clay with the nature of the polymer.

Table 8.1 Classification of the phyllosilicates.

Type of sheet	Interlayer	Group	Subgroup	Species
1:1	Empty or H_2O	Serpentine-kaolin	Serpentines	*Chrysolite, lizardite*
		$X \approx 0$	Kaolins	*Kaolinite, dicktite, nacrite, halloysite*
	Empty	Talc-Pyrophyllite	Talc	*Talc*
		$X \approx 0$	Pyrophyllites	*Pyrophyllite*
		Smectite	Saponites	*Saponite, hectorite, stevensite*
	Interchangeable	$X \approx 0.2$–0.6	Montmorillonites	*Montmorillonite, beidillite, montronite*
	Hydrated Cations	Vermicullite	V.Trioctahedral	*Idem.*
2:1		$X \approx 0.6$–0.9	V.Dioctahedral	*Idem.*
		Mica	Trioctahedral	*Flogopites, biotite, lepidolite*
	Non-hydrated	$X \approx 0.15$–1	Dioctahedral	*Moscovite, illite, glaucomite*
	Cations	Brittle micas	Trioctahedral	*Clintonite*
		$X \approx 2$	Dioctahedral	*Margarite*
			Trioctahedral	*Clinochlore, Chamosite, Nimite, Pennantite*
	Hydroxide layer	Chlorites	Dioctahedral	*Donbassite*
		X variable	Diocta.-Triocta.	*Cookeite*

Besides the use of montmorillonite nanoclays, several studies have been carried out with other non-conventional clays. Kaolinites have been successfully used to obtain nanocomposites with PLA.[2,3] Halloysites (HNT) are currently attracting interest due to their hollow tubular-like shape and to several characteristic features, such as nanoscale lumen and relatively high aspect ratio. HNT are currently considered as promising opponents and cheaper alternatives to both carbon nanotubes and organomodified layered silicates. While organomodified montmorillonites require a high degree of modification to achieve good exfoliation, HNT do not require much, if any, chemical modification to achieve a good favourable dispersion, thus obtaining nanocomposites without the complexity and processing cost associated with montmorillonite.[4] Murariu *et al.*[5] have reported good performance of HNT when incorporated into PLA.

Sepiolite is a fibrillar silicate which has been also used as a nanoclay in PLA,[6] compared with both montmorillonite[7] and halloysite.[8] The use of pillared serpentines as support for ZnO catalyst for PLA polymerization has been recently reported.[9] Zhang *et al.* obtained vermiculite/PLA nanocomposite by *in situ* polymerization of the PLA[10]. The use of different synthetic micas as nanofillers in PLA has also been evaluated.[11]

8.2.1   PLA/Clay Nanocomposite Processing

The main issue in processing of polymer/clay nanocomposites is to achieve sufficient interaction between the nanofiller and the polymer so as to achieve a favourable morphology. Due to the laminar structure of the most common nanoclays, the morphology of nanocomposites can be classified in three different types depending on the structure of the nanoclays and the interaction with the polymer chains: aggregated, intercalated and exfoliated. Figure 8.2 shows the three possible morphologies of a polymer/clay nanocomposite.

The aggregated morphology occurs when the polymer chains are not able to break apart the stacking of the clay and, therefore, small aggregates are found in the composite (Figure 8.2, left). If these aggregates are above the nanometre range, the composite is no longer a nanocomposite, but is now a microcomposite. When the sizes of the aggregates are within the nanometre range they can act almost as single nanofillers, being nanocomposites.

In the intercalated structures, one or several polymer chains are inserted in the interlayer gallery, keeping some kind of parallel order of the clay layers but with a high degree of interaction with the polymer (Figure 8.2, middle). Sometimes, this intercalated structure can be found in combination with aggregated structures of small amounts of layers forming structures with a high aspect ratio, which are frequently called tactoids.

The exfoliated state takes place when the dispersion of the sheets of the clay is fully achieved and, therefore, no interaction between sheets occurs (Figure 8.2, right). The ideal exfoliated state is that in which the clay platelets are homogeneously distributed throughout the polymer matrix. Generally speaking, layered nanocomposites never exhibit a pure structure; on the contrary, their morphology is a combination of all three possibilities with one of them dominant.

When obtaining PLA/clay nanocomposites, three main techniques are frequently used to produce nanocomposites of this material, namely *in situ* intercalative polymerization, solution-casting and melt mixing. Of the three, the *in situ* intercalative polymerization method exhibits the highest performance, since it is the one that results in a higher degree of interaction

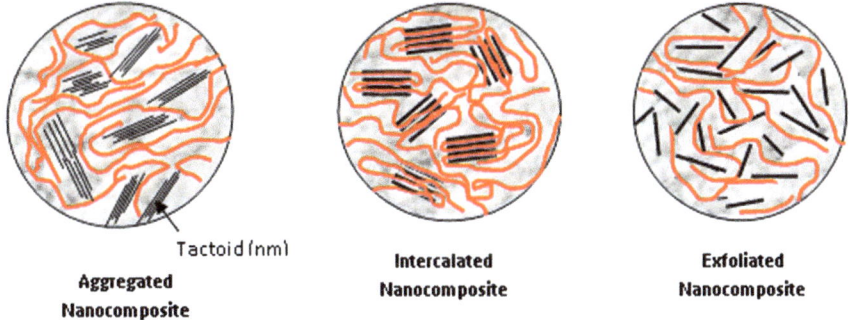

Tactoid (nm)

Aggregated Nanocomposite   **Intercalated Nanocomposite**   **Exfoliated Nanocomposite**

Figure 8.2   Possible morphologies of a polymer/clay nanocomposite.

between the nanoclay and the polymer, yielding a higher dispersion of the clay platelets within the polymer matrix.[4] In this method, the layered silicate is swollen within the liquid monomer or a monomer solution so the polymer formation can occur between the intercalated sheets.

Solution casting has been widely used as a nanocomposite processing method since solvent intercalation is the simplest way to prepare PLA/clay nanocomposites. But, a solvent where polymer is soluble and nanofillers are highly dispersible is necessary. The preparation of PLA-based nanocomposites by this method normally results in good dispersion of the nanofillers within the polymer matrix, and consequently in enhanced properties, as has successfully been demonstrated by some authors.[12,13]

In the melt mixing method, nanoclays are incorporated into the polymer in the molten state. This technique has considerable advantages over either the *in situ* intercalative polymerization or polymer solution intercalation techniques. Firstly, this method is environmentally benign due to the absence of organic solvents. Secondly, melt processing is compatible with current industrial processes, such as extrusion and injection moulding. The melt intercalation method allows the use of biopolymers that were not suitable for *in situ* polymerization. This has been the most widely used method in the literature for obtaining PLA/clay nanocomposites.[14–21]

Other techniques for obtaining PLA/clay nanocomposites have been recently explored, such as masterbatch,[22,23] layer-by-layer[24] and *in situ* intercalative polymerization in supercritical carbon dioxide.[25]

For those systems where montmorillonite clays are used, the final morphology highly depends on the modification of the clay, since it is responsible for the compatibilization of the surface of the clay with the polymer. Krikorian *et al.*[26] studied the effect of three of the most common modifiers for montmorillonites and its effect in the final nanocomposites morphology. However, this modification can also have an unwanted effect when melt processing the nanocomposites: the usual modifiers are based on ammonium quaternary salts and these molecules undergo thermodegradation at temperatures close to the processing melting temperature of PLA.[27]

Besides the change in properties expected from the addition of nanoclays to PLA matrices, some works on the use of the nanoclays as processing agents have also been published. As'habi *et al.*[28] used different nanoclays as compatibilizer agents to develop PLA/LDPE blends. The compatibilizing effect can be explained by the fact that the organoclay is mainly located in the interface of the polymer blends. Nanoclays have also been used as heterogeneous nucleating agents in foaming processes with PLA using montmorillonites[29] and recently also HNT,[30] allowing the size of the foam cell and the cell density to be decreased.

8.2.2   PLA/Clay Nanocomposites Properties

Sinha Ray *et al.*[21] suggested that the barrier properties of non-interacting gases in nanocomposites primarily depend on two factors: one is the dispersed silicate particles aspect ratio and the other is the extent of distribution of these particles within the polymer matrix. When the degree of dispersion and distribution of the nanoclay is maximum, an exfoliated morphology is attained and the barrier properties become dependent on the particles aspect ratio (length/thickness) (see Figure 8.2).

As mentioned above, it is expected that low loadings of well-distributed nanoclays can have a significant impact in enhancing some material properties, such as mechanical properties, thermal stability, UV light protection, conductivity, processability and gas and vapour barrier properties.[2,31–48] Moreover, it is expected that the addition of low nanofiller loadings will not affect to a significant extent the inherently good properties of PLA matrices, such as transparency and toughness. Table 8.2 summarizes some of the works that can be found in the literature regarding improvements in oxygen and water permeability of nanocomposites of PLA.

Sinha Ray *et al.*[15] claimed reductions in oxygen permeability of *ca.* 65% for PLA + 4 wt% of synthetic fluorine mica prepared by melt compounding. Nanocomposites with similar clay contents (4–7 wt%), but with a different type of nanoclay, showed less improvement in oxygen barrier, *i.e.* ranging from 6% to 56%. Moreover, mica base systems have been reported to block

Table 8.2 Reported reductions (%) in oxygen and water vapour permeability for nanocomposites of PLA compared with the virgin resin.

	Nanoclay	Nanoclay content	Reduction in O_2 permeability	Reduction in H_2O permeability
PLA[20]	Organically modified-MMT	4%	12%	
PLA[20]	Organically modified-MMT	5%	15%	
PLA[20]	Organically modified-MMT	7%	19%	
PLA[19]	MMT	4%	14%	
PLA[19]	MMT-modified	4%	12%	
PLA[19]	Saponite	4%	40%	
PLA[19]	Synthetic fluorine mica	4%	65%	
PLA[46]	MMT-layered silicate	5%	48%	50%
PLA[47]	MMT-modified	5%	46%	
PLA[33]	Bentonite	5%	6%	
PLA[48]	Hexadecylamine-MMT	4%	42%	
PLA[48]	Hexadecylamine-MMT	6%	56%	
PLA[48]	Hexadecylamine-MMT	10%	58%	
PLA[48]	Dodecytrimetil ammonium bromide-MMT	4%	41%	
PLA[48]	Dodecytrimetil ammonium bromide-MMT	6%	55%	
PLA[48]	Dodecytrimetil ammonium bromide-MMT	10%	58%	
PLA[48]	Cloisite 25A (Organically modified-MMT)	6%	45%	
PLA[48]	Cloisite 25A (Organically modified-MMT)	10%	56%	
PLA[42]	O2Block™ (Organically modified-MMT)	5%	32%	54%

UV light in addition to enhancing the barrier properties[40] in PLA due to the inherently enhanced scattering and reflection phenomena and absorbing properties of the highly distributed long aspect ratio clay nanolayers (see Figure 8.2). The attained ratio of UV protection and barrier blocking was optimum at 5 wt% of the nanoclay addition in the damaging UV region 280–320 nm. Thus, low nanoclay contents (1 and 5 wt%) added to the transparent PLA matrix led to significant reductions in the UV light transmission and enhanced gas and vapour barrier, while retaining transparency to a significant extent.[40]

Novel uses of nanoclays involve the development of the so-called active nanocomposites that contain functional nanoclays such as oxygen scavenging and antimicrobial nanoclays.[44-45] Thus, a first study discussed PLA nano-biocomposites obtained by solvent casting, containing silver-based antimicrobial layered silicate additives.[44] The composites resulted in transparent films with enhanced water barrier and strong biocide capacity. This study suggested that very low migration levels of silver, *i.e.* within the specific migration levels for silver ions as defined by the European Food Safety Agency (EFSA), can lead to sufficiently strong antimicrobial films, hence supporting the application of this antimicrobial additive in active food packaging applications for maintaining food quality and safety. A subsequent study[45] reported on two PLA nano-biocomposites obtained by melt compounding containing two silver-based biocide montmorillonites. The active clays were different since they contained a different biocide agent oxidation state, *i.e.* in one sample silver was in ionic form and in the other as native nanoparticles. In both cases, composites with enhanced thermal stability and good dispersion and distribution of the antimicrobial compounds were obtained. The biocidal effect was larger for the ionic silver sample. These results support these nano-biocomposites' potential as antimicrobial additives in PLA active packaging applications obtained *via* melt compounding.

Additionally, a zero-valent iron containing montmorillonite and its nano-biocomposite with PLA have also been proposed.[45] The active clay was seen to absorb up to 62 mL of oxygen g^{-1} clay, at 100% RH and 24 °C. The active oxygen absorbing montmorillonite was incorporated by melt mixing into polylactide to obtain a composite with significant oxygen scavenging properties. The thermal stability of the iron composite was reduced by the catalytic effect of iron on polylactide degradation processes; however, other thermal parameters were not affected and the water vapour permeability was slightly improved.

References

1. S. Guggenheim and R. T. Martin, *Clays and Clay Minerals*, 1995, **43**(2), 255–256.
2. L. Cabedo, J. L. Feijoo, J. M. Lagarón, J. J. Saura and E. Giménez, in *Annual Technical Conference–ANTEC*, Conference Proceedings, 2005, **4**, 42.

3. J. Matusik, E. Stodolak and K. Bahranowski, *Appl. Clay Sci.*, 2011, **51**, 102.

4. J. M. Raquez, Y. Habibi, M. Murariu and P. Dubois, *Progr. Polym. Sci.*, 2013, **38**, 1504.

5. M. Murariu, A.-L. Dechief, Y. Paint, S. Peeterbroeck, L. Bonnaud and P. Dubois, *J. Polym. Environ.*, 2012, **20**, 932.

6. K. Fukushima, D. Tabuani, C. Abbate, M. Arena and L. Ferreri, *Polym. Degrad. Stabil.*, 2010, **95**, 2049.

7. K. Fukushima, D. Tabuani and G. Camino, *Mater. Sci. Eng. C*, 2009, **29**, 1433.

8. P. Russo, S. Cammarano, E. Bilotti, T. Peijs, P. Cerruti and D. Acierno, *J. Appl. Polym. Sci.*, 2014, **131**, 397.

9. W. Zhen, J. Li and Y. Xu, *Polym. Compos.*, 2013, **35**(6), 1023.

10. J. H. Zhang, W. Zhuang, Q. Zhang, B. Liu, W. Wang, B. X. Hu and J. Shen, *Polym. Compos.*, 2007, **28**, 545.

11. D. H. S. Souza, S. V. Borges, M. L. Dias and C. T. Andrade, *Polym. Compos.*, 2012, **33**, 555.

12. N. Ogata, G. Jiménez, H. Kawai and T. Ogihara, *J. Polym. Sci. Polym. Phys.*, 1997, **35**, 389.

13. S. Torres-Giner, E. Gimenez and J. M. Lagarón, *Food Hydrocolloids*, 2007, **22**, 601.

14. S. Bandyopadhyay, R. Chen and E. P. Giannelis, *Polym. Mater. Sci. Eng.*, 1999, **81**, 159.

15. S. Sinha Ray, K. Yamada, M. Okamoto and K. Ueda, *Nano Letters*, 2002, **2**, 1093.

16. S. Sinha Ray, P. Maiti, M. Okamoto, K. Yamada and K. Ueda, *Macro-molecules*, 2002, **35**, 3104.

17. S. Sinha Ray, K. Yamada, A. Ogami, M. Okamoto and K. Ueda, *Macromol. Rapid Comm.*, 2002, **23**, 493.

18. S. Sinha Ray, M. Okamoto, K. Yamada and K. Ueda, *Macromol. Rapid Comm.*, 2002, **23**, 943.

19. S. Sinha Ray, K. Yamada, M. Okamoto, A. Ogami and K. Ueda, *Chem. Mater.*, 2003, **15**, 1456.

20. S. Sinha Ray, K. Yamada, M. Okamoto and K. Ueda, *Polymer*, 2003, **44**, 857.

21. S. Sinha Ray, K. Yamada, M. Okamoto, Y. Fujimoto, A. Ogami and K. Ueda, *Polymer*, 2003, **44**, 6633.

22. C. Y. Hung, C. C. Wang and C. Y. Chen, *J. Appl. Polym. Sci.*, 2013, **128**, 2736.

23. C. Y. Hung, D. K. Huang, C. C. Wang and C. Y. Chen, *J. Inorg. Organomet. Polym. Mater.*, 2013, **23**, 1389.

24. A. J. Svagan, A. Åkesson, M. Cárdenas, S. Bulut, J. C. Knudsen, J. Risbo and D. Plackett, *Biomacromolecules*, 2012, **13**, 397.

25. L. Urbanczyk, F. Ngoundjo, M. Alexandre, C. Jérôme, C. Detrembleur and C. Calberg, *Eur. Polym. J.*, 2009, **45**, 643.

26. V. Krikorian and D. J. Pochan, *Chem. Mater.*, 2003, **15**, 4317.

27. L. Cui, D. M. Khramov, C. W. Bielawski, D. L. Hunter, P. J. Yoon and D. R. Paul, *Polymer*, 2008, **49**, 3751.
28. L. As'habi, S. H. Jafari, H. A. Khonakdar, B. Kretzschmar, U. Wagenknecht and G. Heinrich, *J. Appl. Polym. Sci.*, 2013, **130**, 749.
29. B. Lotz, N. Okui, G. Ungar, Y. Ema, M. Ikeya and M. Okamoto, *Polymer*, 2006, **47**, 5350.
30. W. Wu, X. Cao, Y. Zhang and G. He, *J. Appl. Polym. Sci.*, 2013, **130**, 443.
31. V. Cyras, L. Manfredi, M. Ton-That and A. Vazquez, *Carbohydr. Polym.*, 2008, **73**, 55.
32. J. M. Lagaron, E. Gimenez and M. D. Sanchez-Garcia, Thermoplastic nanobiocomposites for rigid and flexible food packaging applications, in *Environmentally Compatible Food Packaging*, ed. E. Chiellini, Woodhead Publishing Ltd, Cambridge, 2008.
33. L. Petersson and K. Oksman, *Compos. Sci. Tech.*, 2006, **66**, 2187.
34. Y. Xu, X. Ren and M. A. Hanna, *J. Appl. Polym. Sci.*, 2006, **99**, 1684.
35. S. I. Marras, K. P. Kladi, I. Tsivintzelis, I. Zuburtikudis and C. Panayiotou, *Acta Biomater.*, 2008, **4**, 756.
36. M. D. Sánchez-García, L. Hilliou and J. M. Lagaron, *J. Agric. Food Chem.*, 2010, **58**, 12847.
37. M. D. Sanchez-Garcia, J. M. Lagaron and S. V. Hoa, *Compos. Sci. Tech.*, 2010, **70**, 1095.
38. C. Wan, X. Qiao, Y. Zhang and Y. Zhang, *J. Appl. Polym. Sci.*, 2003, **89**, 2184.
39. M. D. Sanchez-Garcia and J. M. Lagaron, *Cellulose*, 2010, **17**, 987.
40. M. D. Sanchez-Garcia and J. M. Lagaron, *J. Appl. Polym. Sci.*, 2010, **118**, 188.
41. V. K. Haugaard, A. M. Udsen, G. Mortensen, L. Hoegh, K. Petersen and F. Monahan, Food biopackaging, in *Biobased Packaging Materials for the Food Industry–Status and Perspectives*, ed. C. J. Weber, European Concerted Action Report, The Royal Veterinary and Agricultural University, Copenhagen, 2001.
42. M. D. Sanchez-Garcia, E. Giménez and J. M. Lagaron, *J. Plast. Film Sheet.*, 2007, **23**, 133.
43. C. Bastioli, V. Bellotti, G. F. Del Tredici, R. Lombi, A. Montino and R. Ponti, International Patent Application WO 92/19680, 1992.
44. M. A. Busolo, P. Fernández, M. J. Ocio and J. M. Lagaron, *Food Addit. Contam. A*, 2010, **27**, 1617.
45. M. A. Busolo and J. M. Lagaron, *Innovat. Food Sci. Emerg. Tech.*, 2012, **16**, 211.
46. C. Thellen, C. Orroth, D. Froio, D. Ziegler, J. Lucciarini, R. Farrell, N. A. D'Souza and J. A. Ratto, *Polymer*, 2005, **46**.
47. J. M. Lagaron, L. Cabedo, D. Cava, J. L. Feijoo, R. Gavara and E. Giménez, *Food Addit. Contam.*, 2005, **22**(10), 994–998.
48. J.-H. Chang, Y. U. An and G. S. Sur, *J. Polym. Sci. Polym. Phys.*, 2003, **41**, 94–103.

CHAPTER 9

PLA-nanocellulose Biocomposites

QI ZHOU*[a,b] AND LARS A. BERGLUND[b,c]

[a] School of Biotechnology, Royal Institute of Technology (KTH), AlbaNova University Centre, SE-106 91 Stockholm, Sweden; [b] Wallenberg Wood Science Center, Royal Institute of Technology (KTH), SE-100 44 Stockholm, Sweden; [c] Department of Fibre and Polymer Technology, Royal Institute of Technology (KTH), SE-100 44 Stockholm, Sweden
*Email: qi@kth.se

9.1 Introduction

Bio-reinforced green composites from biodegradable polymers and natural fibres have demonstrated substantially improved mechanical properties, thermal stability and barrier properties, and thus have received increasing scientific and industrial attention in the fields of packaging, medical and automotive sectors. Poly(lactic acid) (PLA) is a biodegradable polyester derived from renewable resources such as corn, wheat and potato starch or sugarcane. It is one of the most promising thermoplastic biopolymers with biocompatibility and good mechanical properties (tensile strength of 50–70 and modulus of 3–4 GPa, equivalent to polystyrene and polyethylene terephthalate). However, the insufficient impact strength, inherent brittleness and low thermal stability of PLA are limiting the scope of its applications. Considerable efforts have been made to improve its properties in order to achieve the compatibility with thermoplastics processing and manufacturing, and compete with commodity polymers. Strategies include co-polymerization,[1] stereo-complexation,[2] polymer blends[3] and addition of

RSC Polymer Chemistry Series No. 12
Poly(lactic acid) Science and Technology: Processing, Properties, Additives and Applications
Edited by Alfonso Jiménez, Mercedes Peltzer and Roxana Ruseckaite
© The Royal Society of Chemistry 2015
Published by the Royal Society of Chemistry, www.rsc.org

plasticizers,[4] nanoscale reinforcing fillers[5,6] and natural fibres.[7] Biocomposites of PLA reinforced with natural and man-made cellulose fibres such as wood pulp fibres, plant fibres from kenaf, flax, hemp, bamboo, ramie, and even agricultural residues, regenerated cellulose fibres and recycled newspaper fibres have been prepared and extensively studied.[7] Such green biocomposites are expected to have a great potential as alternatives for petroleum-based polymers since they are overall renewable and biodegradable without emitting any toxic compounds to the environment. As compared to conventional glass and carbon fibres, natural fibres have the advantages of low density, biodegradability and low environmental impact, while providing composites with good mechanical properties. In particular, cellulose nanofibrils with lateral dimensions three orders of magnitude smaller than the intact plant fibre cells are an ideal reinforcing material for nanocomposites due to their high aspect ratio, large specific surface area and high mechanical strength/weight performance.[8] In this chapter, we review the preparation, structure and properties of PLA/nanocellulose biocomposites with particular focus on the effect of surface modification on nanocellulose dispersion, and the effect of nanocellulose incorporation on crystallization kinetics of PLA matrix and mechanical, thermal and barrier properties of biocomposites.

9.2 Types of Nanocellulose

Nanocelluloses are a new class of nanomaterials having fibrillar structures with one dimension in the nanometre range. They exhibit unique properties compared to micrometre-scale cellulosic fibres. In particular, they are characterized by a very large specific surface area, reactive surface of $-OH$ groups, high strength and modulus and low thermal expansion coefficient and gas permeability. Nanocelluloses are isolated from biological sources that comprise nanosized cellulose fibrils and fibril aggregates. There are basically two types of nanocellulose being widely employed in the PLA biocomposites, *i.e.* cellulose nanofibrils (CNFs) and cellulose nanocrystals (CNCs), as simply categorized by their structure and production methods.

Cellulose nanofibrils (CNFs), also termed microfibrilated cellulose (MFC), nanofibrilated cellulose (NFC) or cellulose nanofibres in the literature, are characterized by their length (usually above 1 micrometre) and ability to form an entangled network structure. CNFs can have a wide width distribution depending on the degree of fibrillation. Methods for their production generally involve enzymatic/chemical/physical methodologies for their isolation from wood and forest/agricultural residues. MFC with a fibril width of 25–100 nm was first isolated from wood pulp by Turbak *et al.* through a repeated homogenization process using a high pressure slit homogenizer.[9] Cellulose nanofibrils with a width of 15–30 nm were then produced by mechanical disintegration of wood pulp fibres that were pretreated with endoglucanase.[10] Furthermore, after pretreatment with a carboxymethylation reaction, cellulose nanofibrils with a width of 5–15 nm were liberated from wood pulp fibres by mechanical homogenization and ultrasonication.[11] Completely individualized cellulose fibrils with a uniform width (*ca.* 3 nm, corresponding to the width of cellulose elementary fibrils) were prepared from wood cellulose fibres by 2,2,6,6-tetramethylpiperidine 1-oxyl radical (TEMPO)-mediated oxidation under moderate aqueous conditions combined with low energy mechanical treatment.[12] As shown in Figure 9.1a, the width of TEMPO-oxidized cellulose nanofibrils (TEMPO-NFC) is 2.4 ± 0.7 nm as measured from the height of AFM image, the lengths of nanofibirls are possibly several micrometres and the nanofibril ends are not apparent. In our recent work, individualized cellulose I nanofibrils with 1.6–2.1 nm width (Figure 9.1b) were mechanically disintegrated from wood pulp that was pretreated through a reaction with glycidyl-trimethylammonium chloride.[13]

Cellulose nanocrystals (CNCs), which have also been named as cellulose nanowhiskers or nanocrystalline cellulose, are characterized by their rod-like or conifer-shaped particles and the production method, strong acid hydrolysis. CNCs are generally prepared by sulfuric acid hydrolysis of native cellulose fibres, resulting in the formation of sulfate esters on the surface.[14] Endoglucanase has also been used for the controlled hydrolysis of microcrystalline cellulose (MCC) or recycled pulp to produce CNCs.[15] The length of CNCs from cotton and wood pulp is typically around 100–300 nm, while the length can be several micrometres for CNCs from tunicate, algal and bacterial cellulose. The width of CNCs is in the range of 5–20 nm depending on the cellulose resources.[16] CNCs with surface carboxylic acid groups have also been prepared by the combination of hydrochloric acid or hydrobromic acid hydrolysis and oxidative carboxylation (TEMPO-mediated oxidation).[17] Subsequent grafting of amine-terminated poly(ethylene glycol) to carboxylated cellulose microcrystals improved the stability of the microcrystal suspensions through steric stabilization.[18] In our recent work, CNCs (Figure 9.1c, width 3.6 ± 1.1 nm, length 288 ± 94 nm) bearing high carboxylate content

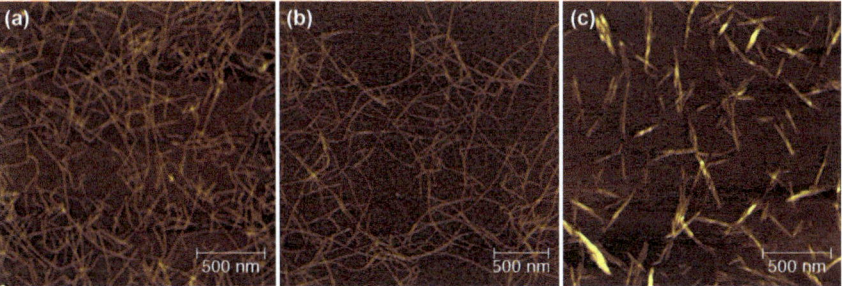

Figure 9.1 AFM height images of a) TEMPO-oxidized cellulose nanofibrils, b) cationically modified cellulose nanofibrils, c) cellulose nanocrystals prepared by direct hydrochloric acid hydrolysis of TEMPO-oxidized cellulose nanofibrils.

(1.5 mmol g^{-1}) have been successfully prepared by direct hydrochloric acid hydrolysis of cellulose nanofibrils obtained from TEMPO-mediated oxidation of softwood pulp.[19]

High-quality CNFs can now be obtained at low cost through mechanical disintegration of wood pulp fibre after enzymatic or chemical pretreatments, and have been widely studied to develop industrially feasible methods for the production of PLA/nanocellulose biocomposites.[20] CNCs from different natural resources have also been extensively used as reinforcement to tailor the properties of PLA particularly in comparison with inorganic nano-particles, such as nanoclays.[21] In addition to the nanocelluloses from wood, CNFs and CNCs from bacterial cellulose have also been utilized for the production of bionanocomposites with PLA.[22–24] Bacterial cells from the genus *Acetobacter* secrete cellulose in the form of entangled nanoribbons that are typically 70–150 nm wide and several micrometres long. This form of cellulose, commonly referred to as "bacterial cellulose" (BC), is highly crystalline, virtually pure and free of any non-cellulosic polymer.

9.3   Surface Modification of Nanocellulose

In nanocomposites, surface properties of the nanoscale building blocks will control their interactions with the matrix, which ultimately dictates the structure and macroscale properties of the composites. In order to obtain well-dispersed hydrophilic reinforcing nanocelluloses in hydrophobic PLA matrix, surface modification of nanocellulose is essential for the preparation of biocomposites. The surface modification chemistries for nanocelluloses in the PLA/nanocellulose biocomposites are illustrated in Figure 9.2. The most used modification method is surface acetylation.[23,25–32] Nanocelluloses in aqueous suspension are generally solvent exchanged to acetone and then to dry toluene or directly to *N,N*-dimethylformamide (DMF), or dried and redispersed in the organic solvents. Acetylation has been carried out using acetic anhydride with perchloric acid or sulfuric acid as catalyst,[25,30,33,34] or with pyridine as an acid acceptor and catalyst.[28,31,32] In the case of TEMPO-oxidized cellulose nanofibrils, acetylation has been performed with acetic anhydride in DMF.[26,27,29] Besides acetic acid, long chain organic acids such as hexanoic acid and dodecanoic acid have also been used to functionalize nanocellulose *via* esterification reaction using *p*-toluenesulfonyl chloride in pyridine.[35] In order to tailor the interfacial energy and adhesion between CNCs and the matrix PLA, homopolyesters poly-ε-caprolactone (PCL) and PLA, and diback copolymers P(CL-*b*-LA) have been successfully grafted from the surface of CNCs by ring-opening polymerization technique.[36–38] Silanization is also a common surface modification method for nanocellulose to improve the interfacial adhesion. Functional trialkoxysilanes bearing various organic moieties (alkyl, (meth)acryloxy and amino groups), *e.g.* 3-aminopropyltriethoxysilane and 3-methacryloxypropyltrimethoxysilane, have been covalently attached onto the surface of nanocelluloses in aqueous or ethanol/water suspensions at pH of *ca.* 5 by adding acetic acid or citric acid

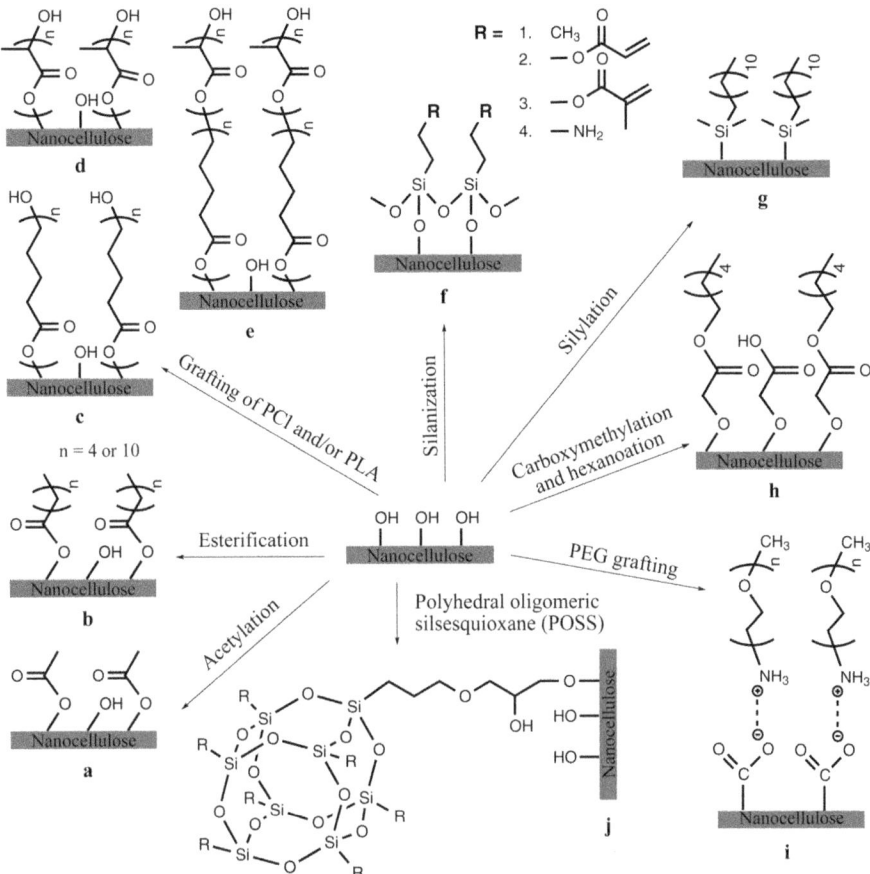

Figure 9.2 Surface modification chemistries of nanocellulose for PLA/nanocellulose biocomposites. a, Acetylation; b, Esterification with various organic acids; c, d, e, Grafting of PCL, PLA, P(CL-*b*-LA); f, Silanization; g, Silylation; h, Carboxymethylation combined with hexanoation; i, PEG grafting; j, Modified with polyhedral oligomeric silsesquioxane (POSS).

at ambient temperature.[39–42] Different from the covalent –Si–O–Si– bond formation during silanization, silylation of the hydroxyl groups on the surface of CNCs with *n*-dodecyldimethylchlorosilane in toluene has also significantly improved their dispersion in PLA.[43] CNFs prepared by carboxymethylation and mechanical disintegration have been further modified by hexanoation, but still exhibited as extensive agglomerates in PLA matrix.[44] Surface grafting of TEMPO-oxidized cellulose nanofibrils with poly(ethylene glycol) (PEG) chains *via* ionic bonds has been achieved by a simple ion-exchange treatment.[45] Interestingly, the PEG-grafted cellulose nanofibrils exhibited nanodispersibility in organic solvents and remarkably improved the mechanical properties of the biocomposites, despite low cellulose addition levels (<1 wt%). Another interesting surface modification for

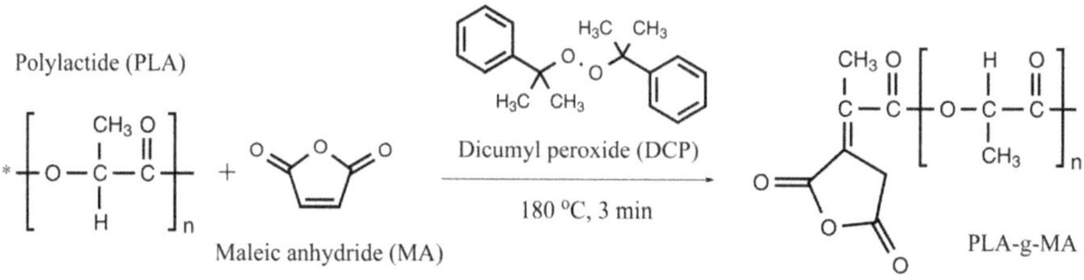

Figure 9.3 Chemical reactions occurring during extrusion of PLA in the presence of maleic anhydride and dicumyl peroxide (reproduced from Figure 1 of Ref. 55 with kind permission from Springer Science and Business Media).

nanocellulose is the introduction of polyhedral oligomeric silsesquioxane (POSS), which consists of a $SiO_{1.5}$ cubic core, surrounded by organic substituents that improve compatibility and/or react with a polymer matrix.[46] POSS-modified nanofibrillated cellulose (PNFC) has been used as a carbon source in intumescing flame retardants for melt-blended poly(lactic acid) (PLA) composites. Apart from the surface chemical modification on nanocellulose, low-molecular-weight poly(ethylene glycol) (PEG)[47–49] and the anionic surfactant, an acid phosphate ester of alkyl phenol ethoxylate,[50–54] have also been used to improve the dispersion of nanocellulose in the matrix PLA. Bacterial cellulose nanofibres networks have also been glyoxalized in order to improve the mechanical properties of BC/PLA biocomposites.[55]

Alternatively, in order to improve the interfacial adhesion between PLA and nanocellulose, PLA has been grafted with maleic anhydride (MA) or γ-methacryloxypropyltrimethoxysilane (MPS). MA-grafted polyolefins are often used to compatibilize non-polar polyolefins and the reinforcing fillers such as glass fibres, talc, clay and cellulose fibres in composites. MA grafted PLA has also been prepared by mixing PLA and MA in a compounder or Brabender at 180 °C using dicumyl peroxide (DCP) as the initiator.[55–58] As shown in Figure 9.3, free radicals on the PLA chains are induced by DCP and reacted with MA. These radicals could also cause the degradation of PLA resulting in PLA-g-MA with a relatively smaller molecular weight. In a similar fashion, PLA has been grafted with MPS, a well-known commonly used silane coupling agent, using xylene as solvent and benzoyl peroxide (BPO) as initiator at 180 °C.[59] PLA-g-MPS has been subsequently used for the surface modification of BC nanofibres.

9.4 Processing/Mixing Strategies

Solvent casting process is the simplest method for laboratory-scale preparation of PLA/nanocellulose biocomposites. Organic solvents for PLA dissolution such as *N,N*-dimethylacetamide (DMAc),[42,48] chloroform[21,26,27,29,32,43,50,51,60–63] and dichloromethane[20,31,38,64,65] have been used for the solution casting and biocomposites are obtained by evaporation of the solvents. Nanocelluloses are dispersed in organic solvents by either solvent exchange or redispersion after freeze-drying using homogenization or ultrasonication. Careful solvent exchange often provides a good dispersion of hydrophilic nanocellulose in non-polar organic solvents, but it is extremely tedious and time consuming. Freeze drying has to be performed from very dilute nanocellulose aqueous suspensions in order to obtain a better redispersion in organic solvents. Chemical modification of nanocelluloses that renders their surfaces hydrophobic is the best way to achieve a nanodispersion in the organic solvent.[32,43] A uniform distribution of relatively low contents of nanocelluloses in the PLA matrix can be obtained with significantly improved mechanical properties. However, this method has drawbacks for the industrial implementation such as difficulties in the elimination of the entrapped solvent and low cost-performance ratio.

The most suitable and practical manufacturing methods of nanocellulose reinforced thermoplastic composites for industrial applications are melt processing techniques. Nanocelluloses including unmodified and chemical modified CNFs and CNCs have been successfully melt blended or compounded with PLA by using an extruder[28,35–37,41,44,49,52,54,66–70] or melting mixer.[8,23,30,39,40,57,71,72] Subsequently, PLA/nanocellulose biocomposites have been prepared by injection or compression moulding, or melt pressing. Instead of feeding dry samples directly into PLA melt,[71] nanocelluloses are often premixed with PLA in an organic solvent, such as acetone,[8,28,30,66] dichloromethane[44] or 1,4-dioxane,[35] to achieve a better dispersion before compounding extrusion. The melt mixing temperature is generally in the range of 165–200 °C. The PLA/nanocellulose-based biocomposites have also been prepared by hand lay-up compression moulding[24,55,73] and spray-up moulding.[74]

In order to prepare PLA nanocomposites with highly dispersed cellulose nanocrystals and porous PLA-based scaffolds with enhanced mechanical properties and thermal stability, CNCs have been incorporated into PLA fibres by electrospinning method.[22,56,75,76] Fibrous biocomposite mats consisting of PLA and CNCs have been electrospun from solvent or solvent mixtures such as 1,1,1,3,3,3-hexafluoro-2-propanol (HFP),[22] DMF/chloroform[56,76] and DMF/tetrahydrofuran (THF).[75] The electrospun PLA/CNCs bionanocomposites have demonstrated rapid *in vitro* biodegradability and cytocompatible properties, and could be potentially suitable in tissue engineering.[56,76]

Another practical approach for the preparation of PLA/CNF biocomposites is the manufacturing process free of toxic organic solvents akin to paper-making. Aqueous suspensions of CNFs were homogenized with PLA latex,[77] fibres[78] or microparticles[79] with the aid of a microfluidizer, Waring blender or Ultra-Turrax. A high-molecular-weight polymer flocculent polyacrylamide was added in the case of PLA fibres, while Polysorbate 80 was used as a surfactant to obtain water suspensions of PLA microparticles. The biocomposites with much higher content of CNFs as compared to solvent casting method have been prepared by compression moulding of the sheets obtained by vacuum filtration of the PLA/CNF water suspensions.

9.5   Properties

Nanocelluloses have been used as a nucleating agent to accelerate the crystallization rate of PLA. PLA/nanocellulose biocomposites frequently exhibit significantly improved thermal and mechanical and various other properties compared to those of pure PLA. The tensile modulus and strength of neat PLA are generally improved with an increase of nanocellulose content. Increased storage modulus at temperatures above T_g is often observed with the incorporation of nanocellulose in PLA, as well as decreased gas permeability.

9.5.1   Crystallization

The slow crystallization of PLA is a major problem industrially, since practical melt processing conditions results in products where the PLA often is in a state of very low crystallinity. An efficient method to increase nucleation density and thus to enhance the crystallization rate is the addition of heterogeneous nucleating agents. Cellulose fibres have been previously used as a nucleating and reinforcing material in the lignocellulosic/thermoplastic composites, particularly in the cellulose/polypropylene systems. Transcrystalline morphology was observed on the surface of the cellulose fibres[80] as well as cellulose nanocrystals.[81] Previously, we have studied the effect of the addition of CNCs as a bio-based nucleating agent on the crystallization behaviour of PLLA for the first time.[43] CNC nanoparticles of 15 nm width and 200–300 nm length were functionalized by partial silylation to improve their dispersion in PLLA. The isothermal crystallization at 125 °C for 5 and 10 min. was observed by polarized optical microscopy. As shown in Figure 9.4, both PLLA and PLLA–SCNC-1 (consisting of 1 wt% silylated CNC in PLLA) showed clean and uniform melt, while microscale aggregates of cellulose nanocrystals were found in the PLLA–CNC-1 (consisting of 1 wt% unmodified CNC in PLLA) before crystallization (0 min.). With the addition of CNC, the nucleus density of PLLA crystallites increased at 5 min. and 10 min. as compared to the pure PLLA. Furthermore, the nucleus density increased significantly with the addition of SCNC, due to the improved dispersion of cellulose nanocrystals in the PLLA matrix. As a result, more crystals were able to nucleate and grow on the increased surface area of the interfaces due to increasing numbers of nucleating particles.

Kose and Kondo studied the size effects of cellulose nanofibres on the crystallization behaviour of PLA.[82] They discovered that the smaller size of cellulose nanofibres on the nanoscale does not necessarily make a better nucleating agent for PLA. Table 9.1 summarizes the Avrami kinetic parameters for the isothermal crystallization of the PLA and PLA biocomposites with different types of nanocelluloses as compared to PLA composites with talc and nanoclay. With the addition of unmodified and silylated CNCs as nucleating agents, the $t_{1/2}$ value increases with increasing T_c, similar to that of nanoclay and corn starch, but opposite to that of talc.[83,84] Comparing the

PLLA   PLLA-CNC-1   PLLA-SCNC-1

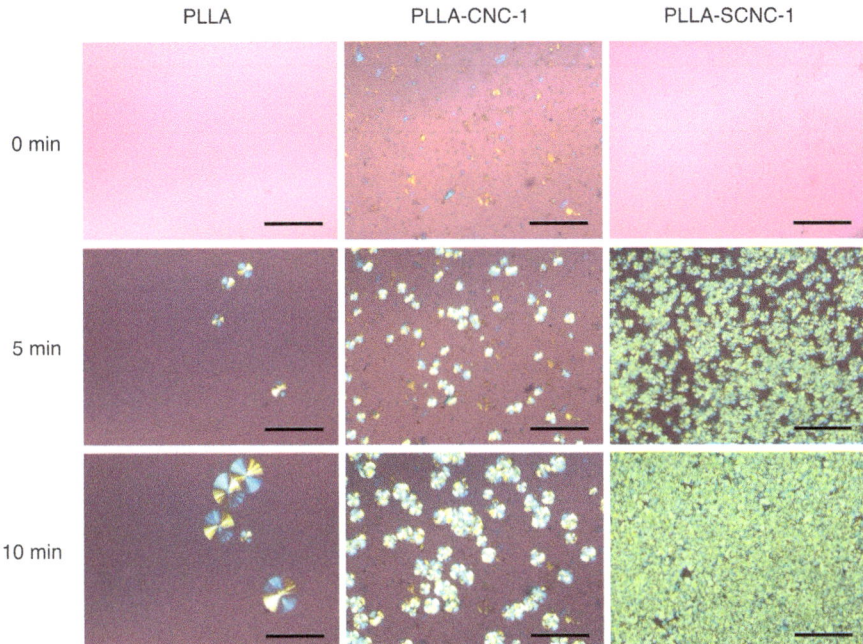

Figure 9.4   Polarized optical microscope images of PLLA, PLLA–CNC-1 and PLLA–SCNC-1 acquired on the 0th, 5th and 10th min. at 125 °C after being quenched from melt at 210 °C. Scale bar, 200 µm (reprinted from Figure 5 of Ref. 43 with permission from Elsevier).

$t_{1/2}$ values of different samples, PLA/NF sample with unmodified cellulose nanofibres of *ca.* 60 nm width has the lowest value of 2.7, only one-fifth of that for pristine PLA, indicating much better acceleration of the crystallization rate of PLA. This value is comparable with that for corn starch. However, it is still lower than that for nanoclay and talc, the most efficient nucleating agents.

9.5.2   Thermal and Mechanical Properties

Crystallinity of PLA has a strong impact on its mechanical properties. Suryanegara *et al.*[64] have prepared PLA/MFC nanocomposites in both fully amorphous and crystallized states. The tensile modulus and strength of pristine PLA were improved with an increase of MFC content in both amorphous and crystallized states. Dynamic mechanical analysis (DMA) has been used to study the effect of MFC reinforcement on the thermo-mechanical properties of PLA in both states and the results are shown in Figure 9.5. In the amorphous state, the storage modulus of pristine PLA below T_g is almost constant at around 3 GPa. Above T_g, the modulus drops to 4 MPa at 80 °C, and then increases to 200 MPa at 100 °C owing to the cold

Table 9.1 Avrami parameters and crystallization half times $(t_{1/2})$ of PLA-based nanocomposites containing 1 wt% of the nucleating agent.

Sample	Width of the nanocellulose	T_c (°C)[a]	n^b	$t_{1/2}$ (min)	Reference
PLA	–	110	2.81	8.6	43
		125	3.28	27.9	
PLA/CNC	15 nm	110	3.47	9.2	
		125	3.24	17.9	
PLA/SCNC	15 nm	110	4.07	4.2	
		125	3.61	7.8	
PLA	–	130	2.8 ± 0.1	13.9 ± 0.2	82
PLA/MCC	11 ± 10 μm		2.6 ± 0.2	14.5 ± 1.2	
PLA/MFC	100 nm to 3 μm		2.8 ± 0.1	11.1 ± 0.7	
PLA/NF	57 ± 22 nm		–	2.7 ± 0.1	
PLA/NC1	35 ± 19 nm		–	3.4 ± 0.1	
PLA/NC2	18 ± 9 nm		–	7.3 ± 0.6	
PLA	–	110	2.1	13.1	83
		125	2.37	10.5	
TacPla1-99	– (Talc)	110	1.87	0.3	
		125	1.67	0.2	
CnPla1-99	– (Corn starch)	110	2.33	2.7	
		125	2.03	20.4	
PLA	–	110.4	2.18	6.2	84
		130.4	1.71	40.2	
PLA/nanoclay	– (Nanoclay)	120.4	2.51	0.9	
		130.5	2.12	1.7	

[a]Isothermal crystallization temperature.
[b]Avrami exponent.

crystallization of the amorphous PLA. In the crystallized state, the storage modulus of PLA decreased gradually with increase in temperature above T_g at around 60 °C due to the relaxation of amorphous structure since PLA is not 100% crystalline. The crystallization significantly increased the storage modulus of PLA, especially at temperatures higher than T_g. The addition of MFC at a fibre content of 20 wt% further improved the storage modulus of crystallized pristine PLA from 505 MPa to 1616 MPa at 80 °C and from 293 MPa to 1034 MPa at 120 °C. Suryanegara *et al.*[20] further studied the deforming behaviour of PLA/MFC composites at an MFC content of 10 wt% with degree of crystallinity (Xc) of 0–43% obtained by annealing at 80 °C with different times. The required annealing time to obtain partially crystallized composite (Xc: 17%) was only around one-seventh of that needed to fully crystallize pristine PLA (Xc: 41%) and, interestingly, above 80 °C the storage modulus of the composite (Xc: 17%) was higher than that of the fully crystallized neat PLA. These results confirm that the combination of MFC reinforcement and crystallization of PLA in the matrix contributes to better heat resistance of the biocomposite.

The mechanical properties of PLA/nanocellulose biocomposites are also strongly affected by the processing strategy and surface chemical modification of nanocellulose. Table 9.2 provides an overview of modulus, tensile

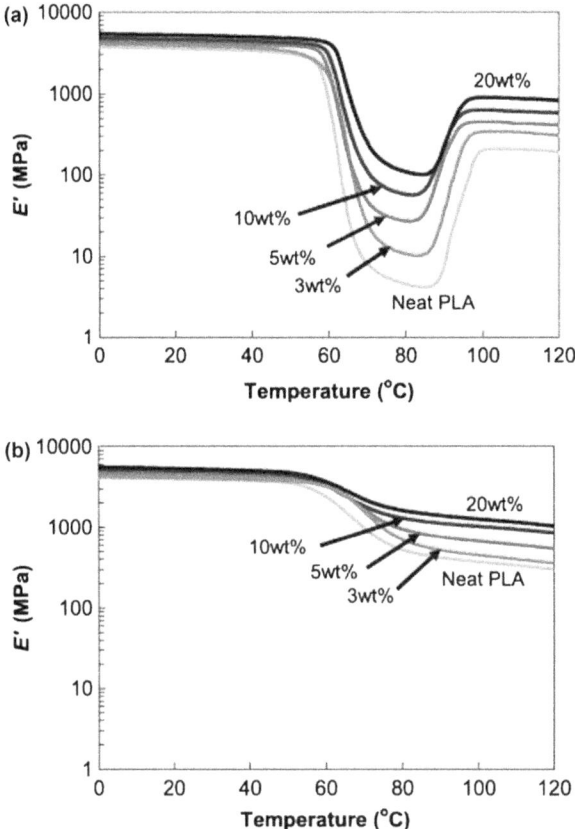

Figure 9.5 Effect of MFC contents (wt%) on the temperature dependency of the storage modulus under (a) amorphous and (b) crystallized states (reprinted from Figure 5 of Ref. 64 with permission from Elsevier).

strength and strain at break of PLA/CNF and PLA/CNC biocomposites prepared by different processing methods with modified and unmodified nanocellulose. Solvent casting method and surface chemical modification lead to better dispersion of nanocellulose in matrix PLA, and thus result in enhancement of mechanical properties. With the addition of 1 wt% well-dispersed acetylated TEMPO-oxidized nanofibrils, the strain-to-failure increased by approximately 30-fold and the work of fracture by an order of magnitude as compared to the pure PLA film.[29] This significant toughening effect is probably due to crazing in the polymer matrix that facilitates plastic deformation. A similar effect has been observed in PLA reinforced with 8 wt% CNC-*g*-PCL, where the nanocomposite had a true strength of 75.5 MPa and elongation at break of 221.6%, 1.8 and 10.7-fold higher than that of pristine PLA, respectively.[38] PLA reinforced with 6 wt% acetylated CNC demonstrated a tensile strength of 71.6 MPa, 61.3% higher than that for pristine PLA.[31] Surface esterification of cellulose nanofibrils from BC

Chapter 9

Table 9.2 Summary of modulus, tensile strength and strain at break of PLA/CNF and PLA/CNC biocomposites.

Filler type	Modification	Processing method	Filler content (wt%)	Modulus (GPa)	Tensile strength (MPa)	Strain at break (%)
TEMPO-NFC[29]	Acetylation (DS = 0.6)	Solvent casting	0	3.2 ± 0.2	41 ± 3	8.4 ± 6.0
			1	0.9 ± 0.2	23 ± 4	258 ± 18.8
			5	3.4 ± 0.2	47 ± 2	1.7 ± 0.1
CNC[31]	Acetylation	Solvent casting	0	0.91	44.4	21
			2	0.97	56.5	12
			6	1.05	71.6	10
			10	1.29	39.0	3
NFC from BC[35]	Esterification[a]	Compression moulding after extrusion	0	1.34 ± 0.04	60.7 ± 0.8	3.59 ± 0.07
			5 (BC)	1.89 ± 0.02	60.9 ± 0.5	2.41 ± 0.08
			5 (C_2-BC)	1.70 ± 0.03	60.1 ± 0.9	2.62 ± 0.07
			5 (C_6-BC)	1.79 ± 0.02	65.1 ± 0.9	2.87 ± 0.03
			5 (C_{12}-BC)	1.98 ± 0.04	68.5 ± 1.5	2.69 ± 0.06
CNC[38]	Grafting PCL	Solvent casting	0	1.55	40.7	20.7
			8	0.8	75.5	221.6
CNF[42]	Silanization	Solvent casting	0	–	39	2.7
			1	–	55.5	3.5
CNC[43]	Silylation[b]	Solvent casting	0	1.1 ± 0.01	48.3 ± 2.9	31.1 ± 3.0
			1 (CNC)	1.0 ± 0.02	49.2 ± 0.4	10.5 ± 2.0
			1 (SCNC)	1.4 ± 0.08	58.6 ± 3.1	8.3 ± 0.6
TEMPO-NFC[45]	PEG grafted	Solvent casting	0	1.23 ± 0.05	40.6 ± 5.7	1.51 ± 0.11
			1	1.72 ± 0.12	51.3 ± 2.6	2.29 ± 0.22

236

Material	Treatment	Method	Loading			
CNW[49]	Mix with PEG and PLA-MA	Twin-screw extrusion and compression moulding	0[c]	2.9 ± 0.1	40.9 ± 3.2	1.9 ± 0.2
			5[d]	3.9 ± 0.3	77.9 ± 6.7	2.7 ± 0.5
			5[e]	2.6 ± 0.2	48.4 ± 3.8	17.8 ± 8.5
CNW[54]	Surfactant	Twin-screw extrusion and compression moulding	0[f]	2.65 ± 0.08	62.8 ± 1.0	19.5 ± 9.7
			5[f]	2.69 ± 0.20	55.5 ± 2.9	9.7 ± 0.9
			0[g]	2.60 ± 0.12	35.1 ± 3.6	1.8 ± 0.3
			5[g]	3.10 ± 0.15	52.4 ± 0.4	3.1 ± 0.2
CNC[56]	PLA-g-MA	Electrospun	0	0.008 ± 0.003	1.6 ± 0.4	51.8 ± 3.8
			5	0.135 ± 0.01	10.8 ± 1.7	44.0 ± 5.1
CNC from BC[22]	Unmodified	Electrospun and melt mixing	0	1.85 ± 0.08	53.77 ± 1.18	4.93 ± 0.45
			1	2.01 ± 0.07	59.04 ± 1.94	5.33 ± 1.66
			2	2.16 ± 0.08	61.36 ± 1.59	3.33 ± 0.58
			3	2.16 ± 0.13	60.89 ± 4.59	4.44 ± 1.08
NFC from wood[77]	Unmodified nanocellulose with PLA latex	Vacuum filtration and compression moulding	0	2.5	28	1.6
			25	5	60	2.1
			60	10	137	4.2
			100	13	220	6.9

[a]BC functionalized with acetic, hexanoic and dodecanoic acid were termed C_2-BC, C_6-BC and C_{12}-BC, respectively.
[b]CNC silylated with n-dodecyldimethylchlorosilane was termed SCNC.
[c]PLA (100).
[d]PLA/PLA-MA/CNW (85/10/5).
[e]PLA/PLA-MA/PEG/CNW (70/10/15/5).
[f]Without surfactant.
[g]With 5 wt% surfactant.

with dodecanoic acid (C$_{12}$-BC) showed an improved reinforcing effect as compared to that esterified with acetic acid (C$_2$-BC).[35] Silanization,[42] silylation[43] and PEG grafting[45] of nanocellulose have also improved the mechanical properties of PLA. By using twin-screw extrusion, the dispersion of CNC in the PLA matrix was improved by the introduction of a surfactant,[54] which resulted in an increase in modulus, but no improvement was observed for tensile strength. This was then improved by the addition of maleic anhydride grafted PLA,[49] and the modulus, tensile strength and strain at break were increased from 2.9 ± 0.1 GPa, 40.9 ± 3.2 MPa and $1.9 \pm 0.2\%$ to 3.9 ± 0.3 GPa, 77.9 ± 6.7 MPa and $2.7 \pm 0.5\%$, respectively. Further addition of PEG as a compatibilizer resulted in an increase of strain to failure to $17.8 \pm 8.5\%$. PLA-*g*-MA has also significantly improved the mechanical properties of electrospun PLA/CNC fibrous scaffolds.[56] These results indicate that tailoring the interface interaction between PLA and nanocellulose is important for property enhancement of the biocomposites. Unmodified BC cellulose nanowhiskers have also been dispersed in PLA through electrospinning in solvent HFP, and subsequently melt compounded with PLA to improve mechanical properties.[22] By using a wet mixing method similar to papermaking, biocomposites with NFC content as high as 60 wt% have been prepared utilizing a PLA matrix in latex form.[77] The modulus was increased from 5 to about 10 GPa and the strength from 60 to 137 MPa going from 25 to 60 wt% NFC, indicating a good dispersion of the NFC and formation of a strong NFC network was obtained at high levels of reinforcement. However, the biocomposites became relatively more sensitive to defects with increasing NFC content.

9.5.3 Barrier Properties

Fukuzumi *et al.*[85] demonstrated that TEMPO-oxidized cellulose nanofibrils coated PLA film has excellent oxygen-barrier properties similar to synthetic polymer films such as poly(vinylidene chloride). The unmodified PLA film (25 μm thick) had an oxygen permeability of about 746 mL m^{-2} day^{-1} Pa^{-1}, which was decreased to 1 by a 0.4 μm thin coating layer of nanocellulose. PLA/nanocellulose biocomposites have also demonstrated water and oxygen barrier properties owing to the incorporation of nanocellulose that alters the tortuous paths for gas transport.[22] Sanchez-Garcia *et al.* reported that the water barrier properties of PLA biocomposites were only reduced by 10% in the sample containing 1 wt% of cellulose microfibres.[60] This was improved by nanofabrication, *i.e.* disintegration of cellulose microfibres to cellulose nanowhiskers (CNWs) with higher levels of crystallinity.[62] As compared to neat PLA, reduction of water permeability of *ca.* 64, 78, 82 and 81% were obtained for the solvent casted PLA biocomposites containing 1, 2, 3 and 5 wt% of CNWs, respectively, while the reductions of oxygen permeability were 83, 90, 90 and 88%. This effect was the result of a reduction in both solubility and diffusion coefficients due to the introduction of well-dispersed nanoscale cellulose crystallites.

9.6   Concluding Remarks

From an application point of view, the addition of nanocellulose to PLA biopolymers is an interesting possibility. Although PLA could be polymerized by ring-opening polymerization in the presence of nanocellulose, melt-processing appears to be the fastest route to commercial application of PLA/nanocellulose biocomposites. The shorter rod-like CNC nanoparticles are more convenient to use than longer swirled CNF nanofibrils, since the increase in viscosity is expected to be more limited for CNC (shorter aspect ratio, less tendency for physical entanglements). There are at least two potential benefits of CNC addition. One is increased crystallization rate, which may improve not only mechanical properties, but also thermal and chemical stability. The other effect is the reinforcement from high-modulus and high-strength CNC nanorods. In research, this is a major challenge since most studies make it impossible to distinguish between the two effects. CNC is added, properties are improved, but the degree of crystallinity is influenced and so are the crystallite morphologies. It thus becomes difficult to understand whether property improvements are primarily due to morphology changes or to CNC reinforcement effects.

In most studies, the nanocellulose content is fairly low, and thus the reinforcement effects can still be much improved. Increased nanocellulose content, and the use of MFC nanofibres rather than CNC are challenges primarily from the point of view of the processing approach. Since nanocellulose is rapidly becoming commercially available in large volumes it is likely that PLA-nanocellulose biocomposites have a bright future. As demonstrated, the properties of PLA can be improved strongly by the use of a biological nanoparticle. The surface characteristics of the nanocellulose in combination with the mixing technique are important in order to avoid nanoparticle agglomerates. This allows for higher nanocellulose content and a high degree of dispersion with corresponding property improvements.

References

1. X. A. Pang, X. L. Zhuang, Z. H. Tang and X. S. Chen, *Biotechnol. J*, 2010, **5**, 1125.
2. H. Tsuji, *Macromol. Biosci.*, 2005, **5**, 569.
3. L. Yu, K. Dean and L. Li, *Prog. Polym. Sci.*, 2006, **31**, 576.
4. N. Ljungberg and B. Wesslen, *Biomacromolecules*, 2005, **6**, 1789.
5. K. Das, D. Ray, I. Banerjee, N. R. Bandyopadhyay, S. Sengupta, A. K. Mohanty and M. Misra, *J. Appl. Polym. Sci.*, 2010, **118**, 143.
6. P. M. Chou, M. Mariatti, A. Zulkifli and S. Sreekantan, *Compos B Eng*, 2012, **43**, 1374.
7. N. Graupner, A. S. Herrmann and J. Mussig, *Compos. Appl. Sci. Manuf.*, 2009, **40**, 810.
8. A. Iwatake, M. Nogi and H. Yano, *Compos. Sci. Tech.*, 2008, **68**, 2103.

9. A. F. Turbak, F. W. Snyder and K. R. Sandberg, *J. Appl. Polym. Sci. Appl. Polym. Symp.*, 1983, **37**, 815.
10. M. Henriksson, G. Henriksson, L. A. Berglund and T. Lindstrom, *Eur. Polym. J.*, 2007, **43**, 3434.
11. L. Wagberg, G. Decher, M. Norgren, T. Lindstrom, M. Ankerfors and K. Axnas, *Langmuir*, 2008, **24**, 784.
12. T. Saito, Y. Nishiyama, J. L. Putaux, M. Vignon and A. Isogai, *Biomacromolecules*, 2006, 7, 1687.
13. A. H. Pei, N. Butchosa, L. A. Berglund and Q. Zhou, *Soft Matter*, 2013, **9**, 2047.
14. J. F. Revol, H. Bradford, J. Giasson, R. H. Marchessault and D. G. Gray, *Int. J. Biol. Macromol.*, 1992, **14**, 170.
15. P. B. Filson, B. E. Dawson-Andoh and D. Schwegler-Berry, *Green Chem.*, 2009, **11**, 1808.
16. S. Beck-Candanedo, M. Roman and D. G. Gray, *Biomacromolecules*, 2005, **6**, 1048.
17. I. Filpponen and D. S. Argyropoulos, *Biomacromolecules*, 2010, **11**, 1060.
18. J. Araki, M. Wada and S. Kuga, *Langmuir*, 2001, **17**, 21.
19. M. Salajkova, L. A. Berglund and Q. Zhou, *J. Mater. Chem.*, 2012, **22**, 19798.
20. L. Suryanegara, A. N. Nakagaito and H. Yano, *Cellulose*, 2010, **17**, 771.
21. L. Petersson and K. Oksman, *Compos. Sci. Tech.*, 2006, **66**, 2187.
22. M. Martinez-Sanz, A. Lopez-Rubio and J. M. Lagaron, *Biomacromolecules*, 2012, **13**, 3887.
23. L. C. Tome, R. J. B. Pinto, E. Trovatti, C. S. R. Freire, A. J. D. Silvestre, C. Pascoal Neto and A. Gandini, *Green Chem.*, 2011, **13**, 419.
24. F. Quero, M. Nogi, H. Yano, K. Abdulsalami, S. M. Holmes, B. H. Sakakini and S. J. Eichhorn, *ACS Appl. Mater. Interfaces*, 2010, **2**, 321.
25. S. Y. Cho, H. H. Park, Y. S. Yun and H.-J. Jin, *Macromol. Res.*, 2013, **21**, 529.
26. M. Bulota, A.-H. Vesterinen, M. Hughes and J. Seppala, *Polym. Compos.*, 2013, **34**, 173.
27. M. Bulota, S. Tanpichai, M. Hughes and S. J. Eichhorn, *ACS Appl. Mater. Interfaces*, 2012, **4**, 331.
28. M. Jonoobi, A. P. Mathew, M. M. Abdi, M. D. Makinejad and K. Oksman, *J. Polym. Environ.*, 2012, **20**, 991.
29. M. Bulota and M. Hughes, *J. Mater. Sci.*, 2012, **47**, 5517.
30. J. Dlouha, L. Suryanegara and H. Yano, *Soft Matter*, 2012, **8**, 8704.
31. N. Lin, J. Huang, P. R. Chang, J. Feng and J. Yu, *Carbohydr. Polym.*, 2011, **83**, 1834.
32. P. Tingaut, T. Zimmermann and F. Lopez-Suevos, *Biomacromolecules*, 2010, **11**, 454.
33. D. Y. Kim, Y. Nishiyama and S. Kuga, *Cellulose*, 2002, **9**, 361.
34. G. Frisoni, M. Baiardo, M. Scandola, D. Lednicka, M. C. Cnockaert, J. Mergaert and J. Swings, *Biomacromolecules*, 2001, **2**, 476.

35. K.-Y. Lee, J. J. Blaker and A. Bismarck, *Compos. Sci. Tech.*, 2009, **69**, 2724.
36. A.-L. Goffin, Y. Habibi, J.-M. Raquez and P. Dubois, *ACS Appl. Mater. Interfaces*, 2012, **4**, 3364.
37. A.-L. Goffin, J.-M. Raquez, E. Duquesne, G. Siqueira, Y. Habibi, A. Dufresne and P. Dubois, *Biomacromolecules*, 2011, **12**, 2456.
38. N. Lin, G. Chen, J. Huang, A. Dufresne and P. R. Chang, *J. Appl. Polym. Sci.*, 2009, **113**, 3417.
39. A. N. Frone, S. Berlioz, J.-F. Chailan and D. M. Panaitescu, *Carbohydr. Polym.*, 2013, **91**, 377.
40. A. N. Frone, S. Berlioz, J. F. Chailan, D. M. Panaitescu and D. Donescu, *Polym. Compos.*, 2011, **32**, 976.
41. J. M. Raquez, Y. Murena, A. L. Goffin, Y. Habibi, B. Ruelle, F. DeBuyl and P. Dubois, *Compos. Sci. Tech.*, 2012, **72**, 544.
42. P. Qu, Y. Zhou, X. Zhang, S. Yao and L. Zhang, *J. Appl. Polym. Sci.*, 2012, **125**, 3084.
43. A. H. Pei, Q. Zhou and L. A. Berglund, *Compos. Sci. Tech.*, 2010, **70**, 815.
44. C. Eyholzer, P. Tingaut, T. Zimmermann and K. Oksman, *J. Polym. Environ*, 2012, **20**, 1052.
45. S. Fujisawa, T. Saito, S. Kimura, T. Iwata and A. Isogai, *Biomacromolecules*, 2013, **14**, 1541.
46. D. M. Fox, J. Lee, C. J. Citro and M. Novy, *Polym. Degrad. Stabil.*, 2013, **98**, 590.
47. P. Qu, L. Bai, Y. Geo, G. Wu and L. Zhang, *Adv. Mater. Sci. Tech.*, 2011, **675–677**, 395.
48. P. Qu, Y. Gao, G.-F. Wu and L.-P. Zhang, *Bioresources*, 2010, **5**, 1811.
49. K. Oksman, A. P. Mathew, D. Bondeson and I. Kvien, *Compos. Sci. Tech.*, 2006, **66**, 2776.
50. E. Fortunati, I. Armentano, Q. Zhou, D. Puglia, A. Terenzi, L. A. Berglund and J. M. Kenny, *Polym. Degrad. Stabil.*, 2012, **97**, 2027.
51. E. Fortunati, M. Peltzer, I. Armentano, L. Torre, A. Jimenez and J. M. Kenny, *Carbohydr. Polym.*, 2012, **90**, 948.
52. E. Fortunati, I. Armentano, Q. Zhou, A. Iannoni, E. Saino, L. Visai, L. A. Berglund and J. M. Kenny, *Carbohydr. Polym.*, 2012, **87**, 1596.
53. L. Petersson, I. Kvien and K. Oksman, *Compos. Sci. Tech.*, 2007, **67**, 2535.
54. D. Bondeson and K. Oksman, *Compos. Interfac.*, 2007, **14**, 617.
55. F. Quero, S. J. Eichhorn, M. Nogi, H. Yano, K.-Y. Lee and A. Bismarck, *J. Polym. Environ.*, 2012, **20**, 916.
56. C. Zhou, Q. Shi, W. Guo, L. Terrell, A. T. Qureshi, D. J. Hayes and Q. Wu, *ACS Appl. Mater. Interfaces*, 2013, **5**, 3847.
57. J. Hong and D. S. Kim, *Polym. Compos.*, 2013, **34**, 293.
58. M. Avella, G. Bogoeva-Gaceva, A. Bularovska, M. E. Errico, G. Gentile and A. Grozdanov, *J. Appl. Polym. Sci.*, 2008, **108**, 3542.
59. Z. Li, X. Zhou and C. Pei, *Int. J. Polym. Anal. Char.*, 2010, **15**, 199.
60. M. D. Sanchez-Garcia, E. Gimenez and J. M. Lagaron, *Carbohydr. Polym.*, 2008, **71**, 235.

61. K. M. Z. Hossain, I. Ahmed, A. J. Parsons, C. A. Scotchford, G. S. Walker, W. Thielemans and C. D. Rudd, *J. Mater. Sci.*, 2012, **47**, 2675.
62. M. D. Sanchez-Garcia and J. M. Lagaron, *Cellulose*, 2010, **17**, 987.
63. J. K. Pandey, C. S. Lee and S.-H. Ahn, *J. Appl. Polym. Sci.*, 2010, **115**, 2493.
64. L. Suryanegara, A. N. Nakagaito and H. Yano, *Compos. Sci. Tech.*, 2009, **69**, 1187.
65. M. Li, D. Li, Q. Deng, D. Lin and Y. Wang, *Appl. Mech. Mater.*, 2012, **174–177**, 885.
66. M. Jonoobi, J. Harun, A. P. Mathew and K. Oksman, *Compos. Sci. Tech.*, 2010, **70**, 1742.
67. K. Okubo, T. Fujii and E. T. Thostenson, *Compos. Appl. Sci. Manuf.*, 2009, **40**, 469.
68. E. Fortunati, I. Armentano, A. Iannoni and J. M. Kenny, *Polym. Degrad. Stabil.*, 2010, **95**, 2200.
69. D. H. Kim, H. J. Kang and Y. S. Song, *Carbohydr. Polym.*, 2013, **92**, 1006.
70. H. Xu, C.-Y. Liu, C. Chen, B. S. Hsiao, G.-J. Zhong and Z.-M. Li, *Biopolymers*, 2012, **97**, 825.
71. E. E. M. Ahmad and A. S. Luyt, *Polym. Compos.*, 2012, **33**, 1025.
72. B. Wang and M. Sain, *Bioresources*, 2007, **2**, 371.
73. A. Lovdal, L. L. Laursen, T. L. Andersen, B. Madsen and L. P. Mikkelsen, *J. Appl. Polym. Sci.*, 2013, **128**, 2038.
74. S. Siengchin, T. Pohl, L. Medina and P. Mitschang, *J. Reinforc. Plast. Compos.*, 2013, **32**, 23.
75. D. Liu, X. Yuan and D. Bhattacharyya, *J. Mater. Sci.*, 2012, **47**, 3159.
76. Q. Shi, C. Zhou, Y. Yue, W. Guo, Y. Wu and Q. Wu, *Carbohydr. Polym.*, 2012, **90**, 301.
77. K. Larsson, L. A. Berglund, M. Ankerfors and T. Lindstrom, *J. Appl. Polym. Sci.*, 2012, **125**, 2460.
78. A. N. Nakagaito, A. Fujimura, T. Sakai, Y. Hama and H. Yano, *Compos. Sci. Tech.*, 2009, **69**, 1293.
79. T. Wang and L. T. Drzal, *ACS Appl. Mater. Interfaces*, 2012, **4**, 5079.
80. D. T. Quillin, D. F. Caulfield and J. A. Koutsky, *J. Appl. Polym. Sci.*, 1993, **50**, 1187.
81. D. G. Gray, *Cellulose*, 2008, **15**, 297.
82. R. Kose and T. Kondo, *J. Appl. Polym. Sci.*, 2013, **128**, 1200.
83. T. Y. Ke and X. Z. Sun, *J. Appl. Polym. Sci.*, 2003, **89**, 1203.
84. M. Day, A. V. Nawaby and X. Liao, *J. Therm. Anal. Calorim.*, 2006, **86**, 623.
85. H. Fukuzumi, T. Saito, T. Wata, Y. Kumamoto and A. Isogai, *Biomacromolecules*, 2009, **10**, 162.

PLA Main Applications

CHAPTER 10

PLA and Active Packaging

RAMÓN CATALÁ, GRACIA LÓPEZ-CARBALLO,
PILAR HERNÁNDEZ-MUÑOZ AND RAFAEL GAVARA*

Packaging lab, Instituto de Agroquímica y Tecnología de Alimentos,
IATA-CSIC, Av. Agustín Escardino 4, 46980 Paterna (Valencia), Spain
*Email: rgavara@iata.csic.es

10.1 Introduction to Active Packaging

The increasing demand for healthier and more nutritional food has prompted the development of new conservation and packaging technologies. Innovation is continuous, as a result of the great efforts and resources that the packaging producer and user sectors invest in research and development. New materials, techniques and production equipment, as well as specific packaging processes, aim to offer solutions to the demands raised by social and legislative changes, of which the consumer is without a doubt the protagonist who also stands at the centre of the system addressed by food production, without forgetting the interests of the food production and commercialization sectors.

The new paradigm for packaging is their active and intelligent behaviour, understanding as such the active participation in maintaining or even improving the quality and shelf-life of packaged food. This is where concepts such as "active packaging" and "intelligent packaging" arise; terms that refer to methods of packaging that go further than the conventional objectives of containing, protecting, presenting and providing information about the food.

RSC Polymer Chemistry Series No. 12
Poly(lactic acid) Science and Technology: Processing, Properties, Additives and Applications
Edited by Alfonso Jiménez, Mercedes Peltzer and Roxana Ruseckaite
© The Royal Society of Chemistry 2015
Published by the Royal Society of Chemistry, www.rsc.org

Active/intelligent packaging attempts to strengthen or take advantage of interactions in the food/packaging/environment system, acting in a co-ordinated manner to improve healthiness and quality of the packaged food and increase its shelf-life. That is, the packaging goes from being a mere container to playing an active part in maintaining or even improving the quality of packaged food. In this way, active/intelligent packaging can be considered an emerging technology in food conservation.[1]

Article 3 of regulation (EC) N° 450/2009 defines *active materials and objects* as those aimed at prolonging shelf-life or improving the condition of packaged food. They are designed to intentionally incorporate components that release substances into the packaged food or its environment, or absorb substances from the food or its environment.

Currently, with the redefinition of the concept of active packaging and its acceptance by international sanitary legislation, designing packaging materials and technologies suited to the consumer market and the different products can be considered, taking into account that all foodstuffs have a specific deterioration mechanism.[2]

Many diverse forms of active packaging to control different deterioration or food quality alteration problems have been proposed, such as controlling gases in the packaging interior–oxygen, carbon dioxide, ethylene *etc.*–regulating humidity, adding chemical preservatives, incorporating aromas, eliminating strange odours and undesired substances or controlling micro-biological contamination. At the same time, systems that are considered intelligent have been developed, such as time/temperature indicators, changes in gaseous composition indicators or data carrier systems that allow for control of the medium and the packaged product, to facilitate decision making that permits extending the shelf-life, increasing safety, improving quality or preventing possible problems.[3]

The diverse active packaging technologies incorporate some kind of agent into the packaging that has to retain, react with or release a specific component whose presence or absence is critical for maintaining the quality of the packaged product. Therefore, it is necessary that this component inter-changes between the food and the packaging with adequate kinetics.

The basis of using polymeric materials for the development of active packaging is to take advantage of the mass transfer processes that occur in these materials.[4] Polymeric materials are not totally inert and interactions take place with the food or the atmosphere that include transport of gas, water vapour and other molecules with low-molecular-weight such as food aroma components, or residues and additives of the packaging material, that can produce chemical, nutritional and sensory changes in the packaged food.[5] These mass transfer processes are considered negative aspects of plastic materials because they can compromise the quality and safety of the packaged food. Nevertheless, it is also possible to capitalize on them in order to benefit the packaged product by obtaining a positive contribution from using these mechanisms as the basis for developing active packaging systems.

Active packaging can be obtained by diverse means, but there are basically two methods for their design:

- Introducing the active agent into the interior of the packaging in an independent device, together with the packaged product;
- Making the active agent a part of the packaging material.

Since the start of the development of these techniques, the way to introduce the agent has been to use small sachets constructed with some kind of polymeric material sufficiently permeable to allow release and/or performance of the active agent, but also to prevent it coming into direct contact with the food. They must therefore be very resistant to breaks, and be suitably labelled to avoid accidents by ingestion of the contents. Nevertheless, current developments head more towards introducing the active agent into the packaging material itself, a more appealing and safer way for the consumer, who does not encounter anything unfamiliar in the inside of the packaging or close to the product that draws his/her attention and/or makes him/her doubt the sanitary quality of the foodstuff. This method also simplifies the packaging technology needed by eliminating the process of introducing elements with the active agent into the inside of the packaging.[6]

Incorporating the active agent directly into the packaging material has the additional advantage of the compound being able to contact directly with the foodstuff on the whole of its surface (direct contact), and not only in a restricted area as is the case with sachets. However, there are times when the active agent is only in contact with the atmosphere surrounding the food (headspace), making the contact between the packaging and the food of the indirect type. For direct contact films, usually films or coatings are used where the active agent can be volatile or not, in case its migration to the foodstuff is desired. By contrast, for indirect contact only volatile active agents can be used, given that, as Appendini and Hotchkiss show,[7] these compounds can vaporize in the headspace after packaging, reach the surface of the foodstuff and be absorbed by it, whereas non-volatile active agents can only migrate to the foodstuff by direct contact.

Figure 10.1 describes the above-mentioned processes. In the case of a non-volatile agent, the substance molecules diffuse through the polymer matrix until they reach the interface food-package where they are partitioned, the extent of which depends on the chemical compatibility of the active molecules with food and polymer. Molecules that are released in the food product diffuse within the bulk of the product by diffusion or dispersion processes depending on the product texture and morphology. No release can be expected through the surface area if there is no direct food/package contact. When the agent molecules are volatile, the release of the food towards the package headspace is also feasible by surface evaporation. The agent can exert its activity from the gas phase or can eventually be solubilized on the food surface and diffuse into the bulk. For indirect contact, active films, conventional films with an active coating or multi-layer films where

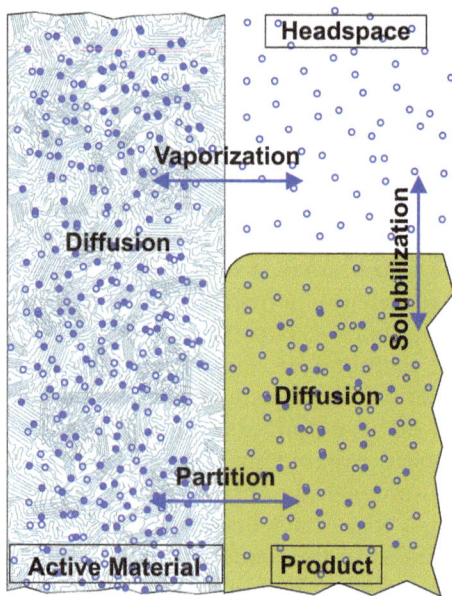

Figure 10.1   Processes involved in mass transport of active agents in a food/
packaging system. (●), non-volatile agent; (○), volatile agent.

one of the layers contains the active agent and the layer in contact with the
food has to be permeable to the compound are normally used.[8]

Polyolefins (polyethylenes and polypropylene) are the most used polymers
for those applications that do not have specific demands for gas or vapour
barriers. When higher impermeability of the packaging material is required,
the alternatives are polyesters and polyamides and above all multi-layer
structures that include high barrier materials, such as ethylene vinyl alcohol
(EVOH) or metalized copolymers. As an alternative to the current conventional polymers, all of them originating from petrol, polymers derived
from renewable sources are receiving increasing attention. Biopolymers
obtained directly from biomass, synthesized from sustainable monomers or
produced by micro-organisms are already found as packaging materials or
food coatings. Edible biopolymers have gone from only being used as
coatings of some fruits to being the carrier vehicle of nutrients, antioxidants,
antimicrobial agents, *etc.*, because diffusion processes are easier to control
in them than in synthetic films. These materials can be biodegradable, and
many of them allow similar or superior physicochemical and microbiological control of the food compared to conventional plastics. Without a
doubt, biopolymers are a good base for the development of active packaging
and coatings, permitting the slow release and incorporation of the active
agents into the food so as to control its deterioration and loss of quality and
lengthen its commercialization time with full quality and safety guarantees.[9]

Many studies have been carried out in which poly(lactic acid) (PLA) homo-
and copolymers are used as a vehicle for the release of molecules of interest

for various applications.[10–13] Among these, a highlight is lactic acid-glycolic acid copolymer capsules for medical and pharmaceutical uses,[14–23] where these copolymers are used for their biocompatibility. These applications are not the focus of this chapter, which is dedicated to active packaging. In this line of research, works have been published demonstrating the possibilities of PLA as a base material for the development of active packaging, particularly for the control of microbial growth or oxidation problems, which are commented on below.

10.2 Antimicrobial Active Packaging

Microbial growth on packaged foods is the primary cause of food spoilage. Diverse yeast, moulds and bacteria can be involved in the deterioration of food products, although their implication in spoilage depends on pH, water activity, headspace composition, temperature and food processing and packaging technology.

Sterilization and pasteurization processes traditionally used for the elimination of pathogenic bacteria are partially being replaced by emerging non-thermal technologies that offer fresh-like products to the consumer. Since these technologies do not individually fully eliminate micro-organisms, their combination (called hurdle technology) can improve food quality and safety. In this sense, antimicrobial active packaging is one of these hurdles. In recent years, most efforts in active packaging development are focused on active antimicrobial packaging systems.

Active antimicrobial packaging can be defined as a packaging system that interacts with the food product or the surrounding headspace to kill micro-organisms or to reduce, inhibit or retard the growth of micro-organisms that might be present in the food product or in the food package.[24] Although initial active packaging systems contain active agents in an independent device (sachet, pad, *etc.*), present trends are focused on their incorporation into the package walls of flexible polymeric packages as described in Figure 10.2. There are diverse activity mechanisms mainly dependent on the active agents and on the polymer matrix. Some present their activity by direct interaction with the micro-organisms from the package wall, that is, agent release is not required; the agent molecules need to be present on the package surface and chemically anchored to the polymer. In others, release at an adequate pace is a requirement. The easiest way to incorporate the agent is by surface spreading of the agent on the package surface by spraying or by wetting with an appropriate agent solution, as such the release is practically immediate and only depends on the agent solubility in the food. Additionally, problems of set-off during film storage are common. In order to obtain some control over the agent release, the agent should be incorporated in the bulk of a polymer matrix. In this case, the release of the agent is partially controlled by agent diffusion within the amorphous regions of the polymer phase. The incorporation into a polymer matrix can be done in a single layer by casting or extrusion processes or in the outermost thin

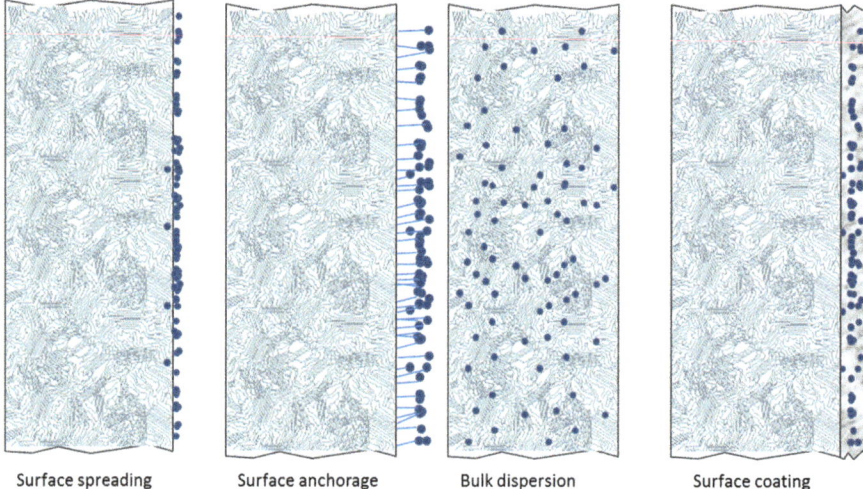

| Surface spreading | Surface anchorage | Bulk dispersion | Surface coating |

Figure 10.2   Methods for the incorporation of active agents into the walls of an active food package.

layer of a structure which can be produced by coextrusion, lamination or printing technologies (gravure or flexography). When the agent is not volatile (as described previously in Figure 10.1), direct contact between the food and the package surface is a requirement.

In comparison to the direct addition to food of antimicrobial substances, where the agent may migrate to the bulk of the food product and may reduce its activity during food processing, the use of active packaging systems that release the agent into the headspace or onto the food surface can improve their efficiency and reduce the amount of agent added to the packaged product.

10.2.1   Antimicrobial Agents

The food-grade condition of an antimicrobial compound is an essential requirement for its incorporation into an active packaging system. Depending on its origin, antimicrobials can be classified into two major groups: chemical agents (triclosan benzoates, sorbates, *etc.*) and natural agents (components of plant extracts and essential oils, bacteriocins and bio-preservatives). Antimicrobials may also be classified by their mode of action in the packaging system as: i) volatile (chlorine dioxide, carvacrol, allyl isothiocyanate, thymol), able to exert their activity from the gas phase and therefore direct surface contact between food and active film is not needed; and ii) non-volatile, including organic acids and salts (benzoates and sorbates), inorganic substances (clays and silver), proteins (enzymes and bacteriocins) and polymers (chitosan). In this class, agents can be released into food by direct contact and increase food resistance to microbial spoilage or they act only from the polymer surface without agent delivery and therefore only act against micro-organisms that contact the package surface. An exhaustive list and description of food antimicrobials can be consulted in the literature.[25]

10.2.2 Incorporation of the Agent into the Polymer Matrix

In order to exert the antimicrobial action by direct contact with the micro-organisms or *via* its release from the walls into the food or headspace, the agent has to be fixed to the polymeric surface or incorporated in the polymeric matrix. Non-volatile antimicrobials can be immobilized onto the polymeric surface, exerting their antimicrobial activity to the target micro-organisms by direct contact with the food product. Common procedures include surface treatment to add anchorage sites and bonding to these active sites of the agent.

To introduce the antimicrobial into the bulk of the polymeric matrix, the selected manufacturing procedure should take into consideration the type of polymer and the characteristics of the antimicrobial agents, especially heat resistance and mechanism of action. From a technological point of view, the agent is mixed with the polymer by dissolution of both into an appropriate solvent followed by evaporation of the solvent (casting or coating technologies), or by polymer melting and incorporation and mixing of the agent in the melt (extrusion). Both techniques are directly applicable to the manufacture of food packages through conventional procedures.

10.2.3 Antimicrobial PLA-based Packaging Developments

Antimicrobial active materials are being studied by using innumerable polymeric materials, including PLA. In the next subsections, a description of the most relevant studies is presented. Table 10.1 lists the antimicrobial agents, the method of incorporation and the application or relevant results in these reports. The review has been divided and organized by considering the processing technology used for the addition of the active agent.

10.2.3.1 Coating onto PLA Films

An antibacterial active film was developed by lauroyl arginine ethyl ester (LAE)-coating onto a previously corona surface activated PLA film. The presence of anchored LAE molecules was confirmed by scanning electron microscopy (SEM). The antibacterial activity of the LAE/PLA films against *Listeria monocytogenes* and *Salmonella typhimurium* was confirmed *in vitro* by the agar diffusion assay and *in vivo* on contaminated cooked ham slices. Films also showed their antimicrobial efficiency in the quantitative JIS Z 2801:2000 method of analysis of antimicrobial surfaces.[26]

Table 10.1 Antimicrobial active materials.

Antimicrobial agent	Incorporation method	Properties and use	Ref.
AIT	Electrospinning	Humidity activated the release of AIT	48
Carvacrol, AIT	Extrusion	Active against *Botrytis cinerea* during storage and in high-pressure treatments	42
Chitosan	Casting	Antifungal activity against *Fusarium proliferatum, Fusarium moniliforme* and *Aspergillus ochraceus*	37
Chlorine dioxide	Casting	Antifungal activity against *Salmonella* spp. and *Escherichia coli*	40
Eugenol and *trans*-cinnamaldehyde	Emulsion evaporation	Active against *Listeria* spp. and *Salmonella* spp.	74
Lactic acid, EDTA, sodium benzoate, potassium sorbate	Polymeric coating	Active against *Escherichia coli* and *Salmonella* Stanley in apples	28
Lauric arginate (LAE)	Surface anchorage after coronas discharge	Active against *Listeria monocytogenes, Salmonella typhimurium*	26
Lemon extract, thymol, lysozyme	Extrusion	Active against *Pseudomonas* spp. and *Micrococcus lysodeikticus*	41
Nisin	Casting	Active against *Listeria monocytogenes, Escherichia coli* and *Salmonella Enteritidis in vitro* and in liquid food	29
Nisin	Polymeric coating	Active against *Listeria monocytogenes* in milk and egg white	30
Nisin	Diffusion	Active against *L. plantarum*	49
Nisin, AIT	Polymeric coating	Active against *Listeria monocytogenes* and *Salmonella spp*	31
Nisin, EDTA,	Extrusion	Active against *Escherichia coli* and *Listeria monocytogenes*	44,45
Nisin, EDTA, sodium benzoate, potassium sorbate	Polymeric coating	Active against *Escherichia coli* and habitual microflora in strawberry fruit	27
Organomodified montmorillonite	Casting	Active against *Listeria monocytogenes*	39
Propolis	Casting	Characterization of release into simulants	36
Silver ions	Casting	Activity against *Salmonella* and feline calicivirus	34
Silver-nanoclay	Casting	Active against *Salmonella spp*	33
Silver nanoparticles	Casting	Silver release increases with the addition of cellulose nanocrystals to PLA	35
Silver nanoparticles	Extrusion	Film characterization	46
Silver zeolites	Casting and extrusion	Cast films were active against pathogen bacteria while extruded films were not	32
Trans-2-hexenal	Extrusion	Thermal degradation of the agent reduces effect on moulds	43
Triclosan	Electrospinning	Active against *Staphylococcus aureus* and *Escherichia coli*	47

10.2.3.2 Polymeric Coatings and Film Casting

Nisin, ethylenediaminetetraacetic acid (EDTA), sodium benzoate, potassium sorbate or combinations of them were incorporated into PLA films by solution casting procedure. Their antimicrobial properties were tested against *E. coli* O157:H7 and habitual microbiota in strawberry puree. The films containing a mixture of all preservatives presented the best results, followed by the combination of the two organic salts. These two films presented higher efficiency than direct addition of the mixed preservatives.[27]

Golden Delicious apples were successfully spray-coated with PLA containing lactic acid, EDTA, sodium benzoate, potassium sorbate or a combination of them. Their antimicrobial properties were tested against *E. coli* O157:H7 and *Salmonella Stanley* previously inoculated in apples. Although this system has no direct application due to the non-edible characteristics of the coating or the inappropriateness of using methylene chloride as a solvent, the final coating showed very efficient antimicrobial protection, not observed in the control coatings, especially coatings containing lactic acid and sodium benzoate.[28]

Nisin was incorporated into a PLA solution in dichloromethane (DCM) and films were obtained by casting. Scanning electron microscopy (SEM) and confocal laser microscopy showed that the agent was evenly distributed in the polymeric matrix. A preliminary study of the release kinetics proved that nisin was able to diffuse through the agar and inhibit the growth of *L. monocytogenes*. Films were tested *in vitro* against *L. monocytogenes*, *E. coli* and *S. enteritidis* and in contaminated orange juice and white liquid egg with good results.[29] This polymeric solution was used to coat the internal surface of glass bottles. Listeria contaminated skimmed milk and white eggs were packaged in the coated bottles. Compared to control samples, the active coating containing 25% of nisin completely inhibited the growth of the pathogen bacteria at 4 and 10 °C.[30]

Equal parts of allyl isothiocyanate (AIT), nisin and PLA were dissolved in DCM and applied as a coating onto diverse extruded materials, including extruded PLA, to manufacture active materials by coating. Samples were immersed in a culture medium inoculated with *L. monocytogenes* or *Salmonella spp.* Results showed that even at these high concentrations, samples with AIT presented bacteriostatic activity while samples containing nisin presented some bacteriostatic effect against *Listeria* but they were not effective against *Salmonella* due to the low activity of this antimicrobial against Gram-negative bacteria.[31] The study also reports on the synergistic effect of both antimicrobials.

Silver ion migration and antimicrobial activity of PLA/silver zeolite composites were investigated. Films prepared by solution-casting/solvent evaporation presented an extended release of silver ions into food simulants and TSB (tryptone soy broth). Antimicrobial activity against *Staphylococcus aureus* and *E. coli* was significant, although the concentration of silver in simulants exceeded the legal limit.[32]

A silver-containing organomodified montmorillonite was incorporated into PLA films by casting procedure from a chloroform (CF) dispersion. The films presented good physical properties and improved barrier properties and when tested against *Salmonella spp*, samples containing concentrations above 1% presented bactericidal activity. The mechanism of action was the release of silver ions as demonstrated in migration studies. In these studies, the food simulants presented silver concentrations above the legal limit.[33] These authors also developed materials by silver ion incorporation into PLA by casting from dimethylformamide (DMF) and tetrahydrofuran (THF) solutions. Film activity was analyzed against *Salmonella* and feline calicivirus *in vitro* and *in vivo* and results showed that the activity of the film is based on the release of silver ions. Successive washing reduced the activity of the films because of silver concentration depletion. The *in vivo* tests with lettuce and paprika showed that the efficiency of the films is significantly reduced by matrix effects and exposure temperature.[34]

Silver nanoparticles were incorporated into PLA-based films which were obtained by casting from a CF solution. To accelerate the release of silver, composites were prepared by adding pure or surfactant-modified cellulose nanocrystals. The TEM images showed that silver and cellulose particles were distributed homogenously, results which were confirmed by the improvement of barrier properties to oxygen and water. However, the migration from the films and the release of silver increased with the presence of cellulose, especially when cellulose was modified with a surfactant. A potential interaction between the surfactant and PLA that resulted in degradation of the polymeric matrix might be the cause of such an effect. This report did not include the antimicrobial activity analysis of the prepared films.[35]

Propolis was incorporated into PLA films by casting. To modify the agent release, a plasticizer (polyethylene glycol, PEG) and a nanoclay (bentonite) were also added. The release of four propolis polyphenols was fully characterized in terms of release extent into food simulants and diffusion within the polymeric matrix. The presence of the modifiers largely increased the release of the agents.[36] No microbiological tests were reported.

Chitosan/PLA biocomposites plasticized with PEG were prepared by casting from a mixture of previous solutions in acetic acid and in CF, respectively. Due to the lack of compatibility between polymers and solvents, films were heterogeneous and presented deficient physical properties. Nevertheless, all developed materials presented inhibitory properties against three mycotoxinogen strains, *Fusarium proliferatum*, *Fusarium moniliforme* and *Aspergillus ochraceus*.[37]

Clay/PLA nanocomposites were prepared by casting method. The tested clays were montmorillonite and two organomodified clays with two quaternary ammonium salts which had exhibited antimicrobial characteristics in chitosan films.[38] In general, the films presented improved physical properties but only films containing Cloisite™30B presented some antimicrobial activity against *L. monocytogenes*. No effect was observed in other micro-organisms. This lack of effect was attributed to the hydrophobic characteristics of PLA.[39]

Sodium chlorite and citric acid powder were incorporated into PLA films at different concentrations and salt ratios to produce antimicrobial films based on the release of chlorine dioxide (ClO_2).[40] Moisture exposure was found to act as triggering mechanism of the activity. The higher the concentration of the salts the higher the gas release with the ratio of salts being an irrelevant factor. Tests carried out on the storability of grape tomato used as a model food showed that the moisture generated by the product was sufficient to trigger the release and that the accumulation of gas in the package was effective in the reduction of *Salmonella spp.* and *Escherichia coli* in previously inoculated tomatoes.

10.2.3.3 Extrusion

Lemon extract, thymol and lysozyme were incorporated into PLA by extrusion at three concentrations and the obtained films were compared to samples obtained through the same process but using poly(ε-caprolactone) (PCL) and low density polyethylene. The antimicrobial activity of the materials containing lemon extract and thymol was assessed against *Pseudomonas spp.*, by turbidimetric tests. The best results were obtained for the PCL-based materials. The antimicrobial activity of materials containing lysozyme was assessed against *Micrococcus lysodeikticus*, by immersing the samples in a phosphate buffer. Data showed that the best results were obtained for the PCL-based materials extruded at 80 °C. Although the actual concentration of the agents in the films was not measured, the authors hypothesized that the loss of agents induced by degradation/evaporation due to the thermal process is the cause of low efficiency observed in the PLA materials.[41]

Carvacrol and AIT were incorporated into PLA by extrusion and the resulting composite was converted into film shape by hot-press moulding. To reduce agent loss by thermal degradation and evaporation, the agents were previously protected by the preparation of inclusion complexes into β-cyclodextrin (βCD). Only the materials prepared with AIT showed activity against *B. cinerea*. The incorporation of AIT as an inclusion complex reduced the release rate of the agent but also increased the activity of the film against the tested mould. According to the report, this proves that the use of βCD complexes reduces AIT degradation and maintains the molecule active, the diffusion rate being less relevant in the protection of a food simulant.[42]

Trans-2-hexenal inclusion complexes in βCD were used as an antifungal agent. These inclusion complexes were extruded, converted into pellets and extruded into flat films. The effect of powder, pellets and films on the growth of several spoilage moulds was analyzed. Powder and pellets proved to inhibit mould growth of *Alternaria solani*, *Aspergillus niger*, *Botrytis cinerea*, *Colletotrichum acutatum* and *Penicillium sp*. However, the extruded pellets did not show any fungistatic effect, indicating that the agent was degraded during extrusion.[43]

Nisin was included in PLA films by extrusion. Since nisin is deactivated at temperatures above 120 °C, Liu *et al.* previously modified PLA with diverse plasticizers and melt-mixed the bacteriocin at 120 °C. Films obtained with lactide and glycerol triacetate maintained the activity of the agent as confirmed by microbiological tests against *Listeria monocytogenes.*[44] Following the same procedure, nisin and EDTA were incorporated into PLA films by extrusion with glycerol triacetate as a plasticizer. The obtained films showed poor compatibility between agents and polymer. The films were tested against *E. coli* O157:H7 by immersion of the samples in a contaminated liquid broth. The films containing the mixture of both agents were the most efficient materials, although their activity was essentially bactericidal.[45]

Silver nanoparticles and microcrystalline cellulose were incorporated into PLA by film extrusion. The obtained composites presented improved mechanical and barrier properties although, optically, the presence of the cellulose reduced transparency. Although the objective is the development of an antimicrobial composite, no tests on antimicrobial activity of the films or monitorization of silver release were reported.[46]

Silver ion migration and antimicrobial activity of PLA/silver zeolite composites were also investigated. Films prepared by extrusion were compared to those prepared by casting. The denser morphology obtained resulted in a large decrease in silver ions release, which eventually caused non-effective antimicrobial activity against *Staphylococcus aureus* and *Escherichia coli*, in contrast with what was observed for cast films.[32]

10.2.3.4   Electrospinning

Triclosan was incorporated into PLA nanofibres by electrospinning methodology. Triclosan was directly added to the fibre-forming PLA solution or a previous cyclodextrin (CD) inclusion complexation was prepared and then added to the PLA solution. The antimicrobial activity of the fibres was analyzed against *S. aureus* and *E. coli* in *in vitro* tests using a disc agar diffusion method. All films presented antimicrobial activity although the inclusion of triclosan encapsulated into βCD and γCD increased its solubility in the agar which increased the inhibition halo.[47]

Similarly, AIT was incorporated in electrospun fibres at diverse concentrations. The study proved that the shape and diameter of fibres were modified with the concentration of AIT. At high concentrations, PLA fibres become thinner while the AIT appeared to be accumulated in spherical particles. The release of AIT was monitored as a function of the relative humidity (RH) and it was observed that humidity activates the release by accelerating the process and increasing its extent.[48]

10.2.3.5   Other Procedures

Nisin was also incorporated by diffusion into a pectin/PLA biopolymer composite by extrusion. The previously extruded pectin and PLA composite had a heterogeneous structure and a high hydrophilic character. Taking advantage of this property, composite films were immersed in a nisin solution and, after drying, submitted to microbiological tests. Composite samples loaded with nisin showed antimicrobial activity against *Lactobacillus plantarum* in both agar diffusion and liquid culture media tests.[49]

10.3 Antioxidant Active Packaging

Oxidation problems are, after microbial growth, the main cause of food alteration. In particular, food with a high content of lipids, and especially those that present a high degree of unsaturation, are susceptible to deterioration by this mechanism. Oxidation can occur spontaneously or be induced by factors like light, temperature, presence of metals or enzymes, *etc*. With oxidation, anomalous odours and flavours appear, the colour and texture of the food are altered and nutritional value is reduced as some vitamins and polyunsaturated fatty acids are lost. Furthermore, the products formed by oxidation can occasionally be harmful to health.[50,51] The addition of antioxidants like dodecyl gallate, butylated hydroxytoluene (BHT) or butylated hydroxyanisole (BHA) to fatty products is common practice when preparing these products, even though their effectiveness is questionable.[9]

To control oxidation of packaged food, the reduction or elimination of the oxygen present is basic, as is that of highly reactive species, such as superoxide, hydroxyl, singlet oxygen, *etc*., which are generated by different mechanisms and are implicated in oxidation reactions of lipids and other food components. Controlling the presence of oxygen in the packaging can be achieved by combining vacuum packaging with the use of high barrier materials, although this may not be sufficient to eliminate all the present oxygen, due to either its residual presence or permeation from the exterior through the packaging wall. To overcome this problem, oxygen-absorbing substances (oxygen scavengers) (iron, ferrous oxide, ascorbic acid, enzymes, *etc*.) are introduced into the packaging, together with the food in sachets or permeable labels that isolate it from the food, or in the packaging wall as part of the material.[52]

Another alternative strategy to the direct addition of antioxidants to the food is to include compounds with antioxidant capacity in the packaging wall so that when they migrate to the food they slow down lipid oxidation processes. These antioxidant compounds can act in a direct way by capturing free radicals, or indirectly by chelating oxidation catalyzing metals like iron and copper for example.[52] In fact, antioxidants are additives commonly used to prevent thermal degradation of polymers during processing. The migration of these additives to the food can also carry out the role of antioxidant in the food itself. With this in mind, the incorporation of antioxidants in packaging polymers with a view to their release into the food is receiving attention and some technologies have already been successfully developed and commercialized.[9]

10.3.1   Antioxidant Agents

In the first published works, synthetic antioxidants like BHT, BHA, PEG, *etc.* were studied. Nevertheless, their presence in food is questionable due to potential risks and they require strict legislative control. An alternative that is being studied widely is the use of phenolic type natural antioxidants from plant species, including both integral extracts obtained by diverse methods and their purified components (such as catechin, quercetin, caffeic acid, *etc.*) obtained by chemical synthesis.

10.3.2   Incorporation of the Agent into the Polymer Matrix

To provide the antioxidant activity, the agent is previously incorporated into the package walls wherefrom it reduces the presence of radical oxygen species from the surrounding headspace, reducing the initiation of peroxidation reactions. Another mechanism of action is the agent release into the headspace and/or food surface where it provides the food with antioxidant protection.[53] The methods of incorporation are similar to those already mentioned in Section 10.1.2 and described in Figures 10.1 and 10.2.

10.3.2.1   *Antioxidant PLA-based Packaging Developments*

All these active systems to control oxidation are based on conventional polymers as base materials, the majority of commercial developments being constituted by polyolefins. Only recent work has started with biopolymers and, although prospects are excellent, results are still scarce. Without doubt, among these PLA is one of the biopolymers that has received most attention and a number of works have been published using the same guidelines and studying the same antioxidants as those developed with conventional polymers. In the next subsections, a description of the most relevant studies is presented. Table 10.2 lists the antioxidant agents, the incorporation method and the application or relevant results reported in these works. As in the antimicrobial section, the review has been divided and organized by the processing technology used for the addition of the agent.

An active packaging system based on PLA was reported by Holm *et al.*[54] This study reports the stability against oxidation of semi-hard cheese packaged in a PLA container that included an oxygen scavenger device. Although this is not an antioxidant active material development, the report showed that the food product improved its stability when stored in active packages.

10.3.2.2   *Polymeric Coatings and Film Casting*

Much closer to the objective of this chapter, Van Aardt *et al.*[55] studied the release of antioxidants in dairy products and liquid food simulants in

Table 10.2 Antioxidants active materials.

Antioxidant agent	Incorporation method	Properties and use	Ref.
Iron-based O$_2$ scavenger	Sachets	Improved stability of hard cheese	54
α-TOC, BHT, BHA	Casting	Only BHT release in water. No antioxidant release in oil	55
α-TOC, BHT	Extrusion	Antioxidant activity, evaluated with DPPH method	61
α-TOC	Casting	No significant effect in mechanical or thermal stability; barrier properties affected by concentration and type of antioxidant	56,57
α-TOC	Extrusion	Protection of soybean oil	62
α-TOC, resveratrol	Compression moulding	Detailed release study in ethanol. Antioxidant concentration is critical in film properties	63,64
Ascorbyl palmitate and α-TOC, BHA, BHT, propyl gallate and TBHQ	Casting	Controlled release into diverse food simulants. Ascorbyl palmitate is not recommended for active packaging	58,75
β-carotene	Casting	Polymer protection against UV oxidation	60
β-carotene	Nanoparticles by oil-in-water dispersion	Protection of the agent against oxidation	71
BHA, BHT, PG and TBHQ	Extrusion	Controlled release into diverse food simulants. No negative effects on film functional properties	66
Cannabidiol and lutein	Melt mixing	PLA/flax fibres composites present biocompatibility	67
Catechin and epicatechin	Extrusion	Antioxidant activity evaluated with the DPPH method	65
Gallic acid	Electrospinning	Controlled release in saline solutions	68
Genistein	Electrospinning	Controlled release in saline solutions	69
Quecitrin	Nanoparticles by oil-in-water dispersion	Fast release in physiological conditions	70
Quercetin	Nanoparticles by oil-in-water dispersion	Fast release in physiological conditions	72

contact with PLGA 50 : 50 films. The studied antioxidants were α-tocopherol (α-TOC) in 2 wt% dispersion or a combination of 1 wt% BHA and 1 wt% BHT added to the PLGA polymer in a methylene chloride solution. Films were obtained by casting and samples were used to study the antioxidants release using aqueous and fat simulants, as well as dry whole milk and dry butter at 4 and 25 °C in the absence and presence of light. They concluded that in an aqueous medium, PLGA showed hydrolytic degradation with BHT release. In the fat simulant there was no degradation of the polymer or release of antioxidant, even after eight weeks at 25 °C. In dry dairy products (water content inferior to 4.6%), antioxidants migration did not occur by hydrolytic degradation of the polymer; their activity only took place by volatilization or surface contact. The authors suggested the use of bio-degradable polymers as an option to prepare active packaging allowing the sustained release of antioxidants to limit oxidization in dairy products with a high fat content.

Gonçalves *et al.*[56] also studied the addition of different quantities of α-TOC to PLA films prepared by casting but focused their attention on the effects on functional properties. No significant changes were found in PLA thermal or mechanical properties by the addition of the antioxidant. Nevertheless, sorption of O_2 increased with higher α-TOC content and the convex shape of the isotherms indicates strong gas-polymer interaction. However, there does not seem to be any significant effect on CO_2 absorption.

Following the same procedure, these authors incorporated α-TOC, BHT and *tert*-butylhydroquinone (TBHQ) into PLA films at concentrations ranging from 0 to 10%.[57] DSC results showed that the polymer glass transition temperature and the crystallinity slightly decreased with the increase of the antioxidant content. Permeability towards water vapour showed a decrease in the wettability of the prepared materials with the increase of the anti-oxidants content. Similar results were obtained for gas permeation. Barrier improvements could not be measured at high concentrations due to phase separation.

The release study of natural antioxidants, such as ascorbyl palmitate and α-TOC, and synthetic phenolic antioxidants such as BHA, BHT, propyl gallate and TBHQ from PLA cast films was studied in 95, 50 and 10% ethanol (used as food simulants) at 40 and 20 °C.[58] In a first report,[59] the addition of these antioxidants at a concentration of 1 wt% did not produce any significant effect on the functional properties of PLA. With the exception of ascorbyl palmitate, which was completely degraded during film preparation, all agents presented Fickian release processes. The diffusion (D) and partition coefficient (K) of this process provided D values ranging between 10^{-9} and 10^{-11} cm^2 s^{-1} and very diverse percentage releases. Antioxidants molecular weight, simulant polarity and temperature were the most influencing factors on the antioxidants release rate from PLA films in contact with food simulants. According to these results, a specific combination of variables could be selected to obtain the needed controlled release of antioxidants.

β-carotene was incorporated into PLA, PCL and polyhydroxybutyrate-*co*-valerate (PHBV) films obtained by casting.[60] The addition of β-carotene resulted in plasticization of the three polymers and reduced their mechanical deterioration when exposed to UV radiation. Therefore, β-carotene can potentially be used as a natural additive to increase the UV stability of these biopolyesters. No release tests were reported.

10.3.2.3 Extrusion

Byun *et al.*[61] prepared PLA films by extrusion with α-TOC and BHT as antioxidants and PEG as plasticizer, studying their mechanical, thermal, optical and gas barrier characteristics as well as antioxidant activity evaluated against the DPPH radical. The presence of PEG significantly reduced the T_g value of the prepared films and PLA films increased their elongation at break and permeability to water vapour while reducing O_2 permeability. PLA films with only BHT showed some antioxidant activity, which significantly increased in films that also incorporated α-TOC; in fact, 1 wt% incorporation of this additive increases radical scavenging activity from 0.84 to 90.4%. Nevertheless, antioxidant films with PLA have the limitation of a noticeable loss of clarity compared to the material without antioxidants.

α-TOC (2.58 wt%) was incorporated into PLA films by blown extrusion with the objective of manufacturing an antioxidant film for the packaging of soybean oil.[62] Optical and thermal properties of the material were studied as well as diffusion kinetics of the antioxidant towards simulants like ethanol and vegetable oil, and the effect of the released antioxidant on the oxidative stability of soybean oil at different temperatures. The diffusion of α-TOC in ethanol showed Fickian behaviour with a 26.9% release. The diffusion of the antioxidant in oil is lower than in ethanol, with only 5.1–12.9%, although this is enough to slow down oxidization of soybean oil at 20 and 30 °C, but not at 40 °C. According to these authors, these results show the potential of the material to protect oily food at room temperature.

PLA films with various concentrations of α-TOC and resveratrol were prepared by compression moulding.[63] The addition of these antioxidant agents resulted in deterioration of optical properties and reduction in T_g and T_m, but a significant improvement of thermostability. Mechanical properties were also improved since this plasticizing effect was accompanied by an increase in the elastic modulus, which is interpreted as a beneficial interaction between polymer and agents. No data about agents' activity or release were reported.

The same authors fabricated PLA/starch blends with various concentrations of two natural antioxidants, α-TOC and resveratrol, by melt blending and compression moulding processes.[64] The sheets showed a yellowish colour and a significant reduction in the glass transition and melting temperatures. The addition of these agents enhanced mechanical properties, an effect which is attributed to a compatibilization effect between PLA and starch chains. A detailed study of the release process showed that blending with starch accelerated the release of resveratrol and α-TOC from PLA into ethanol.

Extruded PLA films containing flavonoids like catechin and epicatechin were prepared as a vehicle for the antioxidant release into food.[65] The release kinetics into ethanol 95% was estimated at 20, 30, 40 and 50 °C, displaying Fickian behaviour with diffusion coefficients in the 10^{-11} cm^2 s^{-1} range. The antioxidant activity of the films was also measured, after extraction with methanol, by using the DPPH test. Results proved that these agents maintain their antioxidant activity after extrusion.

PLA films containing 2% of BHA, BHT, PG and TBHQ were extruded and agent release into 10%, 50% and 95% ethanol at 20 and 40 °C was monitored by HPLC over 60 days.[66] All agents were successfully added with the exception of TBHQ which partially decomposed to 2-tert-butyl-1,4-benzo-quinone during processing and storage. Release presented Fickian diffusion behaviour with diffusion coefficient values varying from 10^{-8} to 10^{-11} cm^2 s^{-1} with variable release. Propyl gallate had the fastest release and BHT the slowest. The release media also influenced the release: the higher the ethanol content the greater extent and the faster release was observed for a specific agent.

PLA composites containing flax fibres rich in poly(hydroxybutyrate) (PHB) as reinforcement were prepared by compression moulding and further analyzed.[67] Biochemical analysis of fibres revealed the presence of several antioxidant compounds, including cannabidiol and lutein. The prepared composites seem to have bacteriostatic, platelet anti-aggregated and non-cytotoxic effect. No tests on compound migration were reported.

10.3.3 Other Packaging Applications

Glucose oxidase was immobilized in PLA fibres manufactured by electrospinning process which were used to activate the lactoperoxidase system in milk.[73] Sorbitan monopalmitate was used as an emulsifier to obtain an aqueous dispersion in an organic solvent mixture. These fibres presented high enzymatic activity compared to that measured when glucose oxidase was incorporated into cast PLA films thanks to the larger surface area of the electrospun fibres. The *in vitro* activation of LP resulted in an increased generation of antimicrobial of hypothiocyanate, which could be used in active food packaging applications.

Acknowledgements

The authors acknowledge the financial support of the Ministry of Economy and Competitiveness (Projects AGL2009-08776 and AGL2012-39920-C03-01) and the technical support of the Associated Unit ITENE. The authors also thank Mr Tim Swillens for correction services.

References

1. M. Ozdemir and J. D. Floros, *Crit. Rev. Food Sci. Nutr.*, 2004, **44**, 185.
2. D. Restuccia, U. G. Spizzirri, O. I. Parisi, G. Cirillo, M. Curcio, F. Iemma, F. Puoci, G. Vinci and N. Picci, *Food Contr.*, 2010, **21**, 1425.
3. A. López-Rubio, E. Almenar, P. Hernández-Muñoz, J. M. Lagarón, R. Catalá and R. Gavara, *Food Rev. Int.*, 2004, **20**, 357.

4. C. Dury-Brun, P. Chalier, S. Desobry and A. Voilley, *Food Rev. Int.*, 2007, **23**, 199.

5. R. Gavara and R. Catalá, in *Engineering and Food for the 21st Century*, CRC Press, Washington D.C., 2002.

6. E. Almenar, P. Hernández-Muñoz, J. M. Lagarón, R. Catalá and R. Gavara, *Advances in postharvest technologies for horticultural crops–Advances in packaging technologies for fresh fruits and vegetables*, Kerala, India, 2006.

7. P. Appendini and J. H. Hotchkiss, *Innovat. Food Sci. Emerg. Tech.*, 2002, **3**, 113.

8. D. Dainelli, N. Gontard, D. Spyropoulos, E. Zondervan van den Beuken and P. Tobback, *Trends Food Sci. Tech.*, 2008, **19**, S99.

9. J. Gómez-Estaca, R. Gavara, R. Catalá, M. P. Balaguer and P. Hernández-Munoz, *Alimentacion Equipos y Tecnologia*, 2011, **63**, 17.

10. K. E. Perepelkin, *Fibre Chemistry*, 2005, **37**, 417.

11. K. E. Perepelkin, *Fibre Chemistry*, 2005, **37**, 381.

12. K. E. Perepelkin, *Fibre Chemistry*, 2005, **37**, 241.

13. G. Mensitieri, E. Di Maio, G. G. Buonocore, I. Nedi, M. Oliviero, L. Sansone and S. Iannace, *Trends Food Sci. Tech.*, 2011, **22**, 72.

14. F. Reno, M. Rizzi and M. Cannas, *J. Biomater. Appl.*, 2012, **27**, 165.

15. E. Aliyev, U. Sakallioglu, Z. Eren and G. Acikgoz, *Biomaterials*, 2004, **25**, 4633.

16. E. A. Simone, T. D. Dziubla, D. E. Discher and V. R. Muzykantov, *Biomacromolecules*, 2009, **10**, 1324.

17. S. S. Feng, L. Mei, P. Anitha, C. W. Gan and W. Zhou, *Biomaterials*, 2009, **30**, 3297.

18. M. Sokolsky-Papkov, L. Golovanevski, A. J. Domb and C. F. Weiniger, *Pharmaceut. Res.*, 2009, **26**, 32.

19. L. F. Miao, J. Yang, C. L. Huang, C. X. Song, Y. J. Zeng, L. F. Chen and W. L. Zhu, *Zhongguo yi xue ke xue yuan xue bao. Acta Academiae Medicinae Sinicae*, 2008, **30**, 491.

20. D. Cun, F. Cui, L. Yang, M. Yang, Y. Yu and R. Yang, *J. Drug Deliv. Sci. Tech*, 2008, **18**, 267.

21. Q. Xu and J. T. Czemuszka, *J. Contr. Release*, 2008, **127**, 146.

22. J. Schnieders, U. Gbureck, R. Thull and T. Kissel, *Biomaterials*, 2006, **27**, 4239.

23. H. Gu, C. Song, D. Long, L. Mei and H. Sun, *Polym. Int.*, 2007, **56**, 1272.

24. J. H. Han, *Food Tech.*, 2000, **54**, 56.

25. P. M. Davidson, J. N. Sofos and A. L. Branen, *Antimicrobials in Food*, Taylor & Francis, 3rd edn, Boca Raton, FL, 2010.

26. P. Theinsathid, A. Chandrachai, S. Suwannathep and S. Keeratipibul, *J. Biobased Mater. Bioenergy*, 2011, **5**, 17.

27. T. Jin, H. Zhang and G. Boyd, *J. Food Protect.*, 2010, **73**, 812.

28. T. Jin and B. A. Niemira, *J. Food Sci.*, 2011, **76**, M184.

29. T. Jin and H. Zhang, *J. Food Sci.*, 2008, **73**, M127.

30. T. Jin, *J. Food Sci.*, 2010, **75**, M83.

31. W. Li, D. R. Coffin, T. Z. Jin, N. Latona, C. K. Liu, B. Liu, J. Zhang and L. Liu, *J. Appl. Polym. Sci.*, 2012, **126**, E361.
32. A. Fernández, E. Soriano, P. Hernández-Munoz and R. Gavara, *J. Food Sci.*, 2010, **75**, E186.
33. M. A. Busolo, P. Fernández, M. J. Ocio and J. M. Lagarón, *Food Addit. Contam. A.*, 2010, **27**, 1617.
34. A. Martínez-Abad, M. J. Ocio, J. M. Lagarón and G. Sánchez, *Int. J. Food Microbiol.*, 2013, **162**, 89.
35. E. Fortunati, M. Peltzer, I. Armentano, A. Jiménez and J. M. Kenny, *J. Food Eng.*, 2013, **118**, 117.
36. E. Mascheroni, V. Guillard, F. Nalin, L. Mora and L. Piergiovanni, *J. Food Eng.*, 2010, **98**, 294.
37. F. Sebastien, G. Stephane, A. Copinet and V. Coma, *Carbohydr. Polym.*, 2006, **65**, 185.
38. J. W. Rhim, S. I. Hong, H. M. Park and P. K. W. Ng, *J. Agr. Food Chem.*, 2006, **54**, 5814.
39. J. W. Rhim, S. I. Hong and C. S. Ha, *LWT Food Sci. Technol.*, 2009, **42**, 612.
40. S. Ray, T. Jin, X. Fan, L. Liu and K. L. Yam, *J. Food Sci.*, 2013, **78**, M276.
41. M. A. Del Nobile, A. Conte, G. G. Buonocore, A. L. Incoronato, A. Massaro and O. Panza, *J. Food Eng.*, 2009, **93**, 1.
42. S. Raouche, M. Mauricio-Iglesias, S. Peyron, V. Guillard and N. Gontard, *Innovat. Food Sci. Emerg. Tech.*, 2011, **12**, 426.
43. M. J. Joo, C. Merkel, R. Auras and E. Almenar, *Int. J. Food Microbiol.*, 2012, **153**, 297.
44. L. Liu, T. Z. Jin, D. R. Coffin and K. B. Hicks, *J. Agr. Food Chem.*, 2009, **57**, 8392.
45. L. Liu, T. Jin, D. R. Coffin, C. K. Liu and K. B. Hicks, *J. Appl. Polym. Sci.*, 2010, **117**, 486.
46. E. Fortunati, I. Armentano, A. Iannoni and J. M. Kenny, *Polym. Degrad. Stabil.*, 2010, **95**, 2200.
47. F. Kayaci, O. C. O. Umu, T. Tekinay and T. Uyar, *J. Agr. Food Chem.*, 2013, **61**, 3901.
48. A. C. Vega-Lugo and L. T. Lim, *Food Res. Int.*, 2009, **42**, 933.
49. L. S. Liu, V. L. Finkenstadt, C. K. Liu, T. Jin, M. L. Fishman and K. B. Hicks, *J. Appl. Polym. Sci.*, 2007, **106**, 801.
50. J. Kanner and I. Rosenthal, *Pure Appl. Chem.*, 1992, **64**, 1959.
51. K. M. Schaich, *Lipids*, 1992, **27**, 209.
52. C. López-de-Dicastillo, *Development and Characterization of Hydrophilic Active Polym.s for Food Packaging*, PhD Thesis, Universidad Politécnica de Valencia, Valencia, 2011.
53. C. López-de-Dicastillo, D. Pezo, C. Nerín, G. López-Carballo, R. Catalá, R. Gavara and P. Hernández-Muñoz, *Packag. Tech. Sci.*, 2012, **25**, 457.
54. V. K. Holm, G. Mortensen, M. Vishart and M. A. Petersen, *Int. Dairy J.*, 2006, **16**, 931.
55. M. van Aardt, S. E. Duncan, J. E. Marcy, T. E. Long, S. F. O'Keefe and S. R. Sims, *Int. J. Food Sci. Tech.*, 2007, **42**, 1327.

56. C. M. B. Gonçalves, L. C. Tome, J. A. P. Coutinho and I. M. Marrucho, *J. Appl. Polym. Sci.*, 2011, **119**, 2468.

57. C. M. B. Gonçalves, L. C. Tome, H. García, L. Brandao, A. M. Mendes and I. M. Marrucho, *J. Food Eng.*, 2013, **116**, 562.

58. M. Jamshidian, E. A. Tehrany and S. Desobry, *Food Bioprocess Technol.*, 2013, **6**, 1450.

59. M. Jamshidian, E. A. Tehrany, F. Cleymand, S. Leconte, T. Falher and S. Desobry, *Carbohydr. Polym.*, 2012, **87**, 1763.

60. A. López-Rubio and J. M. Lagarón, *Polym. Degrad. Stabil.*, 2010, **95**, 2162.

61. Y. Byun, Y. T. Kim and S. Whiteside, *J. Food Eng.*, 2010, **100**, 239.

62. F. Manzanarez-López, H. Soto-Valdez, R. Auras and E. Peralta, *J. Food Eng.*, 2011, **104**, 508.

63. S. W. Hwang, J. K. Shim, S. E. M. Selke, H. Soto-Valdez, L. Matuana, M. Rubino and R. Auras, *Polym. Int.*, 2012, **61**, 418.

64. S. W. Hwang, J. K. Shim, S. Selke, H. Soto-Valdez, L. Matuana, M. Rubino and R. Auras, *J. Food Eng.*, 2013, **116**, 814.

65. F. Iñiguez-Franco, H. Soto-Valdez, E. Peralta, J. F. Ayala-Zavala, R. Auras and N. Gámez-Meza, *J. Agr. Food Chem.*, 2012, **60**, 6515.

66. M. Jamshidian, E. A. Tehrany and S. Desobry, *Food Contr.*, 2012, **28**, 445.

67. M. Wrobel-Kwiatkowska, M. Czemplik, A. Kulma, M. Zuk, J. Kaczmar, L. Dyminska, J. Hanuza, M. Ptak and J. Szopa, *Biotechnol. Progr.*, 2012, **28**, 1336.

68. P. Chuysinuan, N. Chimnoi, S. Techasakul and P. Supaphol, *Macromol. Chem. Phys.*, 2009, **210**, 814.

69. S. Buddhiranon, L. A. DeFine, T. S. Alexander and T. Kyu, *Biomacromolecules*, 2013, **14**, 1423.

70. A. Kumari, S. K. Yadav, Y. B. Pakade, V. Kumar, B. Singh, A. Chaudhary and S. C. Yadav, *Colloids Surf. B*, 2011, **82**, 224.

71. L. Cao-Hoang, H. Phan-Thi, F. J. Osorio-Puentes and Y. Wache, *Food Res. Int.*, 2011, **44**, 2252.

72. A. Kumari, V. Kumar and S. K. Yadav, *Plos One*, 2012, **7**.

73. Y. Zhou and L. T. Lim, *J. Food Sci.*, 2009, **74**, C170.

74. C. Gomes, R. G. Moreira and E. Castell-Perez, *J. Food Sci.*, 2011, **76**, N16.

75. M. Jamshidian, E. A. Tehrany, M. Imran, M. J. Akhtar, F. Cleymand and S. Desobry, *J. Food Eng.*, 2012, **110**, 380.

CHAPTER 11

Biomaterials for Tissue Engineering Based on Nano-structured Poly(Lactic Acid)

ILARIA ARMENTANO,*[a] ELENA FORTUNATI,[a]
SAMANTHA MATTIOLI,[a] NICOLETTA RESCIGNANO[b] AND
JOSÈ MARIA KENNY[a,b]

[a] Materials Engineering Center, UdR INSTM, University of Perugia, 05100, Terni, Italy; [b] Instituto de Ciencia y Tecnología de Polímeros, ICTP-CSIC, Juan de la Cierva, 3 28006, Madrid, Spain
*Email: ilaria.armentano@unipg.it

11.1 Tissue Engineering

Tissue engineering (TE) has been defined as an interdisciplinary field that applies the principles of engineering and the life sciences toward the development of biological substitutes that restore, maintain or improve tissue function.[1] TE is based on a combination of cells, engineering and materials methods, and suitable biochemical and physicochemical factors to improve or replace biological functions.[2,3] While most definitions of tissue engineering cover a broad range of applications, in practice the term is closely associated with applications that repair or replace portions of or whole tissues (i.e. bone, cartilage, blood vessels, bladder, etc.). The tissues involved often require certain mechanical and structural properties for proper function. The term has also been applied to efforts to perform specific biochemical functions using cells within an artificially created support system (e.g. an artificial pancreas, or

RSC Polymer Chemistry Series No. 12
Poly(lactic acid) Science and Technology: Processing, Properties, Additives and Applications
Edited by Alfonso Jiménez, Mercedes Peltzer and Roxana Ruseckaite
© The Royal Society of Chemistry 2015
Published by the Royal Society of Chemistry, www.rsc.org

a bio-artificial liver).[1] TE is a clinically driven field and has emerged as a potential alternative to organ transplantation. The cornerstone of successful tissue engineering rests upon two essential elements: cells and biomaterials. Recently, it was found that stem cells have unique capabilities of self-renewal and multi-lineage differentiation to serve as a versatile cell source, while nanomaterials have lately emerged as promising candidates in producing biomaterials able to mimic the nanostructure in natural extracellular matrices and to efficiently replace defective tissues.[2]

The term regenerative medicine is often used synonymously with tissue engineering, although those involved in regenerative medicine place more emphasis on the use of stem cells to produce tissues. Regenerative medicine is positioned at the cutting edge of the twenty-first century's health systems. Its main purpose is the development of new therapeutic solutions for malfunctioning organs and tissues in the world's increasingly older population, and to fight the continuous rise of health costs, especially in the treatment of chronic illnesses. Recent estimates point out that the USA can save up to 250 billion dollars per year with the adoption of regenerative medicine treatments for chronic illnesses like neurodegenerative diseases (Parkinson's, injury of the spinal marrow, neurovascular accidents).

The tissue engineering approach could be divided into the following steps:

- Appropriate cell sources must be identified, isolated and produced in sufficient numbers.
- Appropriate biocompatible materials that can be used as cell substrates or cell encapsulation.
- Materials isolated or synthesized, manufactured into desired shape and dimensions.
- Cells seeded onto or into materials, maintaining function and morphology.
- Engineered structure placed into appropriate *in vivo* sites.

11.2 Stem Cells

Tissue engineering utilizes living cells as engineering materials. Examples include using living fibroblasts in skin replacement or repair, cartilage repaired with living chondrocytes, or other types of cells used in other ways.[2,3] Stem cells are recognized ideal candidates for tissue regeneration.

Stem cells are undifferentiated cells with the ability to divide in cultures and give rise to different forms of specialized cells. According to their source, stem cells are divided into "adult" and "embryonic", the first class being multipotent and the latter mostly pluripotent, although some cells are totipotent, in the earliest stages of the embryo.[4] While there is still a large ethical debate related to the use of embryonic stem cells, it is thought that stem cells may be useful for the repair of diseased or damaged tissues, or may be used to grow new organs.[3] Over the last few decades, stem cells have emerged as a key player in tissue engineering, for both *in vitro* and *in vivo* tissue regeneration.[5–7]

11.2.1 Embryonic Stem Cells

Embryonic stem cells (ESCs) are cells isolated from the inner mass of blastocysts.[4] The mechanisms by which ESCs maintain self-renewal and pluripotency are still not yet fully understood. Many reports highlight the involvement of miRNAs as crucial players in ESCs regulation and ESCs development. In fact ESCs lose their self-renewal differentiation capacity, following alterations in the machinery involved in miRNAs processing, maintenance and activity. Potential clinical applications of ES cells raise many practical and ethical concerns. Many nations currently have moratoria on either ES cell research or the production of new ES cell lines. Because of their combined abilities of unlimited expansion and pluripotency, embryonic stem cells remain a theoretically potential source for regenerative medicine and tissue replacement after injury or disease.

11.2.2 Adult Stem Cell Sources

Adult stem cells (ASCs) are multipotent stem cells that, under controlled conditions, may differentiate into various cells *in vitro* and *in vivo*.[8–10] ASCs have been isolated from bone marrow, cord blood, skeletal muscle, brain, cornea, tooth and skin among other tissues.[11–13] ASCs can self-renew and undergo multipotential differentiation. Compared to ESCs, adult stem cells exhibit similar self-renewal capacity but show a more restricted differentiation potential that gives rise to a specialized tissue-specific cell type.[14,15] Based on their properties, when ASCs are transplanted they are able to recognize and respond to signals within the host tissue; this phenomenon leads to the preservation of ASCs stem-properties even in tissues of different embryonic origin.[14,16,17] Notably, ASCs respond to microenvironment changes which, in turn, may alter their fate[18–20] suggesting that ASCs are useful for regenerative medicine applications. In this regard, mesenchymal stem cells (MSCs) remain the most promising type of adult stem cells for regenerative medicine in cell therapy and tissue engineering. Their most common sources are bone marrow, fat, amniotic fluid, amniotic membrane and umbilical cord matrix.[21–23]

11.2.3 Induced Pluripotent Stem Cells

Induced pluripotent stem cells (iPSCs) are a new class of stem cells generated *in vitro* from somatic differentiated cells. They hold a great promise for *in vivo* cell therapy and *in vitro* drug screening, and display many features of pluripotency that are typical of embryonic stem cells. Takahashi and Yamanaka demonstrated that the over-expression of specific genes, such as Oct3/4, Sox2, c-Myc and Klf4, in somatic cells can directly generate pluripotent cells with embryonic-like properties.[24] Many laboratories are currently investigating the generation of iPSCs from different cell sources, ASCs and/or somatic differentiated cells, whilst improving the experimental procedure.[3] The clinical potential of these cells resides in their patient-specific nature, given that iPSCs bear the same genotype as the adult donor body. In principle, once generated from a patient, iPSCs can be genetically corrected and re-injected *in vivo* to regenerate a tissue or differentiated *in vitro* towards a specific lineage and used to screen drug functionality.[2] This approach provides an opportunity to explore normal and diseased core blood and bone marrow samples without any limitations associated with virus-based methods.[25]

11.2.4 Differentiation Induction Factors in Tissue Engineering

To induce different lineage commitment, stem cells may require the appropriate extracellular signals to trigger or to promote this process. Differentiation induction factors from protein and chemical origins can constitute an important class of such stimuli. From protein sources, biological growth factors have been widely demonstrated to induce the differentiation of stem cells. For example, bone morphogenetic proteins (BMPs), such as BMP-2 and BMP-7, are known as the most potent growth factors for directing the osteogenesis of stem cells like MSCs and enhancing bone formation,[26] transforming growth factors β1 and 3 (TGF-β1 and TGF-β3) to be utilized to enhance the differentiation of MSCs into chondrocytes.[27] Glial cell line-derived neurotrophic factor (GDNF)[28] and basic fibroblast growth factor (bFGF)[29] are used to induce neural differentiation of NSCs, while epidermal growth factor (EGF) and bFGF can be utilized to promote neural transdifferentiation of MSCs.[30] Hepatocyte growth factor (HGF), EGF, TGF and insulin-like growth factor are employed to help MSCs to trans-differentiate into hepatocytes.[31–33] To support tissue regeneration in an *in vitro* setting, these biochemical agents can be loaded into scaffolds to promote or to induce differentiation of stem cells. However, an increasing number of studies employing only stem cells and scaffolds remain successful in tissue engineering. This outcome represents a development that accentuates the role of the scaffold in substituting some functions of differentiation factors besides just serving as structural support. Therefore, in the following sections, we will survey the recent progress in the application of stem cells and nanomaterials based on PLA for tissue engineering.[2]

11.3 Biomaterials and Nanotechnology

Tissue engineering is generating increasing interest as results obtained in cells associated with engineered materials for the restoration of function to damaged tissues. A crucial point in this field is the complete understanding of cell–material interactions. Biomaterials play a critical role in the success of tissue engineering as they guide the shape and structure of developing tissues, provide mechanical stability and offer opportunities to deliver inductive molecules to transplanted or migrating cells. A critical step of all tissue engineering techniques is the use of a tridimensional structure, which, mimicking the extracellular matrix (ECM), serves as a scaffold that is able to promote and actively guide the tissue regeneration process. A suitable scaffold for tissue engineering should provide:

- Mechanical integrity to tissues by acting as a support for neo-tissue growth.
- Guidance for biological response through the promotion of the dynamic interaction with surrounding tissues.
- A space for host cell survival, enhancing the transport of nutrients and metabolites through maximization of biological and/or pharmaceutical response.
- Adequate biocompatibility/biodegradability, with degradation kinetics suitable to match neo-tissue formation, thus minimizing toxicity in terms of both tissue and systemic response.
- Manufacturing feasibility.

The scaffolds can be considered at different length scales: the macrostructure, the microstructure and the nanostructure. On the macro- and micro-scales it has been shown that surface chemical composition and topography have strong effects on cell behaviour,[34] but less is known about how cells react to nanoscale structures. However, cells are likely to be able to respond to nanostructures, since they live inside the extracellular matrix (ECM) containing nanoscale collagen fibrils and since their own surface is structured on the nanoscale level (receptors and filopodia).

The application of nanoscaled materials and structures, as well as with the manipulation of single atoms and molecules, is an emerging area in tissue engineering.[35–37] The design of multilevel molecular aggregates with novel functional and dynamic properties is desirable for applications in medicine.[2] This approach also offers new possibilities towards the development of personalized medicine, with the development of nanoparticles and nanostructured surfaces as well as nanoanalytical techniques for molecular diagnostics, treatment, follow-up and therapy of diseases (theranostics). Integrated medical nanosystems are also needed that, in the future, may perform monitoring and complex repairs in the body at the cellular level. In fact, nanotechnology considers cells as a complex system of interacting nanomachines. Visionary concepts envisage the construction and control of artificial cells by using engineered nanodevices and nanostructures for medical applications.[38]

11.4 Nanostructured PLA

PLA is widely investigated as a scaffold material to provide two- and three-dimensional matrices for cell adhesion and subsequent tissue development.

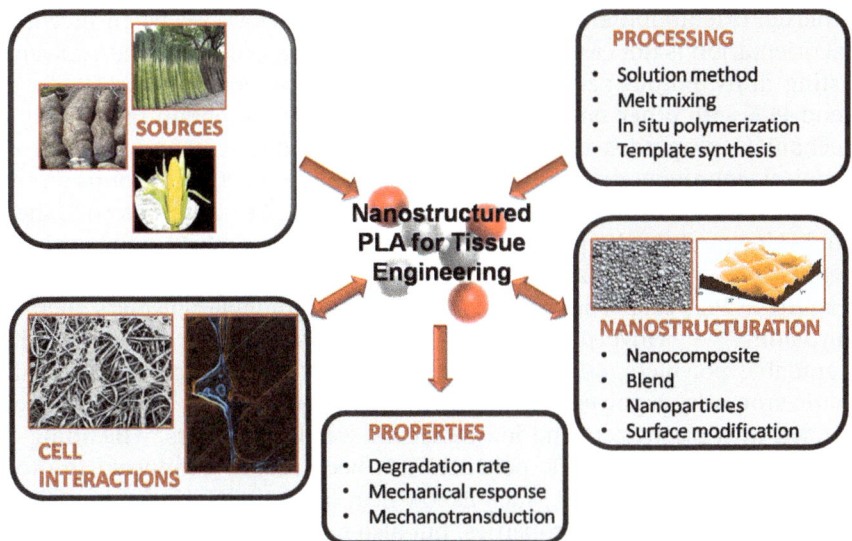

Figure 11.1   Schematic view of the chapter subject of nanostructured PLA for tissue engineering applications.

PLAs have great potential for use in surgical implants, because of their well-known biocompatibility and biodegradability. In the following sub-sections we will present different nanostructured PLAs including blends, nanoparticles, nanocomposites and surface modifications. Figure 11.1 summarizes the main uses of nanostructured PLA for tissue engineering applications, including a scheme of the lactic acid interaction with living systems, starting from synthesis, processing techniques of nanostructured PLAs, blending, nanostructuration approach (blends, nanoparticles, nanocomposites and surface modifications), property modulation and cell interactions.

11.4.1   Blends

Polymer blending is a promising approach for the development of novel biomaterials to be used in tissue engineering and regenerative medicine[39,40] where well-characterized and -tolerated materials are combined to give systems with functionalities that are often inaccessible from the individual components themselves.[40] Polymer blending has been used as the strategy to tailor the mechanical, thermal and chemical properties of polymers but also their degradability and the possibilities to be used as drug release systems.[41,42] The blend compatibility affects the crystallinity and physical properties of the system, consequently it might influence its degradation and controlled release properties. But in many PLA-based systems, bursting strength, elongation and tear strength are also important properties and they can be improved to a certain extent by mechanical drawing, such as biaxial orientation and stretch blow moulding. However, when mechanical orientation is not easily feasible (*e.g.* injection-moulded articles, solvent-casting films, porous scaffolds and mats), the most common strategy is to blend PLA with other synthetic or natural polymers to obtain the desirable mechanical properties and hydrophilicity. In addition, many physical or chemical techniques are developed to bind specific functional groups on the PLA surface to provide the desired functionalities.[43] In general terms, thermoplastic starch is the primary choice for blending with PLA for environmental and biomedical applications. Some other polymers have been used to improve PLA properties, including non-biodegradable polymers such as polyolefins,[44–47] vinyl polymers,[48,49] elastomers,[50] rubbers[51,52] and bio-degradable polymers, such as polyanhydrides,[53] aliphatic polyesters,[54] aliphatic-aromatic copolyesters[55] and bio-elastomers.[56] PLA has also been blended with plasticizers and low-molecular-weight modifiers.[57] Blending of PLA with non-biodegradable polymers has been mainly conducted to overcome the drawbacks of PLA, such as low heat deflection temperature (HDT), fragility and poor barrier properties, but also to create specific morphologies for medical applications. However, the immiscibility of polyolefin/PLA blends due to differences in their chemical structures generally results in weak interfacial adhesion (indicated by poor dispersion, a very broad size distribution and distinct particle interfaces) and poor mechanical properties.

The production and characterization of spun-cast polystyrene (PS)/PLA films were reported by Lim *et al.*[58] and Leung *et al.*[59] Topologies and water contact angles of blend films at various composition ratios were explored to determine their effects on osteoblast adhesion of blend films and it was concluded that it was enhanced by increasing the PLLA weight fraction from 0.7 to 0.9.[58] Leung *et al.*[59] characterized the phase morphology and quantified the chemical composition of spun-cast PS/PLA films by atomic force microscopy (AFM) and synchrotron-based X-ray photoemission electron microscopy (X-PEEM). They reported that a phase inversion of spun-cast PS/PLA films with a 40/60 weight ratio (0.7 wt% solution) occurred with annealing above the PLA glass transition temperature (T_g). PS/PLA blends with a continuous structure are one of the most attractive blends for generating biodegradable controlled-release devices and tissue engineering scaffolds, since after selective removal of the PS phase from the blends, highly structured and completely interconnected porous PLA structures are obtained.[60] The pore size, pore distribution and void volume of the continuous structures can be tailored by controlling the influencing factors or adding compatibilizers.[60,61]

Polyanhydrides, having organic and hydrophobic backbones with anhydride linkages, are gaining interest as biomaterials for drug delivery systems by their biocompatible, non-mutagenic, non-cytotoxic and non-inflammatory character.[62] In addition, the high hydrolytic sensitivity of the anhydride bonds on the blend surface results in rapid degradation, while the hydrophobic backbone prevents the penetration of water into the polymer matrix.[62,63] The hydrolytic degradation of polyanhydrides mainly occurs *via* a surface erosion mechanism. The main shortcomings of these polymers are their poor mechanical strength, high reactivity with amine and nucleophilic groups and high susceptibility to organic solutions, especially at high temperatures, which limited their use in other applications besides drug release devices. Hence, polyanhydrides/PLA blends have been proposed as potential implants for controlled drug delivery systems.

Martin *et al.*[64] proposed to blend thermoplastic starch, polyethylene glycol (PEG), glycerol and oligomeric lactic acid by using a single-screw extruder equipped with a conical-shaped shear element. Glycerol was found to be the least efficient plasticizer, while oligomeric lactic acid and PEG provided substantial increases in elongation at break. The affinity of PLA and thermoplastic starch was poor, leading to blends with weak mechanical properties as compared to the individual polymers. Sheth *et al.*[65] melt-blended PLA and PEG by using a counter-rotating twin-screw extruder at 120–180 °C. They reported that PEG and PLA form partially miscible blends, depending on the relative PEG/PLA concentration. Although PEG is effective in decreasing the PLA stiffness, the use of a low-molecular-weight plasticizer has the disadvantage of some tendency to migrate in the PLA matrix.

Poly(hydroxyalkanoates) (PHAs) are non-toxic, biodegradable and biocompatible bacterial-synthesized polyesters. PHAs show some potential in medical and pharmaceutical applications, such as drug delivery systems and tissue engineering, especially if blended with other biodegradable polymers. Thus, Zhang *et al.*[66] reported that PHB/PDLLA blends prepared by solvent casting were intrinsically immiscible, but some miscibility was observed for 60% PLA and 40% PHB blends if prepared by melt blending. This miscibility was ascribed to possible small-scale trans-esterification reactions. The crystallization rate of PHB decreased with the increase in PLA content, and blends showed some improvement in mechanical properties compared to the neat polymers. Iannace *et al.*[67] studied immiscible PLA/PHBV blends with 20% valerate content prepared by solvent casting. Stress-induced crystallization was observed in neat PHBV films, which was reduced with the addition of PLA to PHBV copolymer.

Poly(ε-caprolactone) (PCL) is widely used as second phase in PLA blends.[68] The non-toxicity, biocompatibility, biodegradability under physiological conditions (pH = 7.4) and high permeability for many drugs of PCL contribute to the significant use in biomedical applications of these blends, since they are bio-compatible materials for hard- and soft-tissue replacement and controlled drug delivery devices.[69–73] However, the high crystallinity (45–60%) and the high hydrophobicity of PCL are responsible for the slow degradation rate compared to other biopolyesters and starch.[74] Blending of PLA with PCL to achieve different performance properties (*i.e.* thermal, mechanical, permeability and/or drug release) and biodegradability is useful to toughen PLA and broaden both polymer applications.[75–77]

PLLA/PCL blends with appropriate physical properties were found to display remarkable bone-like architecture as well as to provide a robust template for skeletal stem cell attachment and bone regeneration. These studies pave the way for the rational design and development of synthetic materials for tissue regeneration and serve as a basis for future exploration of functional cell-based systems with potential applications in hard and soft-tissue repair.[78]

Poly(lactic-*co*-glycolic acid) (PLGA), a biodegradable linear aliphatic copolymer of lactic acid and glycolic acid units, was also used to produce PLGA/PLA blends by the solution method to be used in controlled drug release applications. One successful approach is to use PLGA and PLA as microspheres for drug support. However, there is an initial bursting problem for some drugs with high toxicity, which can cause a problem for patients. Rahman and Mathiowitz[79] prepared PLGA/PLA microspheres with different ratios by solution blending in methylene chloride and they were then phase-separated to provide drug encapsulation in the internal wall, resulting in the prevention of the initial burst. Some of these blends were found to be immiscible and phase-separated. In addition, phase separation occurred at blend concentrations higher than 30 wt% in the solvent, regardless of the PLGA/PLA ratio. For effective drug administration, a higher ratio of PLGA and pre-encapsulation step would provide the internal encapsulation of the drug and a lower rate of drug release over time compared to neat PLA, preventing the initial burst.

PLLA exhibits good mechanical properties and easy processing. However, as a major drawback it can elicit undesirable inflammatory and allergenic reactions[80] by the decrease in the local pH as a consequence of hydrolytic degradation. The combination of a natural polymer, such as chitosan, with PLLA might overcome some of these drawbacks and lead to blends with interesting properties to be used in the biomedical field. The production of homogeneous chitosan/PLLA blends is very difficult to achieve, but some solutions were proposed by Jiao *et al.*,[81] who reported the fabrication of PLLA/chitosan composite sponges with potential applications in tissue engineering. Their porosity is greatly influenced by the polymer concentration in solution and on their relative ratio. The presence of a much higher concentration of PLLA in the blend produces structures exhibiting larger and more heterogeneous pores.[82]

11.4.2 Nanoparticles and Nanoshells

11.4.2.1 *Nanoparticles*

Nanoparticles (NPs) are entities with at least one dimension in the nanoscale or solid colloidal particles ranging in size from 10 nm to 1000 nm.[83] They can be used in optical, electronic and magnetic devices due to their properties, while formulation of different types of stimuli-responsive polymers for controlled drug delivery in various diseases, such as cancer, diabetes, filarial, viral infections and bacterial infections is proposed.[84] These structures have received attention in the biomedical field due to their numerous advantages, including large-specific surface areas, such as controlled[85] and targeted[86] drug delivery. Drugs can be encapsulated[87] or conjugated[88] on the surface.

The particle size is essential to control the drug release and *in vivo* bio-distribution,[89,90] but limited studies have been performed on biodegradable nanoparticles with narrow size distribution for tissue engineering. Typically, nanoparticles are usually formed by high-speed homogenization or ultra-sonication.[91] Some reports have shown that cellular uptake is correlated with particle size, but results could be ambiguous and even contradictory due to the broad size distribution of nanoparticles.[92]

The advantages of using nanoparticles as drug delivery systems include:

1. Particle size and surface characteristics of nanoparticles can be easily manipulated to achieve both passive and active drug targeting.[93]
2. Controlled release and particle degradation characteristics can be readily modulated by the choice of the matrix constituents. Drug loading is relatively high and drugs can be incorporated into the system without any chemical reaction; this is an important factor for preserving the drug activity.
3. The system can be used for various administration routes, including oral, nasal, parenteral, intra-ocular *etc.*
4. The nanoparticles size for crossing different biological barriers is dependent on the tissue, target site and circulation. For the cellular internalization of nanoparticles, surface charge is important in determining whether these nanoparticles would cluster in blood flow or would adhere to or interact with oppositely charged cells membranes.

Release mechanisms can be modulated by the molecular weight of the used biopolymer. The higher molecular weight of the polymer decreases the *in vitro* drug release.[94,95]

The development of biodegradable polymers for drug delivery has been largely empirical; few polymers have been developed specifically for the purpose of drug delivery. A case in point is the widespread use of PLA and copolymers, such as poly (DL-lactide-*co*-glycolide) and PCL for the preparation of degradable nanocomposites.[96] These polymers were first used in the production of biodegradable sutures and later found to have properties desirable for controlled release devices. It is not surprising that these materials are by far the most widely studied class of biodegradable polymers in biomedical applications.

One serious challenge for the application of PLA in the biomedical field is associated with hydrophobicity. For drug delivery applications, the PLA hydrophobicity and copolymers enhance the uptake of drug-loaded nanoparticles (NPs) through mononuclear phagocyte system (MPS), resulting in short residence times. This effect causes a decrease in the *in vivo* drug efficiency. One promising approach to permit the long circulation of hydrophobic PLA-based NPs in the blood is to coat their surface with hydrophilic polymers by preparation of well-controlled PLA containing amphiphilic block copolymers (ABPs) for drug delivery applications.

11.4.2.2 Nanoshells

Nanoshell particles (NSs) are highly functional materials with tailored properties, which are quite different from either the core or the shell material. Therefore, nanoshells are preferred over nanoparticles since their properties can be modified by changing either the constituting materials or the core-to-shell ratio.[97] These particles are synthesized for a variety of purposes like providing chemical stability to colloids,[98] band structures,[99] biosensors,[100] drug delivery,[101] *etc.*

The most promising core-shell nanostructures can be considered those comprising a polymeric core and/or a polymeric shell dispersed in a matrix of any material whose property is to be modified or enhanced. Alginate, chitosan, gelatin, hyaluronic acid, poly(glycolic acid) (PGA), PLA and their co-polymers are some examples of biodegradable polymers intensively investigated in the formulation of polymeric-based core-shell nanoparticles, mainly as drug, protein and gene delivery vehicles.[102]

The combination of enzymatic degradation and stimuli-responsive "smart" behaviour of magnetic nanoparticles is a viable approach to fabricate a drug carrier.[103,104] These nanoshells will serve as effective drug delivery systems with targeting and controlled drug release. Monodisperse magnetite (Fe_3O_4) nanoparticles constituting the core are encapsulated in biodegradable polymers, such as PLA, as drug carriers. The magnetic drug-targeting carrier has some advantages, since the magnetic drug carrier can be targeted to a selected site through the use of external magnetic fields and the encapsulated drug can be subsequently released.[105]

11.4.3 Nanocomposites

In the large field of nanotechnology, polymer-based nanocomposites have become a prominent area of current research and development for biomedical applications.[106–112] In principle, nanocomposites are an extreme case of composite materials in which interface interactions between the two phases, the matrix and the reinforcement, are maximized. In the literature, the term nanocomposite is used for polymers with sub-micrometre dispersions. In polymer-based nanocomposites, nanometre sized particles of inorganic or organic materials are homogeneously dispersed as separate particles in a polymer matrix. Researchers have tried a variety of processing techniques to obtain dense polymer nanocomposite films. The incorporation of nanostructures into polymers can generally be done in different ways as indicated below:[113]

(1) Solution method, involving dissolution of polymers in adequate solvents with nanoscale particles and evaporation of solvent or precipitation.
(2) Melt mixing, where the polymer is directly melt-mixed with nanoparticles.

(3) *In situ* polymerization, where nanoparticles are first dispersed in liquid monomers or monomer solutions. Polymerization is performed in the presence of nanoscale particles.

(4) Template synthesis by using polymers as template, the nanoscale particles are synthesized from precursor solutions.

The use of nanocomposites for biomedical and clinical applications requires the selection of the appropriate biopolymer matrix, since it can have a profound impact on the quality of the newly formed tissue. Given that few biomaterials possess all the necessary characteristics for such application, researchers have pursued the development of hybrid or composite biomaterials to get synergies from the beneficial properties of multiple materials. The combination of biopolymers with nanostructured materials, including the use of nanoparticles, nanofibres and other nanoscaled features, has demonstrated the ability to enhance cellular interaction, to encourage integration into host tissue and to provide tunable material properties and degradation kinetics.[114] Materials with nanometre scaled dimensions, also known as nanophase or nanostructured materials, can be used to produce nanometre features on the surface of three-dimensional substrates for scaffolds.

Results from our laboratories demonstrated that adult, embryonic and induced pluripotent stem cells modify the biomaterial properties and change their fate. In particular we explored the response of human bone marrow MSCs, iPSCs and ESCs to nanocomposite PLLA fibrous mats containing 1 wt% to 8 wt% calcium deficient hydroxyapatite (d-HAp) producing a set of materials with similar architecture and tunable mechanical properties.[115] Results indicated that PLLA/d-HAp nanocomposite mats induce osteogenic differentiation in both multipotent and pluripotent stem cells in the absence of soluble differentiating agents and highlight the direct interaction of stem cell-polymeric nanocomposites as the key step for the osteogenic differentiation process. This mechano-sensitive feedback modulates cellular functions, such as proliferation, differentiation, migration and apoptosis, and it is crucial for organ development and homeostasis. Cells can act as mechano-sensitive units responding to the mechanical stimulation of the extracellular matrix through focal adhesions and changes in cytoskeletal organization.

11.4.4   Surface Modification

Polymer surface engineering may potentially be used to create materials to control cellular adhesion and maintain differentiated phenotypic expressions, merely by the introduction of nanoparticles within the polymer generating surface modifications that may affect both cell behaviour and the antibacterial character of the material.[115,116]

Surface modification of polymeric biomaterials is becoming an increasingly popular method to improve material multifunctional, biological and mechanical properties, as well as biocompatibility of artificial devices while obviating the need for large expense and long time to develop brand new materials. The nanostructure of a polymeric surface controls properties such as charge, conductivity, roughness, porosity, wettability, friction, physical and chemical reactivity and compatibility with the organism. There is a growing need for smart surfaces showing high biocompatibility, particularly in the area of artificial organs and prosthetics.[117–122]

Different modification methods have been recently developed to modulate the PLA surface properties. These methods can be classified as non-permanent (non-covalent attachment of functional groups) or permanent (covalent attachment).[123] The non-permanent approach includes surface coating that involves deposition/adsorption of molecules onto the polymer surface. Some authors reported that PLA can be coated with modifying species to control polymer-cell interactions. Eid *et al.*[124] covered PLA with arginine-glycine-aspartic acid (RGD) peptides; Chen *et al.*[125] used biomimetic apatite to produced scaffold and demonstrated that Saos-2 osteoblast-like cell compatibility of the scaffolds was greatly enhanced with these coatings; extracellular matrix proteins can also be used as reported by Atthof *et al.*[126] Although coating is a simple and convenient surface modification protocol, passive adsorption could induce competitive adsorption of other materials in the system and change the configuration of adsorbed species.

Another technique is the biomacromolecules entrapment, as reported by Quirk *et al.*[127] They used PEG and poly(L-lysine) (PLL) to improve biomaterial-cell interaction, while Cai *et al.*[128] modified PLA surfaces by entrapping poly(aspartic acid) to enhance their cell affinity. Rat osteoblasts were seeded onto the modified surfaces to examine their effects on cell adhesion and proliferation. Results showed that modified PLA surfaces may enhance the cell surface.[128] Moreover, migratory additives are blended with polyester as a way to tailor surface properties.[129] Yu *et al.*[130] blended poly(D,L-lactic acid)-block-poly(ethylene glycol) (PLE) copolymer and RGD derivatives with PLA to engineer the surface properties of the resultant blend to promote chondrocyte attachment and growth.

Plasma techniques can be easily used to introduce desired functional groups or chains onto the material surface, and they have special applications to improve the cell affinity of scaffolds. It has been reported that plasma treatment is a unique and powerful method to modify polymeric materials without altering their bulk properties. Cell affinity is the most important factor to be considered when biodegradable polymeric materials, are utilized as cell scaffolds in tissue engineering. Plasma treatment was used to improve the PLA surface properties.[131–133] This technique modifies PLA hydrophilicity, topography, roughness and cell affinity without affecting the degradation rate.[131] Fortunati and co-workers treated PLGA/silver nanocomposite surfaces with oxygen plasma to enhance antimicrobial and cell adhesion properties,[133] while Yang *et al.*[134] used anhydrous ammonia plasma treatment to create reactive amine groups on porous PLA scaffolds that anchored collagen through polar and hydrogen bonding interactions. These surface modified scaffolds showed enhanced cell adhesion.

The effects of plasma treatment were also evaluated by using oxygen on the surface of PLLA and the influence thereof on protein adsorption and on bone-cell behaviour was studied.[135] Results indicated that the PLLA surface became hydrophilic and its roughness increased with the treatment time having a major influence on the protein adsorption process. The outcome of the plasma treatment on various PLLA surfaces has shown to be the up-regulator of the cell-adhesive proteins expression and consequently the improvement of cell adhesion and growth. Oxygen-treated PLLA promoted higher adhesion and proliferation of the mesenchymal stem cells in comparison to untreated samples. It was concluded that following plasma treatment, PLLA samples show enhanced affinity for osteoprogenitor cells.[135]

The permanent surface modification methods include chemical conjugation using wet chemistry. Cai *et al.*[136] covalently attached chitosan to the PLA surface through alkaline surface hydrolysis followed by acid-chitosan conjugation, demonstrating that this treatment improved rat osteoblast attachment and proliferation. Liquid phase photografting was also used to immobilize useful chains onto PLA surfaces by the *"grafting to"* approach[137] or to introduce hydrophilic groups by *"grafting from"*.[138] Vapour phase photografting was used to avoid detrimental solvent effects.[139,140]

11.5 Conclusions

The potential applications of nanostructured poly(L-lactic acid) in bio-medicine are very broad. In this chapter we provided an overview on some fascinating developments on PLA modifications in the area of tissue engineering research and applications. Our aim was mainly to demonstrate the highly trans-disciplinary character of these topics and to give an overview on developments and research topics in chemistry, biology, physics and engineering that can lead to a revolution in clinical therapies and diagnostics. As with any emerging technology, a significant effort with extensive and interdisciplinary collaborations and partnerships is required to move concepts from research and development to fruitful commercialization and important clinical applications.

Acknowledgements

The authors wish to thank Prof. A. Bianco, Prof. S. Martino, Prof. C. Emiliani, Prof. L. Visai, Prof. H. Arzate, Prof. F. Elisei, Prof. L. Latterini, Prof. P. Locci and Prof. E. Becchetti for fundamental collaborations in biological and material science fields.

References

1. R. Langer and J. Vacanti, *Science*, 1993, **260**, 920.

2. C. Zhao, A. Tan, G. Pastorin and H. K. Ho, *Biotechnol. Adv.*, 2013, **31**, 654.
3. S. Martino, F. D'Angelo, I. Armentano, J. M. Kenny and A. Orlacchio, *Biotechnol. Adv.*, 2012, **30**, 338.
4. M. Richards, S. P. Tan, J. H. Tan, W. K. Chan and A. Bongso, *Stem Cells*, 2004, **22**, 51.
5. N. Carlesso and A. A. Cardoso, *Curr. Opin. Hematol.*, 2010, **17**, 281.
6. A. Orlacchio, G. Bernardi and S. Martino, *Discov. Med.*, 2010, **9**, 546.
7. A. Orlacchio, G. Bernardi and S. Martino, *Curr. Med. Chem.*, 2010, **17**, 595.
8. Z. Jiang, G. B. Adams, A. M. Hanash, D. T. Scadden and R. B. Levy, *Biol. Blood Marrow Transplant*, 2002, **8**, 588.
9. P. Bossolasco, T. Montemurro, L. Cova, S. Zangrossi, C. Calzarossa, S. Buiatiotis, D. Soligo, S. Bosari, V. Silani, G. L. Deliliers, P. Rebulla and L. Lazzari, *Cell Res.*, 2006, **16**, 329.
10. S. Ilancheran, A. Michalska, G. Peh, E. M. Wallace, M. Pera and U. Manuelpillai, *Biol. Reprod.*, 2007, 77, 577.
11. S. Martino, F. D'Angelo, I. Armentano, R. Tiribuzi, M. Pennacchi, M. Dottori, S. Mattioli, A. Caraffa, G. G. Cerulli and J. M. Kenny, *Tissue Eng. Part A*, 2009, **15**, 3139.
12. S. Martino, R. Tiribuzi, E. Ciraci, G. Makrypidi, F. D'Angelo, I. di Girolamo, A. Gritti, G. M. C. de Angelis, G. Papaccio, M. Sampaolesi, A. C. Berardi, A. Datti and A. Orlacchio, *Int. J. Biochem. Cell Biol.*, 2011, **43**, 775.
13. A. L. Vescovi and E. Y. Snyder, *Brain Pathol.*, 1999, **9**, 569.
14. G. Cossu and P. Bianco, *Curr. Opin. Genet. Dev.*, 2003, **13**, 537.
15. R. McKay, *Phil. Trans. R. Soc. B*, 2004, **359**, 851.
16. P. H. Lee and H. J. Park, *J. Clin. Neurol.*, 2009, **5**, 1.
17. X. R. Ortiz-Gonzalez, C. D. Keene, C. M. Verfaillie and W. C. Low, *Curr. Neurovasc. Res.*, 2004, **1**, 207.
18. I. R. Lemischka and K. A. Moore, *Nature*, 2003, **425**, 778.
19. A. A. Kiger, D. L. Jones, C. Schulz, M. B. Rogers and M. T. Fuller, *Science*, 2001, **294**, 2542.
20. A. Spradling, D. Drummond-Barbosa and T. Kai, *Nature*, 2001, **414**, 98.
21. A. Parolini and M. Caruso, *Placenta*, 2011, **32**, S186.
22. Y. Jiang, H. Lv, S. Huang, H. Tan, Y. Zhang and H. Li, *Neurol. Res.*, 2011, **33**, 331.
23. A. J. Marcus and D. Woodbury, *J. Cell. Mol. Med.*, 2008, **12**, 730.
24. K. Takahashi and S. Yamanaka, *Cell*, 2006, **126**, 663.
25. K. Hu, J. Yu, K. Suknuntha, S. Tian, K. Montgomery, K. D. Choi, R. Stewart, J. A. Thompson and I. I. Slukvin, *Blood*, 2011, **117**, e109.
26. E. Wang, D. Israel, S. Kelly and D. Luxenberg, *Growth Factors*, 1993, **9**, 57.
27. J. S. Park, H. N. Yang, D. G. Woo, S. Y. Jeon and K. H. Park, *Biomaterials*, 2011, **32**, 1495.
28. Y. C. Chen, D. C. Lee, T. Y. Tsai, C. Y. Hsiao, J. W. Liu, C. Y. Kao, H. K. Lin, H. C. Chen, T. J. Palathinkal and W. F. Pong, *Biomaterials*, 2010, **31**, 5575.

29. W. Li, Y. Guo, H. Wang, D. Shi, C. Liang, Z. Ye, F. Qing and J. Gong, *J. Mater. Sci. Mater. Med.*, 2008, **19**, 847.
30. M. P. Prabhakaran, J. R. Venugopal and S. Ramakrishna, *Biomaterials*, 2009, **30**, 4996.
31. A. Banas, Y. Yamamoto, T. Teratani and T. Ochiya, *Dev. Dynam.*, 2007, **236**, 3228.
32. K. D. Lee, T. K. C. Kuo, J. Whang-Peng, Y. F. Chung, C. T. Lin, S. H. Chou, J. R. Chen, Y. P. Chen and O. K. S. Lee, *Hepatology*, 2004, **40**, 1275.
33. M. Chivu, S. O. Dima, C. I. Stancu, C. Dobrea, V. Uscatescu, L. G. Necula, C. Bleotu, C. Tanase, R. Albulescu and C. Ardeleanu, *Transl. Res.*, 2009, **154**, 122.
34. K. Anselme, *Biomaterials*, 2000, **21**, 667.
35. L. Zhang and T. J. Webster, *Nano Today*, 2009, **4**, 66.
36. T. Dvir, B. P. Timko, D. S. Kohane and R. Langer, *Nature Nanotechnol.*, 2010, **6**, 13.
37. A. Curtis and C. Wilkinson, *Mater. Today*, 2001, **4**, 22.
38. K. Riehemann, S. W. Schneider, T. A. Luger, B. Godin, M. Ferrari and H. Fuchs, *Angew. Chem. Int. Ed.*, 2009, **48**, 872.
39. Q. Tran-Cong-Miyata, S. Nishigami, T. Ito, S. Komatsu and T. Norisuye, *Nat. Mater.*, 2004, **3**, 448.
40. C. G. Simon Jr, N. Eidelman, S. B. Kennedy, A. Sehgal, C. A. Khatri and N. R. Washburn, *Biomaterials*, 2005, **26**, 6906.
41. Y. Cha and C. G. Pitt, *Biomaterials*, 1990, **11**, 108.
42. M. Yasin and B. J. Tighe, *Biomaterials*, 1992, **13**, 9.
43. Y. Wan, Y. Fang, H. Wu and X. Cao, *J. Biomed. Mater. Res. A*, 2007, **80A**, 776.
44. Y. Wang and M. A. Hillmyer, *J. Polym. Sci. Polym. Chem.*, 2001, **39**, 2755.
45. K. S. Anderson, S. H. Lim and M. A. Hillmyer, *J. Appl. Polym. Sci.*, 2003, **89**, 3757.
46. Y. Kim, C. Choi, Y. Kim, K. Lee and M. Lee, *Fiber Polym.*, 2004, **5**, 270.
47. K. S. Anderson and M. A. Hillmyer, *Polymer*, 2004, **45**, 8809.
48. G. Biresaw and C. J. Carriere, *J. Polym. Sci. Polym. Phys.*, 2002, **40**, 2248.
49. L. Zhang, S. H. Goh and S. Y. Lee, *Polymer*, 1998, **39**, 4841.
50. C. H. Ho, C. H. Wang, C. I. Lin and Y. D. Lee, *Polymer*, 2008, **49**, 3902.
51. H. J. Jin, I. J. Chin, M. N. Kim, S. H. Kim and J. S. Yoon, *Eur. Polym. J.*, 2000, **36**, 165.
52. S. Ishida, R. Nagasaki, K. Chino, T. Dong and Y. Inoue, *J. Appl. Polym. Sci.*, 2009, **113**, 558.
53. X. Chen, S. L. McGurk, M. C. Davies, C. J. Roberts, K. M. Shakesheff, S. J. B. Tendler, P. M. Williams, J. Davies, A. C. Dawkes and A. Domb, *Macromolecules*, 1998, **31**, 2278.
54. D. Newman, E. Laredo, A. Bello, A. L. Grillo, J. L. Feijoo and A. J. Müller, *Macromolecules*, 2009, **42**, 5219.
55. F. Signori, M. B. Coltelli and S. Bronco, *Polym. Degrad. Stabil.*, 2009, **94**, 74.

56. W. Zhang, L. Chen and Y. Zhang, *Polymer*, 2009, **50**, 1311.
57. V. P. Martino, A. Jiménez and R. A. Ruseckaite, *J. Appl. Polym. Sci.*, 2009, **112**, 2010.
58. J. Y. Lim, J. C. Hansen, C. A. Siedlecki, R. W. Hengstebeck, J. Cheng, N. Winograd and H. J. Donahue, *Biomacromolecules*, 2005, **6**, 3319.
59. B. O. Leung, A. P. Hitchcock, J. L. Brash, A. Scholl and A. Doran, *Macromolecules*, 2009, **42**, 1679.
60. Z. Yuan and B. D. Favis, *Biomaterials*, 2004, **25**, 2161.
61. D. Yao, W. Zhang and J. G. Zhou, *Biomacromolecules*, 2009, **10**, 1282.
62. N. Kumar, R. S. Langer and A. J. Domb, *Adv. Drug Deliv. Rev.*, 2002, **54**, 889.
63. L. S. Nair and C. T. Laurencin, *Progr. Polym. Sci.*, 2007, **32**, 762.
64. O. Martin and L. Avérous, *Polymer*, 2001, **42**, 6209.
65. M. Sheth, R. A. Kumar, V. Davé, R. A. Gross and S. P. McCarthy, *J. Appl. Polym. Sci.*, 1997, **66**, 1495.
66. L. Zhang, C. Xiong and X. Deng, *Polymer*, 1996, **37**, 235.
67. S. Iannace, L. Ambrosio, S. J. Huang and L. Nicolais, *J. Appl. Polym. Sci.*, 1994, **54**, 1525.
68. R. Pucciariello, L. Tammaro, V. Villani and V. Vittoria, *J. Polym. Sci. Polym. Phys.*, 2007, **45**, 945.
69. J. M. Raquez, R. Narayan and P. Dubois, *Macromol. Mater. Eng.*, 2008, **293**, 447.
70. M. Sun and S. Downes, *J. Mater. Sci. Mater. Med.*, 2009, **20**, 1181.
71. J. T. Yeh, C. J. Wu, C. H. Tsou, W. L. Chai, J. D. Chow, C. Y. Huang, K. N. Chen and C. S. Wu, *Polym. Plast. Tech. Eng.*, 2009, **48**, 571.
72. V. R. Sinha, K. Bansal, R. Kaushik, R. Kumria and A. Trehan, *Int. J. Pharm.*, 2004, **278**, 1.
73. X. J. Xu, J. C. Sy and V. Prasad-Shastri, *Biomaterials*, 2006, **27**, 3021.
74. G. Chouzouri and M. Xanthos, *Acta Biomater.*, 2007, **3**, 745.
75. D. Wu, Y. Zhang, M. Zhang and W. Yu, *Biomacromolecules*, 2009, **10**, 417.
76. D. Wu, Y. Zhang, M. Zhang and W. Zhou, *Eur. Polym. J.*, 2008, **44**, 2171.
77. S. S. Sabet and A. A. Katbab, *J. Appl. Polym. Sci.*, 2009, **111**, 1954.
78. F. Khan, R. S. Tare, J. M. Kanczler, R. O. C. Oreffo and M. Bradley, *Biomaterials*, 2010, **31**, 2216.
79. N. A. Rahman and E. Mathiowitz, *J. Contr. Release*, 2004, **94**, 163.
80. X. Zhang, H. Hua, X. Shen and Q. Yang, *Polymer*, 2007, **48**, 1005.
81. Y. Jiao, Z. Liu and C. Zhou, *J. Biomed. Mater. Res. A*, 2007, **80A**, 820.
82. A. R. C. Duarte, J. F. Mano and R. L. Reis, *J. Supercrit. Fluids*, 2010, **54**, 282.
83. W. H. De Jong and P. J. Borm, *Int. J. Nanomed.*, 2008, **3**, 133.
84. D. Yadav, S. Suri, A. A. Chaudhary, M. N. Beg, V. Garg, M. Asif and A. Ahmad, *Pol. J. Chem. Tech.*, 2012, **14**, 57.
85. A. Kumari, S. K. Yadav and S. C. Yadav, *Colloids Surf. B*, 2010, **75**, 1.
86. M. L. Hans and A. M. Lowman, *Curr. Opin. Solid State Mater. Sci.*, 2002, **6**, 319.

87. T. Jung, W. Kamm, A. Breitenbach, G. Klebe and T. Kissel, *Pharmaceut. Res.*, 2002, **19**, 1105.

88. L. Nobs, F. Buchegger, R. Gurny and E. Allémann, *Int. J. Pharmaceut.*, 2003, **250**, 327.

89. I. Gutierro, R. M. Hernández, M. Igartua, A. R. Gascón and J. L. Pedraz, *Vaccine*, 2002, **21**, 67.

90. J. Cheng, B. A. Teply, I. Sherifi, J. Sung, G. Luther, F. X. Gu, E. Levy-Nissenbaum, A. F. Radovic-Moreno, R. Langer and O. C. Farokhzad, *Biomaterials*, 2007, **28**, 869.

91. A. Budhian, S. J. Siegel and K. I. Winey, *Int. J. Pharmaceut.*, 2007, **336**, 367.

92. M. Gaumet, R. Gurny and F. Delie, *Int. J. Pharmaceut.*, 2007, **342**, 222.

93. V. Mohanraj and Y. Chen, *Trop. J. Pharmaceut. Res.*, 2007, **5**, 561.

94. Y. Y. Yang, T. S. Chung and N. P. Ng, *Biomaterials*, 2001, **22**, 231.

95. J. Panyam, M. M. Dali, S. K. Sahoo, W. Ma, S. S. Chakravarthi, G. L. Amidon, R. J. Levy and V. Labhasetwar, *J. Contr. Release*, 2003, **92**, 173.

96. V. Lassalle and M. L. Ferreira, *Macromol. Biosci.*, 2007, 7, 767.

97. S. J. Oldenburg, R. D. Averitt, S. L. Westcott and N. J. Halas, *Chem. Phys. Lett.*, 1998, **288**, 243.

98. J. N. Smith, J. Meadows and P. A. Williams, *Langmuir*, 1996, **12**, 3773.

99. S. Kim, B. Fisher, H. J. Eisler and M. Bawendi, *J. Am. Chem. Soc.*, 2003, **125**, 11466.

100. C. Loo, A. Lowery, N. Halas, J. West and R. Drezek, *Nano Lett.*, 2005, **5**, 709.

101. M. Filippousi, S. A. Papadimitriou, D. N. Bikiaris, E. Pavlidou, M. Angelakeris, D. Zamboulis, H. Tian and G. Van Tendeloo, *Int. J. Pharmaceut.*, 2013, **448**, 221.

102. Y. Y. Yang, Y. Wang, R. Powell and P. Chan, *Clin. Exp. Pharmacol. Physiol.*, 2006, **33**, 557.

103. A. Ramzi, D. Schwahn, W. E. Hennink and C. F. van Nostrum, *J. Phys. Chem. B*, 2008, **112**, 784.

104. M. Ballauff and Y. Lu, *Polymer*, 2007, **48**, 1815.

105. J. L. Zhang, R. S. Srivastava and R. D. K. Misra, *Langmuir*, 2007, **23**, 6342.

106. D. W. Schaefer and R. S. Justice, *Macromolecules*, 2007, **40**, 8501.

107. H. Liu and T. J. Webster, *Biomaterials*, 2007, **28**, 354.

108. A. R. Boccaccini and V. Maquet, *Compos. Sci. Tech.*, 2003, **63**, 2417.

109. S. Ramakrishna, J. Mayer, E. Wintermantel and K. W. Leong, *Compos. Sci. Tech.*, 2001, **61**, 1189.

110. S. D. McCullen, D. R. Stevens, W. A. Roberts, L. I. Clarke, S. H. Bernacki, R. E. Gorga and E. G. Loboa, *Int. J. Nanomed.*, 2007, **2**, 253.

111. M. O. Montjovent, L. Mathieu, H. Schmoekel, S. Mark, P. E. Bourban, P. Y. Zambelli, L. A. Laurent-Applegate and D. P. Pioletti, *J. Biomed. Mater. Res. A*, 2007, **83**, 41.

112. E. Nejati, H. Mirzadeh and M. Zandi, *Compos. Appl. Sci. Manuf.*, 2008, **39**, 1589.

113. I. Armentano, M. Dottori, E. Fortunati, S. Mattioli and J. M. Kenny, *Polym. Degrad. Stabil.*, 2010, **95**, 2126.
114. I. Armentano, N. Bitinis, E. Fortunati, S. Mattioli, N. Rescignano, R. Verdejo, M. A. Lopez-Manchado and J. M. Kenny, *Progr. Polym. Sci.*, 2013, **38**, 1720.
115. F. D'Angelo, I. Armentano, I. Cacciotti, R. Tiribuzi, M. Quattrocelli, C. Del Gaudio, E. Fortunati, E. Saino, A. Caraffa, G. G. Cerulli, L. Visai, J. M. Kenny, M. Sampaolesi, A. Bianco, S. Martino and A. Orlacchio, *Biomacromolecules*, 2012, **13**, 1350.
116. S. Montesano, E. Lizundia, F. D'Angelo, E. Fortunati, S. Mattioli, F. Morena, I. Bicchi, F. Naro, M. Sampaolesi, J. Sarasua, J. M. Kenny, A. Orlacchio, I. Armentano and S. Martino, *ISRN Tissue Engineering*, 2013, **13**, 8.
117. F. Zeifang, M. Grunze, G. Delling, H. Lorenz, C. Heisel, G. Tosounidis, D. Sabo, H. G. Simank and J. H. Holstein, *Med. Sci. Monit.*, 2008, **14**(2), BR35.
118. U. Schmelmer, A. Paul, A. Küller, M. Steenackers, A. Ulman, M. Grunze, A. Gölzhäuser and R. Jordan, *Small*, 2007, **3**, 459.
119. A. Welle, M. Grunze and D. Tur, *J. Colloid Interface Sci.*, 1998, **197**, 263.
120. J. P. Spatz and B. Geiger, *Methods Cell. Biol.*, 2007, **83**, 89.
121. C. Mohrdieck, F. Dalmas, E. Arzt, R. Tharmann, M. M. Claessens, A. R. Bausch, A. Roth, E. Sackmann, C. H. Schmitz and J. Curtis, *Small*, 2007, **3**, 1015.
122. T. Steinberg, S. Schulz, J. P. Spatz, N. Grabe, E. Mussig, A. Kohl, G. Komposch and P. Tomakidi, *Nano Lett.*, 2007, 7, 287.
123. R. M. Rasal, A. V. Janorkar and D. E. Hirt, *Progr. Polym. Sci.*, 2010, **35**, 338.
124. K. Eid, E. Chen, L. Griffith and J. Glowacki, *J. Biomed. Mater. Res.*, 2001, 57, 224.
125. Y. Chen, A. F. T. Mak, M. Wang, J. Li and M. S. Wong, *Surf. Coating. Tech.*, 2006, **201**, 575.
126. B. Atthoff and J. Hilborn, *J. Biomed. Mater. Res. B Appl. Biomater.*, 2007, **80B**, 121.
127. R. A. Quirk, M. C. Davies, S. J. Tendler, W. C. Chan and K. M. Shakesheff, *Langmuir*, 2001, **17**, 2817.
128. K. Cai, K. Yao, X. Hou, Y. Wang, Y. Hou, Z. Yang, X. Li and H. Xie, *J. Biomed. Mater. Res.*, 2002, **62**, 283.
129. D. J. Irvine, A. V. G. Ruzette, A. M. Mayes and L. G. Griffith, *Biomacromolecules*, 2001, **2**, 545.
130. G. Yu, J. Ji, H. Zhu and J. Shen, *J. Biomed. Mater. Res. B Appl. Biomater.*, 2006, **76B**, 64.
131. S. Mattioli, J. M. Kenny and I. Armentano, *J. Appl. Polym. Sci.*, 2012, **125**, E239.
132. T. Hirotsu, K. Nakayama, T. Tsujisaka, A. Mas and F. Schue, *Polym. Eng. Sci.*, 2002, **42**, 299.

133. E. Fortunati, S. Mattioli, L. Visai, M. Imbriani, J. L. G. Fierro, J. M. Kenny and I. Armentano, *Biomacromolecules*, 2013, **14**, 626.

134. J. Yang, J. Bei and S. Wang, *Biomaterials*, 2002, **23**, 2607.

135. I. Armentano, G. Ciapetti, M. Pennacchi, M. Dottori, V. Devescovi, D. Granchi, N. Baldini, B. Olalde, M. J. Jurado, J .I .M. Alava and J. M. Kenny, *J. Appl. Polym. Sci.*, 2009, **114**, 3602.

136. K. Cai, K. Yao, Y. Cui, S. Lin, Z. Yang, X. Li, H. Xie, T. Qing and J. Luo, *J. Biomed. Mater. Res.*, 2002, **60**, 398.

137. A. Zhu, M. Zhang, J. Wu and J. Shen, *Biomaterials*, 2002, **23**, 4657.

138. Z. Ma, C. Gao, Y. Gong and J. Shen, *Biomaterials*, 2003, **24**, 3725.

139. M. Källrot, U. Edlund and A. C. Albertsson, *Biomaterials*, 2006, **27**, 1788.

140. M. Källrot, U. Edlund and A. C. Albertsson, *Biomacromolecules*, 2007, **8**, 2492.

Degradation and Biodegradation of PLA

CHAPTER 12

Abiotic-hydrolytic Degradation of Poly(lactic acid)

KIKKU FUKUSHIMA* AND GIOVANNI CAMINO

Polytechnic of Turin, Alessandria Branch, INSTM Research Unit, Viale Teresa Michel 5, 15100 Alessandria, Italy
*Email: kikku.fukushima@gmail.com

12.1 Introduction

Research on biodegradable polymers has gained considerable interest in recent years due to their increasingly attractive environmental, biomedical and agricultural applications.[1] Among all biodegradable polymers, polyesters play a key role because of their highly hydrolyzable ester bonds, making this class of polymers highly subject to degradation in humid environments, which is favourable for biodegradation mechanisms.

Poly(lactic acid) (PLA) is one of the most frequently used polyesters in biomedical applications because of its bio-resorbability and biocompatible properties in the human body. The main reported examples on medical or biomedical PLA applications are fracture fixation devices like screws, sutures, delivery systems and micro-titration plates.[2,3]

Understanding the hydrolytic degradation of PLA is essential for the plastics industry to meet the current strict environmental regulations. Furthermore, evaluation and control of its hydrolytic degradation and biodegradation has been essential for medical applications. In this way, hydrolysis of PLA has been reported[1] to proceed either at the surface (homogeneous) or within the bulk (heterogeneous) and it is controlled by a wide variety of variables, *e.g.* matrix morphology, chain orientation,

RSC Polymer Chemistry Series No. 12
Poly(lactic acid) Science and Technology: Processing, Properties, Additives and Applications
Edited by Alfonso Jiménez, Mercedes Peltzer and Roxana Ruseckaite
© The Royal Society of Chemistry 2015
Published by the Royal Society of Chemistry, www.rsc.org

chemical composition, stereochemical structure, sequence distribution, molecular weight distribution, the presence of residual monomers and the environmental degradation conditions. Most of the studies on PLA degradation have been concentrated on abiotic hydrolysis,[4,5] and have found that, most of the time, an unfavourable hydrolytic degradation rate of PLA matrix limits its applications.[6] Consequently, considerable efforts have been made to control its hydrolytic degradation rate (usually to accelerate it), so that PLA can have wider biomedical or ecological applications without a compromise of the product properties prior to degradation.

In this context, numerous hydrolytic degradation tests have been performed on PLA to simulate its degradation process in the human body ($T \sim 37\,°C$)[7–13] and in natural media as soil or compost ($25\,°C < T < 58\,°C$)[14–16], all reporting that PLA can be hydrolyzed to give low-molecular-weight water-soluble oligomers.

The hydrolytic degradation of PLA has been reported to take place mainly in the bulk of the material rather than on the surface[6,17,18] and has been assumed as an autocatalytic hydrolysis which occurs homogeneously along sample cross-section.[6,17,19] The formation of lactic acid oligomers, which follows the chain scission, increases the carboxylic acid end-groups concentration in the degradation medium, making the hydrolytic degradation of PLA a self-catalyzed and self-maintaining process thanks to the catalytic action of carboxyl end-groups.

In parallel, the physical structure of PLA has been reported to affect its hydrolytic degradation mechanism. For example, it has been found that the hydrolytic chain cleavage proceeds preferentially in the amorphous regions, leading to the increase in polymer crystallinity.[1,19,20] The rate of the degradation reaction seems also to be affected by the shape of the specimen and by the chemical structure of the polymer, as well as by the conditions under which the hydrolysis is performed, including pH and temperature.[21,22]

Thus, considering the high importance of the appropriate hydrolytic degradation rate of PLA when this is utilized for its recycling to lactic acid or as biomedical materials with an optimal degradation rate in the human body (as pharmaceutical matrices), the aim of this chapter is to review the recent studies reported in the literature on the mechanisms, affecting parameters and possible applications of the abiotic hydrolytic degradation of PLA.

12.2 Molecular Hydrolytic Degradation Mechanisms of PLA

Poly(L-lactide) [poly(L-lactic acid) (PLLA)], poly(D-lactide) [poly(D-lactic acid) (PDLA)] and their copolymers belong to the family of aliphatic polyesters and therefore their ester groups can be hydrolytically degraded in the presence of water according to the following reaction:[23]

$$-COO- + H_2O \rightarrow -COOH + HO- \tag{1}$$

The position of the ester group (chain-end or within the chain) seems not to affect the hydrolytic degradability of PLA-based materials, as long as the polymers are composed of one monomer unit. Hakarainen *et al.*[24] indicated that PLLA was hydrolytically (abiotically) degraded *via* random cleavage while a chain-end cleavage was observed for biotic hydrolytic degradation. An exception is the report on the acidic hydrolysis of amorphous poly(DL-lactic acid) (PDLLA) in the solid state by Shih,[22,25] which stated that the ester group adjacent to the chain-terminal was more susceptible to hydrolytic degradation than those in the middle of the chain. Degradation at the chain-end and in the middle of the chain was called exo-chain and endo-chain cleavage, respectively.

In parallel, the hydrolytic degradation of semicrystalline PLA has been reported to occur in two stages.[26] The first stage starts with water diffusion into the amorphous regions, which are less organized and allow water to diffuse more easily. The second stage starts when most of the amorphous regions are degraded. The hydrolytic attack then proceeds from the edge towards the center of the crystalline domains. This explains the reported much faster hydrolysis rate of amorphous PLA compared to semicrystalline PLA.

In general, the hydrolytic degradation of PLA seems to take place mainly in the bulk of the material rather than in its surface.[26] A bulk erosion mechanism takes place when the hydrolytic degradation rate of the material surface is not so high compared to the diffusion rate of water, and therefore hydrolytic degradation takes place homogeneously, irrespective of the depth from the material surface (Figure 12.1b). According to Tsuji,[23] this degradation mechanism can be divided into at least three stages: (1) initial hydration or water absorption; (2) gradual decrease in molecular weight without weight loss; (3) weight loss through the formation and dissolution of water-soluble oligomers and monomers. In contrast, a surface erosion mechanism occurs when the hydrolytic degradation rate of the material surface in contact with water (containing catalytic substances such as alkalis and enzymes) is much higher than the diffusion rate of water molecules or of catalytic substances within the material. In such a case, the hydrolytic degradation seemingly occurs solely on the material surface because the hydrolytic degradation rate is much higher on the surface than at the core (Figure 12.1a).[27]

Some structural or external factors can determine the predominant hydrolytic degradation mechanism. For example, when the material thickness becomes higher than some determined value (critical thickness) the hydrolytic degradation mechanism can change from bulk to surface erosion.[28,29] On the other hand, in some cases PLA and some of its copolymers[30–32] can be hydrolytically degraded *via* a core-accelerated bulk erosion when the material is thicker than 0.5–2 mm. In such cases, hydrolysis-formed oligomers and monomers with high catalytic effect are trapped and accumulated in the core part of the materials,[33,34] resulting in an accelerated hydrolytic degradation in the core part (core-accelerated bulk erosion) as shown in Figure 12.1c.

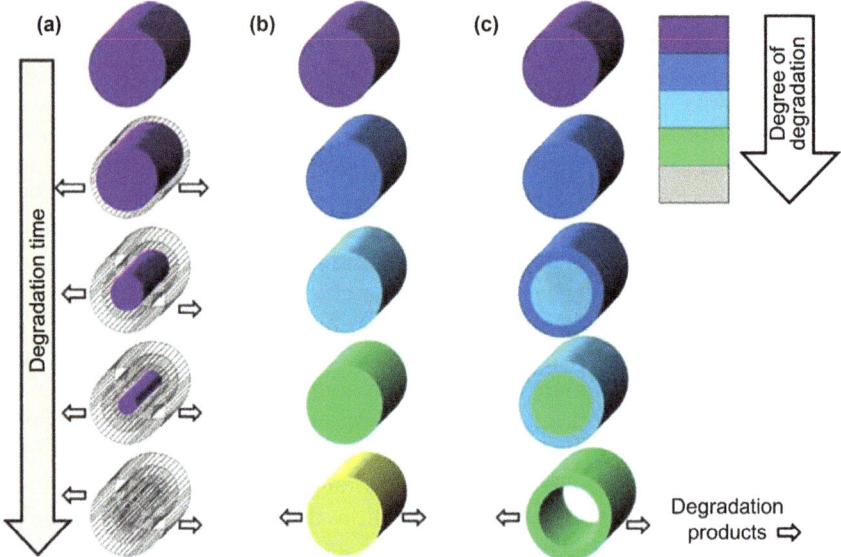

Figure 12.1   Degradation modes for biopolymers under (a) surface erosion, (b) bulk
degradation and (c) core-accelerated bulk erosion.
Taken from Sin et al.,[34] with permission from Elsevier.

The study of PLA hydrolysis has been performed in aqueous media, such
as phosphate-buffered solutions or water, at 37 °C, to simulate its degradation in body fluids at the appropriate temperature. Studies have also been
done at higher temperatures, in acidic, alkaline or buffered solutions, to
determine the hydrolytic effects of PLA under severe and accelerated conditions. Simultaneously, different sample structural parameters have been
studied during the hydrolytic degradation of PLA, such as shape and history
of the specimen, chemical structure, molecular weight, tacticity, crystallinity
and molar distribution, purity and morphology. We summarize and discuss
in detail the above-mentioned parameters in the next parts of this chapter.

12.3   Factors Controlling Hydrolytic Degradation of PLA

12.3.1   Degradation Medium Conditions

12.3.1.1   pH

Hydrolytic degradation of PLA-based materials involves the cleavage of ester
groups, and therefore the rate of reaction ester splitting through catalysis
can be influenced by the presence of hydronium and hydroxide ions.[35] By
comparing poly(glycolic acid) and poly(lactide-co-glycolide) sutures, Chu[36]
reported that the breaking strength of an entire suture depended on the
medium pH, especially at high and low pH values. Under acidic and basic

conditions, an ion exchange occurred to promote a stable condition when chain cleavage took part.

De Jong *et al.*[37] reported that the cleavage mechanism of PLA in solution depended on the media pH. As shown in Figure 12.2, in acidic solutions hydrolytic degradation seems to proceed *via* a chain-end scission mechanism in a lactyl monomer unit, forming lactic acid, whereas in alkaline solutions hydrolytic degradation takes place *via* back-biting to form lactyl dimer (lactide) units, which are further hydrolyzed to give lactoyl lactic acid. Hydrolytic degradation under controlled neutral pH conditions (similar to *in vivo* degradation experiments in the human body) were studied by van Nostrum *et al.*[38] utilizing acetylated and non-acetylated (hydroxy-terminated)

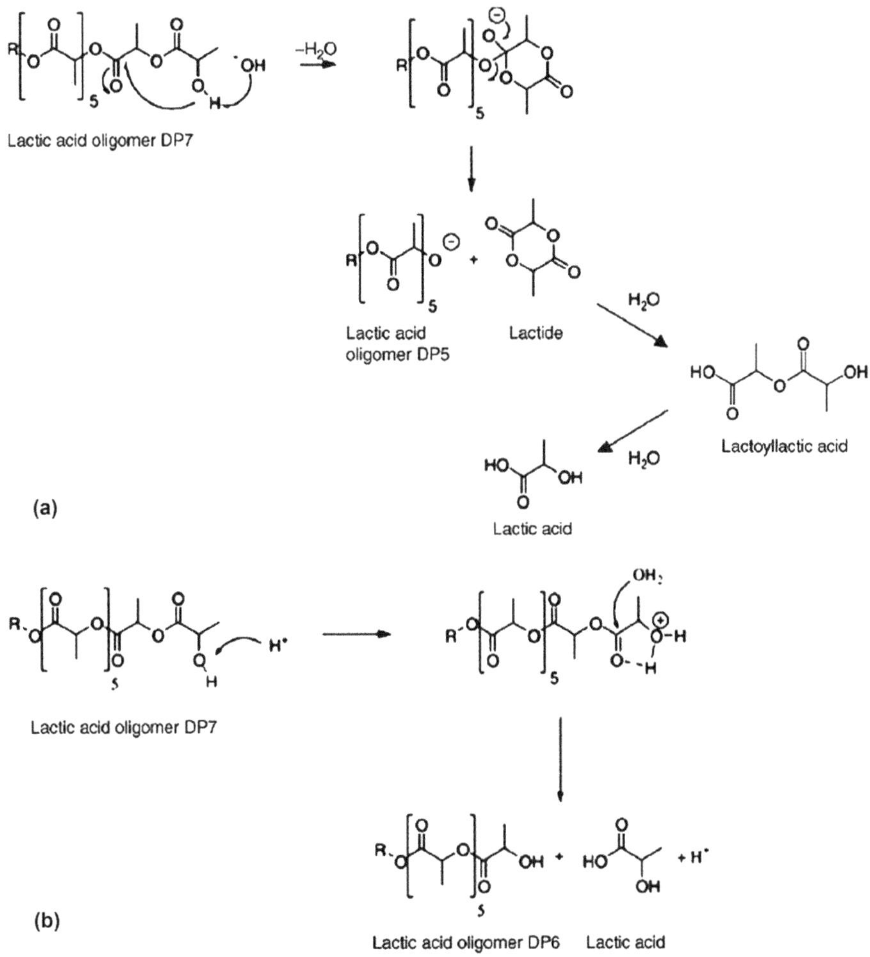

(a)

Lactic acid oligomer DP7

Lactic acid oligomer DP5

Lactide

Lactoyllactic acid

Lactic acid

(b)

Lactic acid oligomer DP7

Lactic acid oligomer DP6   Lactic acid

Figure 12.2 Hydrolytic chain cleavage mechanisms of PLA in alkaline (a) and acidic (b) solutions.
Taken from Tsuji,[27] with permission from John Wiley & Sons, Inc.

L-lactic acid oligomers esterified with *N*-(2-hydroxypropyl) methacrylate and showed that the hydrolytic degradation of hydroxy-terminated L-lactic acid oligomers seemed to proceed *via* a back-biting cleavage, which resulted in dimer formation, whereas that of acetylated L-lactic acid oligomers took place *via* random chain cleavage.

Concerning the PLA hydrolytic degradation rate, it has been reported that the high concentration of hydroxide ions in alkaline media strongly accelerates the process.[36,39,40] For instance, Makino *et al.*[39] indicated that the alkaline hydrolytic degradation of PDLLA and PLLA increased with ionic strength. Therefore, alkaline media could be used for accelerated tests for hydrolyzable polymeric materials. On the other hand, Kulkarni *et al.*[41] indicated that the PDLLA and poly(DL-lactic acid-*co*-glycolic acid) (P(DLLA-*co*-GA)) degradation in an alkaline solution proceeded *via* chain-end cleavage and that it was enhanced at low and high pH. They found that the degradation rates initially increased with time, followed by a maximum value, and finally decreased, according to molecular weight analyses.

Acidic media can also accelerate the hydrolytic degradation of PLA-based materials.[39] However, when the molecular weight of PLA is higher than 1×10^5 g mol^{-1}, hydronium ions seem to have an insignificant or a very low catalytic effect on the hydrolytic degradation of PLA-based materials.[42]

12.3.1.2 Temperature

It is well recognized that high temperatures can accelerate the hydrolytic degradation process of PLA.[43] Hyon *et al.*[44] found that PLA fibres remained intact for 6 months at 37 °C in phosphate-buffered saline solution, pH 7.4, whereas a 50% weight loss was detected after 30 h at 100 °C. Fukushima *et al.*[45] also found that in phosphate-buffered solution, pH 7.0, crystalline PLA films were considerably faster degraded at 58 °C than at 37 °C due to the higher occurrence of micro-structural changes and molecular rearrangements at temperatures close to or higher than PLA glass transition temperature (T_g), resulting in high water absorption into the polymer. Similar conclusions were obtained by Karjomaa *et al.*,[46] reporting increases of the water absorption level into the polymer matrix with temperature.

Interestingly enough, the degradation temperature seems to affect not only the PLA degradation rate, but also its degradation mechanism. In this way, conflicting works on the type of degradation mechanism of PLA as related to temperatures above or below T_g have been reported. For example, Weir *et al.*[47] suggested that above or below T_g, PLLA degradation proceeded by the same mechanism. However, Agrawal *et al.*[48] found that activation energies for the degradation reaction in 50:50 poly(lactic-*co*-glycolic acid) (PLGA) copolymers at temperatures below and above T_g were different and recommended that tests performed at $T > T_g$ should not be used to predict the polymer degradation trend at $T < T_g$. Reed and Gilding[49] revealed that in neutral conditions the PLA hydrolytic degradation rate dramatically increased when the degradation temperature was over T_g. They found that the hydrolytic degradation mechanism of crystallizable PLA-based materials varied when the degradation temperature became higher than the melting temperature (T_m), since crystalline regions melted and disappeared at temperatures exceeding T_m, and hydrolytic degradation in the melt took place homogeneously.

In a recent work, Fukushima *et al.*[45] found that in neutral conditions the hydrolytic degradation rate of crystalline PLA was considerably increased from 37 °C to temperatures close to PLA's T_g ($\sim$58 °C), observing at $T \sim T_g$ that no signals of PLA crystallization were detected after only 3 weeks under degradation, according to DSC measurements. At 37 °C, only some decrease in the PLA cold crystallization temperature (T_{cc}) and melting were distinguished after 8 weeks as the result of the crystallization of shorter polymer chains, leading to a lower T_{cc}. On the other hand, the order of the crystalline zones decreased at low molecular weights, owing to the disorder introduced by chain ends, thus leading to a decrease of T_m.[6]

12.3.2 Structure and Properties of PLA-based Materials

12.3.2.1 Shape and Geometry

12.3.2.1.1 Thickness. The hydrolytic degradation mechanism of PLA seems to change from the bulk erosion to the surface erosion mechanism when the material thickness becomes higher than a critical value.[28] For instance, Jie *et al.*[29] found that poly(DLLA-*co*-1,3-trimethylene carbonate (TMC)) cylindrical specimens with relatively large diameters of 2.5 mm were hydrolytically degraded in a neutral medium *via* a surface erosion mechanism. On the other hand, other researchers have found that in neutral conditions the hydrolytic degradation of PLLA,[31] PDLLA,[30] poly(L-lactic acid-*co*-glycolic acid) (P(LLA-*co*-GA)) and P(DLLA-*co*-GA)[32] and PDLLA-*b*-PEO[49,50] proceeds *via* a core-accelerated bulk erosion even at thicknesses 2 mm–7.4 cm. They attributed this phenomenon to the accumulation in the specimen core of highly catalytic oligomers and monomers formed during hydrolysis, resulting in an accelerated hydrolytic degradation by a core-accelerated bulk erosion, as shown in Figure 12.1c. They concluded that the hydrolytic degradation of PLA can proceed *via* bulk erosion, core-accelerated erosion and surface erosion mechanism for material thicknesses lower than 0.5–2 mm, between 2 mm and 7.4 cm and higher than 7.4 cm, respectively.

12.3.2.1.2 Porosity and Surface Area. It has been reported that pore formation decreases the degradation rate of PLA-based materials.[51] Lee *et al.*[52] found that higher sample thickness and lower porosity values in PLLA increased hydrolytic degradation rate in neutral conditions because the large thickness and small surface area could lead to the slow outward diffusion rate and high bulk concentration of the oligomeric degradation products, resulting in highly catalyzed bulk acidic degradations.

However, Chen and Ma[53] found that nanofibrous foams based on PLA presented higher hydrolytic degradation rate in neutral conditions than solid-wall foams, according to molecular weight measurements. Huang et al.[54] also indicated that under neutral conditions, porous P(DLLA-*co*-GA) had higher degradation rate than dense P(DLLA-*co*-GA). Additionally, Tsuji et al.[55] found that surface pores could enhance the alkaline and enzymatic surface hydrolytic degradation of aliphatic polyesters by the increased surface area per unit mass of porous materials. Similarly, Shirahase et al.[56] reported that pores formed by the removal of water-soluble polymers from PLLA/water-soluble polymer blends accelerated the alkaline hydrolytic degradation of PLLA. All these results indicate that the hydrolytic degradation rate of PLA should be determined by considering the combined effects of water diffusion rate on the material and the concentration of hydrolysis-formed catalytic products.

12.3.2.2 *Molecular Weight*

Molecular weight is one of the most crucial factors for the hydrolytic degradation rate of PLA-based materials. Saha et al.[57,58] studied the effect of different molecular weight values on the hydrolytic degradation of PLLA in neutral conditions. They found that in the range of 8×10^4–4×10^5 g mol^{-1} the effect of molecular weight on the degradation rate of PLA was insignificant, observing similar decreases in the number average molecular weight (M_n) of PLLA matrix against degradation time. However, at M_n values lower than 4×10^4 g mol^{-1} the hydrolytic degradation was considerably accelerated by decreasing molecular weight. This last phenomenon was explained considering that at low-molecular-weight values higher molecular mobility, higher density of hydrophilic terminal carboxyl- and hydroxyl groups and higher probability of formation of water-soluble oligomers and monomers could take place, thus increasing the water diffusion rate and content, and catalyzing PLA degradation.

Similarly, Wu et al.[59] reported that the rate of neutral hydrolytic degradation of PLLA co-polymers with glycolide (GA) (P(DLLA-*co*-GA)) and weight average molecular weight (M_w) 1.1×10^4–1.7×10^5 g mol^{-1} decreased at high molecular weights. A similar trend was also reported by Arvanitoyannis et al.[60,61] in branched PLLA samples co-initiated with glycerol (3-arm) or sorbitol (5-arm) with $M_n = 8 \times 10^4$ g mol^{-1} under alkaline conditions.

12.3.2.2.1 Chemical Structure

12.3.2.2.1.1 Tacticity. It has been found that low optical purity (isotacticity) in PLLA by incorporation of D-configured lactyl-units can significantly accelerate the neutral and alkaline hydrolytic degradation of PLA,[16,57] given that the chain heterogeneity of low optical purity PLA can cause a higher degree of disordered chain packing, leading to higher water absorption rates onto the amorphous regions of the polymer matrix. Saha et al.[57] reported that even the addition of 1% of D-lactyl units could significantly catalyze the neutral hydrolytic degradation of PLA samples. Similarly, Malin *et al.*[62] found that copolymers of P(LLA-*co*-CL) retained higher molecular weight than P(DLLA-*co*-CL) during their neutral hydrolytic degradation in buffered solutions.

By using molecular modelling techniques, Karst and Yang[63] investigated the effect of the L-/D- units ratio and of their block or random arrangements on the hydrolysis of stereo-copolymers of poly(L-lactide-*co*-D-lactide) P(LLA–DLA). They found that among the stereo-copolymers with the longest blocks of L- and D-lactyl units, those containing 50% L-lactyl units were more hydrolytically resistant compared to stereo-copolymers with 26% or 74% of L-lactyl units. They related this last phenomenon to the fact that the amount of stable stereo-complexes was higher for block stereo-copolymers than for random stereo-copolymers and was highest for block stereo-copolymers with 50% L-lactyl units.

12.3.2.2.1.2 Branching Level. Higher branching levels are expected to accelerate the PLA hydrolytic degradation due to the higher number per unit mass of hydrophilic and/or catalytic terminal groups and of terminal groups where LA is formed by back-biting.[23,64,65] Li and Kissel[66] found that the neutral hydrolytic degradation of prepared linear block copolymers of LLA and GA was slower than in the case of 4-arm/8-arm branched block copolymers of LLA and GA co-initiated with branched PEO. Star-block copolymers possess on average shorter chain lengths of poly(L-lactide-*co*-glycolide) (PLLGA) block due to the star architecture in comparison to the linear block copolymers. Water-soluble breakdown products, such as PLLGA oligomers and monomers, can be produced faster in star-block copolymers than in linear block copolymers after fewer hydrolytic cleavage steps, leading to faster erosion of the matrix.

Numata *et al.*[67] reported that 22 branches of polyglycerine co-initiated PDLLA enhanced the alkaline hydrolytic degradation of the polymer matrix. They also found that six branches of myo-inositol-*co*-initiated PDLLA had no effect on the alkaline hydrolytic degradation of PLLA, according to gel permeation chromatography results.

12.3.2.2.1.3 Crosslinking. It is expected that crosslinking reduces the hydrolyzable regions of PLA considering that crosslinked rigid structures are thought to reduce the number of terminal groups in the specimen and to disturb the diffusion of water molecules into the bulk material. In this way, Younes *et al.*[68] reported relatively lower neutral hydrolytic degradation rates of P(DLLA-*co*-CL) co-polymers initiated with glycerol and crosslinked with 2,2-bis(CL-4-yl)-propane as compared to analogous samples without any crosslinking.

12.3.2.2.2   Orientation of Polymer Chains.   It is usually reported that the orientation of polymer chains has relatively small effect on the neutral and alkaline hydrolytic degradation of PLA, in contrast to the enzymatic hydrolytic degradation of PLLA.[69] In general, polymer crystallinity, rather than bi-axial orientation of polymer chains, seems to determine at higher extent the alkaline and neutral hydrolytic degradation rate of PLLA. However, Hyon *et al.*[70] indicated that hydrostatically extruded PLLA rods could retain longer their bending strength and modulus upon hydrolysis as compared to uniaxially stretched PLLA rods prepared by drawing in silicon oil, due to their superior molecular orientation and packing able to delay the hydrolytic deterioration of their mechanical properties.

12.3.2.2.3 Crystallinity. PLA and copolymers, having sufficiently long stereo-regular sequence length, are in theory able to crystallize.[23] In this manner, upon crystallization some chains can assemble to form crystalline regions, as depicted in Figure 12.3a and it was discussed in chapters 2 and 3 in this volume. The chains in the crystalline regions have been reported to be hydrolysis-resistant compared to those in the amorphous regions because the access of water molecules to chains inside the rigid crystalline regions is highly restricted. This causes selective or predominantly hydrolytic cleavage of chains in the amorphous regions and removal of hydrolysis-formed water-soluble oligomers and monomers, with only some residual crystalline regions, or "crystalline residues" remaining (see Figure 12.3b).[23] This degradation trend has indeed been observed in a previous work by Fukushima *et al.*,[71] who reported considerably higher decreases in the sample weight of fully amorphous PDLLA as compared to those reported in the literature for semicrystalline PLA under similar conditions of hydrolytic degradation (phosphate-buffered solution pH = 4.0 at 37 °C). They reported that after 18 weeks, amorphous PDLLA presented a weight decrease by *ca.* 14% with respect to its initial mass, whereas Tsuji *et al.*[72,73] obtained similar weight losses for a semicrystalline PLA by hydrolysis in phosphate buffered saline with pH = 3.4 at 37 °C, but only after 20 months. These results indicate that the hydrolysis level obtained for the amorphous PDLLA matrix is reasonably high and attributed to the easier hydrolytic attack of ester bonds in amorphous polymer matrices as compared to crystalline polymers. The authors also studied the hydrolytic degradation of amorphous PDLLA, PCL and their blends in a phosphate-buffered solution of pH = 4.0 at 37 °C and reported a considerably faster degradation of PDLLA as compared to PCL due to the hydrophobic and semicrystalline nature of the PCL matrix, able to partially prevent water diffusion into the bulk specimen.

Thus, an increase in polymer crystallinity should generally decrease polymer degradation rates, as extensively reported for PLA matrices under alkaline hydrolytic degradation conditions.[20,74] Other researchers[19,75] reported accelerated hydrolysis in neutral media with increasing polymer crystallinity for PLLA[13,76] and PLLA/PDLA blends.[77] A possible explanation to this behaviour has been related to the consideration that, upon crystallization of PLLA, hydrophilic terminal groups (-OH and -COOH) can be condensed in the amorphous area between the crystalline regions (see

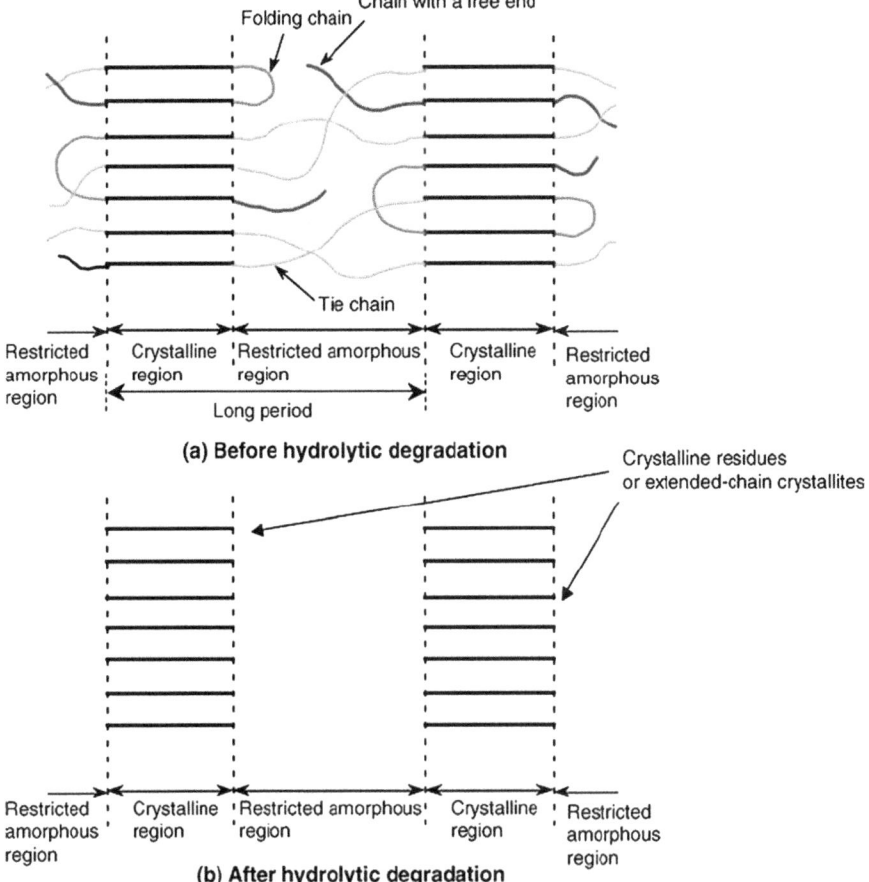

Figure 12.3 Schematic representation of structures of crystallized PLA material before and after hydrolytic degradation, and of the formation of crystalline residues, taken from Tsuji,[27] with permission from John Wiley & Sons, Inc.

Figure 12.3a).[23,76] The high density of the hydrophilic terminal groups can cause loosening of chain packing in the amorphous area between the crystalline regions compared to the chain packing in completely amorphous regions, enhancing the diffusion of water molecules into the bulk material. The high level of absorbed water and the catalytic effect by the high density of carboxylic groups can subsequently accelerate the hydrolytic degradation of crystallized PLLA specimens. Moreover, the terminal groups are sites for the formation of lactoyl-lactic acid and lactic acid, which could catalyze the hydrolytic degradation in a restricted area of the PLA amorphous regions.

12.3.2.2.4 Copolymerization. A great number of PLA-based copolymers have been investigated to improve the thermal, mechanical and physical properties, as well as to tune the PLA degradation rate. In general, with the incorporation of a co-monomer or of a co-monomer polymeric block, selective degradation takes place. The most representative example is the selective cleavage of ester groups connecting glycolyl units in copolymers of LA and GA, and the predominant removal of glycolyl units resulting in decreasing content of GA units in the copolymer during hydrolytic degradation.[23]

Several researchers have extensively studied the effect of GA and CL units on the hydrolytic degradation rate of LLA-based copolymers,[58,78] and found catalytic effect in alkaline and neutral media by the use of both co-monomers, especially by the incorporation of hydrophilic GA units as compared to the hydrophobic CL. In neutral conditions the degradation rate of P(LLA-*co*-GA) seems to increase with GA unit content.[13,49] Similar evidence was obtained for P(DLLA-*co*-GA) copolymers under neutral[79] and acidic[59] media, and they were ascribed to the increase in the bound reactive water content in the specimens at higher content of GA units. The effect of addition of CL units on the hydrolytic degradation of LLA- or DLLA-based copolymers is still not clear. Some researchers reported higher hydrolytic degradation rates in neutral media for nanoparticles based on P(LLA-*co*-GA-*co*-CL) (63/27/20) copolymers as compared to nanoparticles based on P(LLA-*co*-GA) (70/30) copolymers.[80] Conversely, other researchers reported lower neutral hydrolytic degradation rates for block copolymers of PDLLA-*b*-PCL, P(DLLA-*co*-GA-*co*-CL) and P(LLA-*co*-GA-*co*-CL) with the increase in the PCL content.[81–83]

The polymerization of LLA, DLLA, LA/GA and other combinations of LA with co-monomers in the presence of hydrophilic poly(ethylene glycol) or poly(ethylene oxide) (PEO) segments is also frequently used to synthesize block PLA copolymers, given their high potential as matrices for hydrophilic drug release systems. Concerning their degradation, it was reported that the incorporation of PEO units accelerates the hydrolytic degradation of PLA-based homopolymers and copolymers. During the hydrolytic degradation of copolymers, the high hydrophilicity of PEO seems to accelerate the diffusion of water into the specimen, increasing the water content in the bulk material and resulting in enhanced cleavages of the ester linkages, while the release of water-soluble PEO causes a rapid increase in weight loss.[84,85] Chaubal *et al.*[86] reported faster hydrolytic degradation of PLLA-*b*-PEO-*b*-PLLA and PDLLA-*b*-poly(ethylene propylene) (PEP) copolymers at an early stage of the process in neutral media, and postulated the occurrence of an initial cleavage of the ethylphosphate-lactyl linkage with a subsequent cleavage of the lactyl-lactyl linkages.

Other PLLA-based copolymers cited in the literature are those containing hydrophilic β-malic acid units and α-L-malic acid [poly(L-lactide-*co*-β-malic acid)], generally showing in neutral hydrolytic conditions that most of the hydrophilic β-malic acid units can break down from the copolymer within 1 week and the ester bond between L-lactide and β-malic acid hydrolyzes prior to the inner ester bond of PLLA.[87]

Zhu and Lei[88] reported that the neutral degradation rate of poly(DLLA-*co*-adipic anhydride) increased at higher content of adipic anhydride. Nakayama *et al.*[89] reported that the addition of 20 mol% of β-methyl-valerolactone (VL) considerably increased the degradation rate of P(LLA-*co*-β-methyl-VL) copolymers. However, high contents of δ-valerolactone units in PLLA copolymers were found to decrease the hydrolytic degradation rate in neutral media.

PLA-based copolymers with amino acid units have also been explored. For instance, Helder *et al.*[90] investigated the neutral hydrolytic degradation of P(DLLA-glycine), and found that the degradation rate of the copolymers was highly increased at high -DLLA contents (80 mol%). Simultaneously, Barrera *et al.*[91] found that the presence of lysine units in PLLA-based copolymers accelerated the neutral hydrolytic degradation of the PLLA specimens. Similar results were also found by Tarvainen *et al.*[92] reporting higher degradation rates for 2,2-bis(2-oxazoline)-PDLLA copolymers than for pristine PDLLA. Other researchers[93,94] reported analogous increases in the hydrolytic degradation rate of PLLA and PDLLA-based copolymers after incorporation of aspartic acid units.

12.3.2.2.5 Blending with other Polymers. Polymer blending is commercially advantageous to prepare PLA-based materials with a wide variety of physical and hydrolytic degradation properties.[23] Generally, the addition to PLLA of hydrophilic polymers, such as PEO,[95] PGA[96] and oligomeric PDLLA chains[97] has shown increases in hydrolytic degradation rate. For instance, Sheth *et al.*[98] investigated the alkaline hydrolytic degradation of pure PLLA and PLLA/PEG blends to obtain faster degradation in blends as compared to pristine PLLA. They observed that when PEG content was 30 wt% or lower, the weight loss of samples was found to occur as a combination of PLA degradation and PEG dissolution.

Shirahase *et al.*[99] found that the addition of 30 wt% poly(methyl methacrylate) (PMMA) enhanced the alkaline hydrolytic degradation of PLLA/PMMA blends, given that hydroxyl ions could probably diffuse into the interface between the PLLA and PMMA-rich domains and catalyze the surface hydrolytic degradation of PLLA domains.

Mallardé *et al.*[100] reported that oligomeric PDLLA ($M_n = 1100$ g mol^{-1}) enhanced the alkaline hydrolytic degradation of poly(3-hydroxyoctanoate). Tsuji *et al.*[72] investigated the non-enzymatic hydrolysis of blend films from PLLA and PCL in a phosphate-buffered solution of pH 7.4 as a model of phase-separated aliphatic polyester blends. They found that the PLLA hydrolysis was much faster than that of PCL, and that a small amount of PCL could accelerate the PLLA hydrolysis in blends, probably because of the increased concentration of the terminal carboxyl groups in the film by addition of low-molecular-weight PCL molecules. However, the PLLA hydrolysis in the blend films with high PCL content was greatly retarded, probably due to the prevention of water diffusion into the PLLA phase dispersed in the PCL matrix. Similarly, Fukushima *et al.*[71] studied the hydrolytic

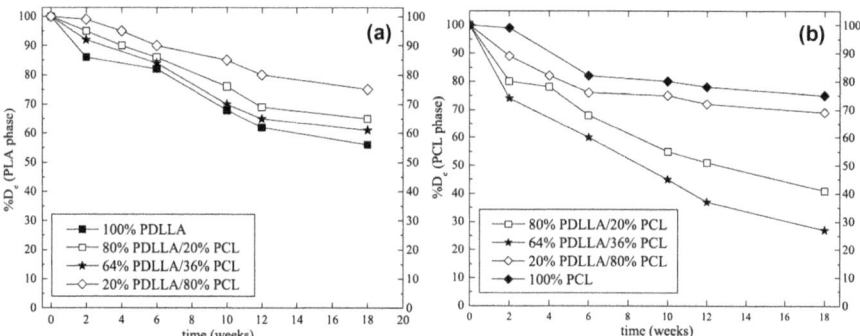

Figure 12.4 Average percentage of PDLLA and PCL ester group loss (%D_e) for PDLLA phase (a) and PCL phase (b) for pristine PDLLA, PCL and their blends as function of degradation time. %D_e was calculated based on the values of normalized absorbances of ester groups obtained for each polymer phase in all materials before and after degradation.
Adapted from Fukushima *et al.*[71] with permission from Elsevier.

degradation of amorphous PDLLA, PCL and their partially miscible blends in a phosphate-buffered solution of pH 4.0 at 37 °C and found that the PCL phase in compositions rich in PCL was very stable against hydrolysis, but high PDLLA contents (80 wt% and 64 wt%) in the blends seemed to catalyze the hydrolytic degradation of the PCL phase. This should probably be associated to the easy diffusion of water into the PCL phase due to the presence of PDLLA amorphous regions, being less organized than the PCL crystalline domains, allowing water to penetrate more easily into the bulk material (Figure 12.4). In general, they found that the degradation of pristine PDLLA proceeded faster than in PDLLA/PCL blends, indicating that PCL molecules partially delayed the hydrolysis of PDLLA molecules, probably due to a higher barrier effect of water diffusion into the PDLLA matrix (see Figure 12.4). This water diffusion was proportional to the PCL content in the blends, except for a blend with a particular PDLLA/PCL interphase structure (64% PDLLA/36% PCL), which showed that not only did PCL molecules not significantly impede hydrolysis of PDLLA chains in this blend, but they could also catalyze the PDLLA hydrolytic degradation and partially attenuate the delaying effect of PCL molecules on water diffusion into the bulk material given a considerable increase of the end-hydroxyl groups concentration in the PDLLA/PCL interphase.

12.3.2.2.6 Blending with Additives/(Nano)particles. Degradation rates can be controlled by blending PLA with additives, inorganic fillers and nanoparticles. Schwach *et al.*[101] found that zinc metal particles used during the polymerization of DL-lactide could, for instance, increase the water absorption level in PDLLA bulk samples, resulting in higher degradation rates. Shikinami and Okuno[102] found a catalytic hydrolytic degradation for PLLA by addition of hydroxyapatite. They ascribed this higher hydrolytic degradation rate to a surface pore formation along the whole sample as the consequence of the removal of hydroxyapatite particles during degradation.

Conversely, other researchers[103–105] reported that the addition of MgO particles, calcium compounds and tri-calcium phosphate can neutralize the catalytic carboxyl groups formed by the hydrolytic degradation of P(LLA-*co*-DLA), P(DLLA-*co*-GA) and PLLA-b-poly(ethylene/hexamethylene–sebacate), and delayed their hydrolytic degradation rate in neutral media. Simultaneously, Zhang *et al.*[106] indicated that the addition of alkaline metal salts with low water solubility reduced the hydrolytic degradation rate of P(DLLA-*co*-GA) (50/50) because these salts might disrupt the autocatalytic effect caused by the carboxylic groups. Similar observations were reported by Niemelä *et al.*[107] on the hydrolytic degradation of P(LLA-*co*-DLA) (70/30) containing bioactive glass particles (spherical bioactive glass 13–93 particles with a composition of 6 wt% Na_2O, 12 wt% K_2O, 5 wt% MgO, 20 wt% CaO, 4 wt% P_2O_5 and 53 wt% SiO_2 and particle size distribution 50–125 µm). They found that the addition of bioactive glass decreased the degradation rate according to molecular weight analysis.

Concerning the use of nanoparticles, Armentano *et al.*[108] reported interesting results regarding the effect of carbon nanotubes on the PLA hydrolytic degradation. Indeed, they described a correlation between the hydrophilicity level of carbon nanotubes with the hydrolytic degradation rate of PLA in neutral conditions.

Montmorillonite (MMT), a smectite clay, is probably the most extensively studied nanomaterial in terms of mechanical, thermal, fire retardant or crystallization behavior of polylactide,[6] especially when these nanoparticles are organically modified allowing the achievement of intercalated and exfoliated nanocomposites.[109] These nanocomposites show enhanced properties as compared to microcomposites and pristine polymer. However, biodegradation and hydrolytic degradation of PLA in the presence of nanoclays has been investigated to a small extent.

Several authors found the catalytic effect of montmorillonites on the biodegradation and hydrolytic degradation of different aliphatic polyesters, due to the high hydrophilicity of these nanoparticles.[14,110,111] Indeed, in a previous work,[112] we studied the degradation of PLA nanocomposites based on organically modified montmorillonites at 5 wt% loading in compost at 40 °C and neutral pH. We found that the addition of nanoclays increased the PLA hydrolytic degradation rate because of the presence of hydroxyl-groups belonging to the silicate layers. This phenomenon was particularly evident for the highest dispersed clay in the polymer matrix. Nonetheless, other authors reported that nanoclays might retard the degradation of aliphatic polyesters during either biodegradation or hydrolytic degradation, due to the enhancement in the barrier properties of layered silicate nanocomposites.[113,114] Consistently, in a previous work,[115] we also evidenced that the PLA hydrolytic degradation can be partially delayed upon addition of 5 wt% of unmodified sepiolites (SEPS9), but only if the degradation temperature is close to the PLA T_g. We found that even if PLA and PLA/SEPS9 presented a significant level of degradation in compost at 58 °C by a preferential mechanism of bulk degradation, the presence of sepiolite particles partially delayed the PLA matrix hydrolytic degradation probably due to the preventing effect of these particles on polymer chain mobility at this relatively high temperature close to $T_g \sim 55$–60 °C.

Zhou *et al.*[6] studied the effects of the filler type combined with the polymer crystallinity and degradation temperature on the hydrolytic degradation of semicrystalline and amorphous PLA-based materials, containing unmodified and organically modified montmorillonites. They showed that the effective pH and hydrophilicity of the organically modified nanofillers promoted significantly higher degradation rates for nanocomposites. In contrast, the degradation rate for microcomposites (based on pristine nanoclays) was lower than that for the unfilled polymers due to the reduction of the carboxyl-group catalytic effect through neutralization with the hydrophilic alkaline filler. They found that bulk hydrolytic degradation started apparently from the interface between polymer and fillers, resulting in significant differences between nanocomposites, microcomposites and the neat polymer. In a similar way, Luo *et al.*[116] found that the hydrolysis of PLA/TiO$_2$ nanocomposites in phosphate buffer solution of pH = 7.4 at 37 °C occurred at the interface of the PLA matrix and nanofillers, and stated that the dispersion of nanofillers in the polymer matrix as well as the crystallinity of nanocomposites were important parameters affecting the water absorption level and degradation rate of the PLA-based nanocomposites.

Subsequently, Fukushima *et al.*[45] studied the hydrolytic degradation at 37 and 58 °C in a pH 7.0 phosphate-buffered solution of PLA and nanocomposite films based on modified montmorillonite (C30B), fluorohectorite (SOMMEE) and unmodified sepiolite (SEPS9) at 5 wt% clay loading. They found that the addition of C30B and SEPS9 delayed the PLA degradation at 37 °C, especially for C30B, due to the induced PLA crystallization effect. Although SOMMEE also induced the PLA crystallization, it was found that PLA/SOMMEE presented higher decreases in molecular weight as compared to neat PLA after 8 weeks, indicating certain inducing effect of SOMMEE particles on polymer degradation (see Figure 12.5 and Table 12.1). Differences in the degradation trend of PLA/SOMMEE as compared to PLA/C30B and PLA/SEPS9 were related to the probable higher water uptake of C30B and SEPS9 in the nanocomposites as compared to neat PLA and SOMMEE. Water segregation between C30B layers and SEPS9 pores could be responsible for the delayed polymer degradation due to the reduced amount of water available for hydrolysis. This effect was probably reduced in the case of PLA/SOMMEE due to the low hydrophilic properties of SOMMEE. Concerning hydrolysis at 58 °C, the presence of nanoparticles did not significantly affect the PLA degradation trend, achieving similar molecular weight decrease for all materials. This was related to the easy access to water molecules in the amorphous and semicrystalline regions at this temperature, initiating the hydrolytic chain scission in both phases. In general, they concluded that PLA

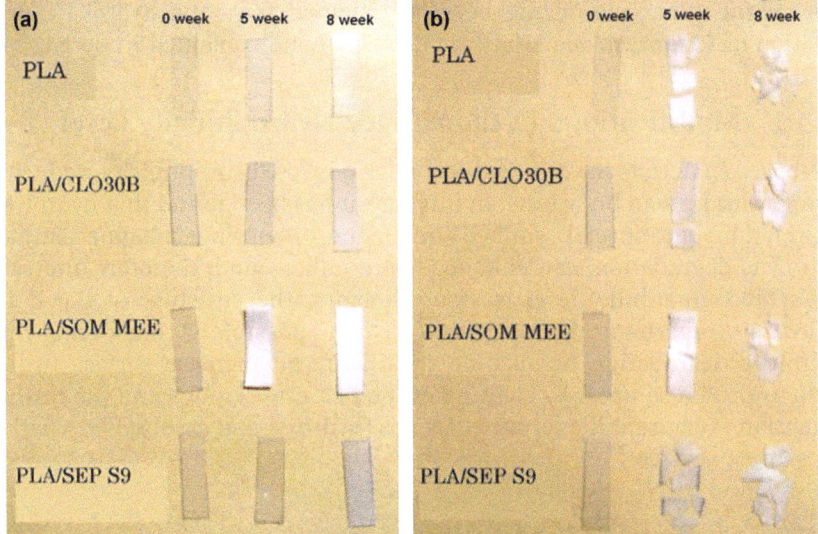

Figure 12.5 PLA and nanocomposites before and after degradation in neutral phosphate buffered solution at 37 °C (a) and 58 °C.
Taken from Fukushima *et al.*[45] with permission from Elsevier.

Table 12.1 Size Exclusion Chromatography data on PLA and nanocomposites before and after degradation in neutral phosphate buffered solution at 37 °C and 58 °C. Data extracted from Fukushima *et al.*,[45] with permission from Elsevier.

| | Week 0 | | | Week 8 | | | | | |
| | | | | 37 °C | | | 58 °C | | |
Sample	M_n (g mol^{-1})	M_w (g mol^{-1})	M_w/M_n	M_n (g mol^{-1})	M_w (g mol^{-1})	M_w/M_n	M_n (g mol^{-1})	M_w (g mol^{-1})	M_w/M_n
PLA	36000	77000	2.1	21000	50000	2.4	1800	2500	1.4
PLA + CLO30B	27000	69000	2.6	22000	60000	2.7	1800	2400	1.4
PLA + SOMMEE	22000	56000	2.5	12000	34000	3.0	1300	2300	1.8
PLA + SEPS9	37000	103000	2.8	27000	94000	3.5	2500	34300	1.4

hydrolysis in the presence of nanoparticles is a complex phenomenon depending on the degradation temperature, nature and dispersion of filler, and polymer crystallinity.

12.4 Possible Benefits of Hydrolytic Degradation of PLA

In addition to the importance of an effective and controlled hydrolytic degradation of PLA for its environmental and biomedical applications, in this section other interesting benefits obtained from the hydrolytic degradation of PLA-based materials are identified and explained in detail.

12.4.1   Modifications in the Surface Hydrophilicity Level

PLA-based materials have generally a relatively low hydrophilicity, reducing affinity with human body cells. In this way, it has been found that hydrolytic degradation, in special surface treatments involving alkaline surface hydrolytic degradation, can enhance hydrophilicity and, therefore, increase PLA's biocompatibility level without changing the structure of the bulk, maintaining mechanical properties.[117] This is possible if considering that hydrolytic degradation or alkaline treatment can increase surface hydrophilicity of PLA samples by cleaving of the PLA ester groups and the further formation of hydrophilic terminal groups (hydroxyl and carboxyl groups) on the sample surface.[23]

12.4.2   Structural Modification

Specific desired polymeric structures can be attained after removal of PLA chains by hydrolytic degradation. For instance, Ho *et al.*[118] developed novel thin polystyrene (PS) films with perpendicularly and cylindrical nanopatterns through the selective hydrolytic degradation and removal of PLLA segments from spin coated PLLA-*b*-PS copolymers. Rzayev and Hillmyer[119] prepared novel nanochannel array triblock copolymers of PS-poly(dimethylacrylamide) (PDMA)-PLA after removal of PLA segments from the aligned materials.

Other researchers[32,54,120] have studied the possibility to change the monomer ratio in copolymers of P(LLA-*co*-GA) and P(DLLA-*co*-GA) by their hydrolytic degradation in neutral media. The authors reported that the GA fraction can be reduced during hydrolytic degradation of copolymers without altering their relative ratio of L- and D-lactyl units.

12.4.2.1   Variations in Crystallization and Thermal Properties

The PLA crystallization kinetics has been extensively studied and found to be rather slow, as in the case of poly(ethylene terephthalate) (PET).[121] Nevertheless, it has been reported that it can be increased with a decrease in its molecular weight. As described in different research works,[13,23] crystallization of PLA-based materials can take place during their hydrolytic degradation. Fully amorphous PLLA matrices crystallize during hydrolytic degradation in phosphate-buffered solution.[122] Similarly, L-lactide copolymers of P(LLA-*co*-GA) and P(LLA-*co*-CL) are able to crystallize in higher extent during the course of their hydrolytic degradation.[58]

In a previous work we studied the effect of the hydrolytic degradation on crystallization of PLA and nanocomposites based on organically modified montmorillonite (PLA/C30B) by DSC analysis.[45] Results corresponding to cooling and second heating scans for neat PLA and PLA/C30B degraded in phosphate buffered solution at 37 °C indicated that their crystallization temperature (T_c) during cooling slightly decreased by hydrolysis within 8 weeks, but their crystallization enthalpy (ΔH_c) increased (Table 12.2). This was related to significant molecular weight decreases in both materials, since shorter polymer chains with higher mobility could crystallize during cooling at a larger extent than the original PLA chains. Second heating scans of these materials degraded at 37 °C showed that T_{cc} and T_m shifted to lower values after 8 weeks as the result of the molecular weight decrease. However, melting enthalpy (ΔH_m) of both materials increased due to the additional polymer crystallization attributed to PLA plasticization by water molecules as well as by lactic acid oligomers that would give sufficient mobility to the polymer chains to organize and further crystallize.[14] Moreover, the crystallinity (χ) of PLA and nanocomposites after 8 weeks at 37 °C increased, especially for nanocomposites due to a considerable nucleating effect of nanoparticles on PLA crystallization (see Table 12.2).

On the other hand, it is important to highlight that when the degradation of PLA and PLA/C30B was carried out at 58 °C, T_c of neat PLA and PLA/C30B decreased after 1 week.[45] Furthermore, no sign of crystallization was detected after 3 weeks. In parallel, thermograms corresponding to the second heating scan for both materials degraded at 58 °C showed a decrease of T_{cc} and T_m after 1 week, with their subsequent disappearance after 3 weeks (as reported in Table 12.2). These results were explained by considering the significant polymer degradation in these materials (in agreement with size exclusion chromatography analysis). In this way, chain scission due to 1-week hydrolysis could lead to shortening of the polymer macromolecules,

Table 12.2 DSC data on PLA and PLA/CLO30B obtained prior to and after hydrolytic degradation at 37 °C and 58 °C, by cooling and second heating curves. Data taken from Fukushima *et al.*,[45] with permission from Elsevier.

| Sample | Cooling | | Second heating | | | | | |
	$T_c{}^a$ (°C)	ΔH_c (J g^{-1})	$T_{cc}{}^a$ (°C)	ΔH_{cc} (J g^{-1})	$T_m{}^a$ (°C)	ΔH_m (J g^{-1})	χ (%)	T_g (°C)
PLA	93	3	101	20	163	29	10	58
PLA – week 6 at 37 °C	94	3	101	38	163	50	12	58
PLA – week 8 at 37 °C	92	5	97	40	160	50	11	55
PLA – week 6 at 58 °C	–	–	–	–	–	–	–	50
PLA – week 8 at 58 °C	–	–	–	–	–	–	–	44
PLA + CLO30B	94	3	98	25	164	39	21	58
PLA + CLO30B – week 6 at 37 °C	94	12	94	25	162	43	27	57
PLA + CLO30B – week 8 at 37 °C	92	17	91	24	160	47	27	54
PLA + CLO30B – week 6 at 58 °C	–	–	–	–	–	–	–	45
PLA + CLO30B – week 8 at 58 °C	–	–	–	–	–	–	–	43

^{a}Temperatures quoted are peak temperatures.

which were still long enough to reach crystalline organization although of lower order than pristine material. But, after 3 weeks of hydrolysis, polymer chains were too short to crystallize. It is possible to assume that interesting increases in thermal properties of PLA-based samples can be achieved, but certain experimental parameters, such as degradation temperature and time are crucial to avoid the contrary effect.

12.4.2.2   *Improvements of Particular Mechanical Properties*

At late stages of the hydrolytic degradation of PLA-based materials, some decrease in the mechanical properties occur as the result of important decreases in their molecular weight. However, at early stages of their hydrolytic degradation, some mechanical properties can increase as the consequence of the improved packing of polymer chains in the amorphous regions due to the polymer chain annealing in the presence of water molecules which can act a plasticizer.[19,76] Such kinds of phenomena have been, for instance, observed by Karjalainen *et al.*[123] during the neutral hydrolytic degradation of P(LLA-*co*-CL) (45/55) copolymers at 23 and 37 °C.

12.4.2.3   *Surface Reactive Sites for Grafting*

The generation of surface reactive sites in PLA-based materials can be very useful for grafting processes with other polymer matrices. For example, Croll *et al.*[124] obtained reactive carboxyl- or amine groups on alkaline hydrolytically degraded P(DLLA-*co*-GA) copolymers, using NaOH, ethylenediamine or an *N*-aminoethyl-1,3-propanediamine solution for their further grafting with chitosan. Tsuji *et al.*[125] obtained PLLA films grafted with PCL chains by surface alkaline hydrolytic degradation. In another work, Tsuji *et al.*[126] prepared block copolymers by ring-opening polymerization of DLLA co-initiated with PLLA crystalline residues obtained from the neutral hydrolytic degradation of crystalline PLLA films. They found that not only could crystalline residues act as macro-co-initiators by their many hydroxyls as terminal-groups on their surface, but also that the chains inside these crystalline residues were inert and expected to be protected from interseg-mental trans-esterification between block chains synthesized in the first step (L-lactide chains) and the second step (DL-lactide chains) due to the rigidity of the crystalline lattice.

12.5   Conclusions

The control of the PLA hydrolytic degradation is necessary for medical applications and for the plastics industry. The mechanism of abiotic-hydrolytic degradation as well as the rate of degradation reaction for PLA-based materials seems to be affected by a wide variety of compositional and property variables, as reported in this chapter. Consequently, knowing and understanding all parameters could bring the possibility to control in a certain way the rate and mechanism of the PLA hydrolytic degradation, so that these materials could find their most optimal final use. Some possible advantages of the hydrolytic degradation of PLA-based materials to modify or prepare novel PLA compositions with improved surface, thermal, crystalline or mechanical properties were identified widening their possible applications in the industrial, biomedical, pharmaceutical and environmental sectors.

References

1. M. Hakkarainen, *Adv. Polym. Sci.*, 2002, **157**, 113.
2. P. Gruber and M. O'Brien, *Biopolymers, Polyesters III – Applications and Commercial Products*, ed. Y. Doi and A. Steinbüchel, Wiley-VCH, Weiheim, Germany, 2002, vol. 4, pp. 235–250.
3. R. Auras, B. Harte and S. Selke, *Macromolecular Bioscience*, 2004, **4**, 835.
4. S. M. Li, H. Garreau and M. Vert, *J. Mater. Sci. Mater. Med.*, 1990, **1**, 198.
5. M. Hakkarainen, A. C. Albertsson and S. Karlsson, *Polym. Degrad. Stabil.*, 1996, **52**, 283.
6. Q. Zhou and M. Xanthos, *Polym. Degrad. Stabil.*, 2008, **93**, 1450.
7. H. Pistner, R. Gutwald, R. Ordung, J. Reuther and K. Muhling, *Biomaterials*, 1993, **14**, 671.
8. H. Pistner, D. R. Bendix, J. Muhling and J. F. Reuther, *Biomaterials*, 1993, **14**, 291.
9. H. Pistner, H. Stallforth, R. Gutwald, J. Muhling, J. Reuther and C. Michel, *Biomaterials*, 1994, **15**, 439.
10. C. Migliaresi, L. Fambri and D. Cohn, *J. Biomater. Sci. Polym. Ed.*, 1994, **5**, 591.
11. A. Pegoretti, L. Fambri and C. Migliaresi, *J. Appl. Polym. Sci.*, 1997, **64**, 213.
12. A. Sodergard, J. F. Selin and J. Nasman, *Polym. Degrad. Stabil.*, 1996, **51**, 351.
13. E. A. Duek, C. A. C. Zavaglia and W. D. Belangero, *Polymer*, 1999, **40**, 6465.
14. M. A Paul, C. Delcourt, M. Alexandre, P. Degée, F. Monteverde and P. Dubois, *Polym. Degrad. Stabil.*, 2005, **87**, 535.
15. S. Sinha Ray and M. Bousmina, *Progr. Mater. Sci.*, 2005, **50**, 962.
16. H. Tsuji and Y. Tezuka, *Macromolecular Bioscience*, 2005, **5**, 135.
17. H. Tsuji, *Polylactides, Biopolymers for Medical and Pharmaceutical Applications*, Wiley-VCH, USA, 2005, pp. 183–219.
18. I. Grizzi, H. Garreau, S. Li and M. Vert, *Biomaterials*, 1995, **16**, 305.
19. H. Tsuji and Y. Ikada, *Polym. Degrad. Stabil.*, 2000, **67**, 179.
20. H. Tsuji and S. Miyauchi, *Polym. Degrad. Stabil.*, 2001, **71**, 415.
21. M. Vert, G. Schwach and J. Coudane, *J. Macromol. Sci. Pure Appl. Chem.*, 1995, **32**, 787.
22. C. Shih, *J. Contr. Release*, 1995, **34**, 9.

23. H. Tsuji, *Poly(lactic acid). Synthesis, Structures, Properties, Processing, and Applications*, ed. R. Auras, L. T. Lim, S. E. M. Selke and H. Tsuji, John Wiley & Sons, USA, 2010, pp. 345–382.

24. M. Hakarainen, S. Karlsson and A. C. Albertsson, *Polymer*, 2000, **41**, 2331.

25. C. Shih, *J. Pharmaceut. Sci.*, 1995, **12**, 2036.

26. C. C. Chu, *J. Appl. Polym. Sci.*, 1981, **15**, 1727.

27. H. Tsuji, *Polyesters 3, Biopolymers*, ed. Y. Doiand and A. Steinbuechel, Wiley-VCH, Weinheim, 2002, vol. 4, p. 129.

28. F. von Burkersroda, L. Schedl and A. Goepferich, *Biomaterials*, 2002, **23**, 4221.

29. C. Jie and J. Zhu, *Polym. Int.*, 1997, **42**, 373.

30. S. M. Li, H. Garreau and M. Vert, *J. Mater. Sci. Mater. Med.*, 1990, **1**, 123.

31. G. Schwach and M. Vert, *Int. J. Biol. Macromol.*, 1999, **25**, 283.

32. S. M. Li, H. Garreau and M. Vert, *J. Mater. Sci. Mater. Med.*, 1990, **1**, 131.

33. S. Li and M. Vert, *Biodegradable Polymers: Principles and Applications*, ed. G. Scott and D. Gilead, Chapman & Hall, Cambridge, 1995, pp. 43–87.

34. L. T. Sin, A. R. Rahmat and W. A. W. A. Rahman, *Polylactic Acid. PLA Biopolymer Technology and Applications*, ed. L. T. Sin, A. R. Rahmat and W. A. W. A. Rahman, Elsevier, Amsterdam, 2013, ch. 7, pp. 247–299.

35. M. Vert, *J. Mater. Sci. Mater. Med.*, 2009, **20**, 437.

36. C. C. Chu, *Ann. Surg.*, 1982, **195**, 55.

37. S. J. de Jong, E. R. Arias, D. T. S. Rijkers, C. F. van Nostrum and J. J. Kettenes-van den Bosch, *Polymer*, 2001, **42**, 2795.

38. C. F. van Nostrum, T. F. J. Veldhuis, G. W. Bos and W. E. Hennink, *Polymer*, 2004, **45**, 6779.

39. K. Makino, H. Ohshima and T. Kondo, *J. Microencapsul.*, 1986, **3**, 203.

40. X. Yuan, A. F. T. Mak and K. Yao, *Polym. Degrad. Stabil.*, 2003, **79**, 45.

41. A. Kulkarni, J. Reiche and A. Lendlein, *Surf. Interface Anal.*, 2007, **39**, 740.

42. H. Tsuji and K. Nakahara, *J. Appl. Polym. Sci.*, 2002, **86**, 186.

43. M. Itavaara, S. Karjomaa and J. F. Selin, *Chemosphere*, 2002, **46**, 879.

44. S. H. Hyon, K. Jamshidi and Y. Ikada, *Polymers as Biomaterials*, Plenum, New York, 1984, pp. 51–65.

45. K. Fukushima, D. Tabuani, M. Dottori, I. Armentano, J. M. Kenny and G. Camino, *Polym. Degrad. Stabil.*, 2011, **96**, 2120.

46. S. Karjomaa, T. Suortti, R. Lempiainen, J. F. Selin and M. Itavaara, *Polym. Degrad. Stabil.*, 1998, **59**, 333.

47. N. A Weir, F. J. Buchanan, J. F. Orr, D. F. Farrar and G. R. Dickson, *J. Eng. Med.*, 2004, **218**, 321.

48. C. M. Agrawal, D. Huang, J. P. Schmitz and K. A. Athanasiou, *Tissue Eng.*, 1997, **3**, 345.

49. A. M. Reed and D. K. Gilding, *Polymer*, 1981, **22**, 494.

50. M. Stefani, J. Coudane and M. Vert, *Polym. Degrad. Stabil.*, 2006, **91**, 2853.

51. Q. Cai, G. Shi, J. Bei and S. Wang, *Biomaterials*, 2003, **24**, 629.
52. I. C. Lee, L. P. Cheng and T. H. Young, *J. Biomed. Mater. Res.*, 2005, **76A**, 842.
53. V. J. Chen and P. X. Ma, *Biomaterials*, 2006, **27**, 3708.
54. J. Huang, M. S. Lisowski, J. Runt, E. S. Hall, R. T. Kean, N. Buehler and J. S. Lin, *Macromolecules*, 1998, **31**, 2593.
55. H. Tsuji and T. Ishizaka, *J. Appl. Polym. Sci.*, 2001, **80**, 2281.
56. T. Shirahase, Y. Komatsu, H. Marubayashi, Y. Tominaga, S. Asai and M. Sumita, *Polym. Degrad. Stabil.*, 2007, **92**, 1626.
57. S. K. Saha and H. Tsuji, *Polym. Degrad. Stabil.*, 2006, **91**, 1665.
58. S. K. Saha and H. Tsuji, *Macromol. Mater. Eng.*, 2006, **291**, 357.
59. X. Wu and N. Wang, *J. Biomater. Sci. Polym. Ed.*, 2001, **12**, 21.
60. I. Arvanitoyannis, A. Nakayama, E. Psomiadou, N. Kawasaki and N. Yamamoto, *Polymer*, 1995, **36**, 2947.
61. I. Arvanitoyannis, A. Nakayama, E. Psomiadou, N. Kawasaki and N. Yamamoto, *Polymer*, 1996, **37**, 651.
62. M. Malin, M. Hiljanen-Vainio, T. Karjalainen and J. Seppälä, *J. Appl. Polym. Sci.*, 1996, **59**, 1289.
63. D. Karst and Y. Yang, *Macromol. Chem. Phys.*, 2008, **209**, 168.
64. Y. Zhao, X. Shuai, C. Chen and F. Xi, *Chem. Mater.*, 2003, **15**, 2836.
65. S. H. Kim and Y. H. Kim, *Biodegradable Plastics and Polymers*, ed. Y. Doi and K. Fukuda, Elsevier, Amsterdam, 1994, pp. 464–469.
66. Y. Li and T. Kissel, *Polymer*, 1998, **39**, 4421.
67. K. Numata, R. K. Srivastava, A. Finne-Wistrand, A. C. Albertsson, Y. Doi and H. Abe, *Biomacromolecules*, 2007, **8**, 3115.
68. H. M. Younes, E. Bravo-Grimaldo and B. G. Amsden, *Biomaterials*, 2004, **25**, 5261.
69. H. Tsuji, M. Ogiwara, S. K. Saha and T. Sakaki, *Biomacromolecules*, 2006, 7, 380.
70. S. H. Hyon, F. Jin, K. Jamshidi, S. Tsutsumi and T. Kanamoto, *Macromol. Symp.*, 2003, **197**, 355.
71. K. Fukushima, J. L. Feijoo and M. C. Yang, *Polym. Degrad. Stabil*, 2012, **97**, 2347.
72. H. Tsuji and Y. Ikada, *J. Appl. Polym. Sci.*, 1998, **67**, 405.
73. H. Tsuji and T. Ishizaka, *Int. J. Biol. Macromol.*, 2001, **29**, 83.
74. H. Tsuji and S. Miyauchi, *Polymer*, 2001, **42**, 4463.
75. S. Chye, J. Loo, C. Ooi, S. Wee, Y. Chiang and F. Boey, *Biomaterials*, 2005, **26**, 2827.
76. H. Tsuji, A. Mizuno and Y. Ikada, *J. Appl. Polym. Sci.*, 2000, 77, 1452.
77. H. Tsuji and C. A. Del Carpio, *Biomacromolecules*, 2003, **4**, 7.
78. M. Vert, S. M. Li and H. H. Garreau, *J. Contr. Release*, 1991, **16**, 15.
79. M. Hakarainen, S. Karlsson and A. C. Albertsson, *Polymer*, 2000, **41**, 2331.
80. H. Zhang, W. Cui, J. Bei and S. Wang, *Polym. Degrad. Stabil.*, 2006, **91**, 1929.
81. Q. Cai, J. Bei and S. Wang, *Polym. Adv. Tech.*, 2002, **13**, 105.

82. A. S. Sawhney and J. A. Hubbell, *J. Biomed. Mater. Res.*, 1990, **24**, 1397.
83. W. P. Ye, F. S. Du, W. H. Jin, J. Y. Yang and Y. Xu, *React. Funct. Polymer*, 1997, **32**, 161.
84. E. Cellikkaya, E. B. Denkbas and E. Piskin, *J. Appl. Polym. Sci.*, 1996, **61**, 1439.
85. L. Youxin, C. Volland and T. Kissel, *J. Contr. Release*, 1994, **32**, 121.
86. M. V. Chaubal, G. Su, E. Spicer, W. Dang, K. E. Branham, J. P. English and Z. Zhao, *J. Biomater. Sci. Polym. Ed.*, 2003, **14**, 45.
87. B. He, J. Bei and S. Wang, *Polym. Adv. Tech.*, 2003, **14**, 645.
88. K. J. Zhu and Y. Lei, *Polym. Int.*, 1997, **43**, 210.
89. A. Nakayama, N. Kawasaki, Y. Maeda, I. Arvanitoyannis, S. Aiba and N. Yamamoto, *J. Appl. Polym. Sci.*, 1997, **66**, 741.
90. J. Helder, P. J. Dijkstra and J. Feijen, *J. Biomed. Mater. Res.*, 1990, **24**, 1005.
91. D. A. Barrera, E. Zylstra, P. T. Lansbury and R. Langer, *Macromolecules*, 1995, **28**, 425.
92. T. Tarvainen, T. Karjalainen, M. Malin, S. Pohjolainen, J. Tuominen, J. Seppälä and K. Järvinen, *J. Contr. Release*, 2002, **81**, 251.
93. H. Shinoda, Y. Asou, A. Suetsugu and K. Tanaka, *Macromolecular Bioscience*, 2003, **3**, 34.
94. Y. S. Huang and F. Z. Cui, *Curr. Appl. Phys.*, 2005, **5**, 546.
95. A. J. Nijenhuis, E. Colstee, D. W. Grijpma and A. J. Pennings, *Polymer*, 1996, **37**, 5849.
96. Y. You, S. W. Lee, J. H. Youk, B. M. Min, S. J. Lee and W. H. Park, *Polym. Degrad. Stabil*, 2005, **90**, 441.
97. J. Mauduit, E. Pérouse and M. Vert, *J. Biomed. Mater. Res.*, 1996, **30**, 201.
98. M. Sheth, R. A. Kumar, V. Dáve, R. A. Gross and S. P. McCarthy, *J. Appl. Polym. Sci.*, 1997, **66**, 1495.
99. T. Shirahase, Y. Komatsu, Y. Tominaga, S. Asai and M. Sumita, *Polymer*, 2006, **47**, 4839.
100. D. Mallardé, M. Vailiére, C. David, M. Menet and P. Guérin, *Polymer*, 1998, **39**, 3387.
101. G. Schwach, J. Coudane, R. Engel and M. Vert, *Biomaterials*, 2002, **23**, 993.
102. Y. Shikinami and M. Okuno, *Biomaterials*, 1999, **20**, 859.
103. S. A. T. van der Meer, J. R. de Wijn and J. G. C. Wolke, *J. Mater. Sci. Mater. Med.*, 1996, 7, 359.
104. M. Ara, M. Watanabe and Y. Imai, *Biomaterials*, 2002, **23**, 2479.
105. Y. Imai, M. Nagai and M. Watanabe, *J. Biomater. Sci. Polym. Ed.*, 1999, **10**, 421.
106. Y. Zhang, S. Zale, L. Sawyer and H. Bernstein, *J. Biomed. Mater. Res.*, 1997, **34**, 531.
107. T. Niemelä, H. Niiranen and M. Kellomäki, *Acta Biomater.*, 2008, **4**, 156.
108. I. Armentano, M. Dottori, D. Puglia and J. M. Kenny, *J. Mater. Sci. Mater. Med.*, 2008, **19**, 2377.
109. B. Pukanszky, *Eur. Polym. J.*, 2005, **41**, 645.

110. S. Ray, K. Yamada, M. Okamoto and K. Ueda, *Macromol. Mater. Eng.*, 2003, **288**, 203.

111. P. Maiti, C. Batt and E. Giannelis, *Polym. Mater. Sci. Eng.*, 2003, **88**, 58.

112. K. Fukushima, C. Abbate, D. Tabuani, M. Gennari and G. Camino, *Polym. Degrad. Stabil.*, 2009, **94**, 1646.

113. Y. Someya, N. Kondo and M. Shibata, *J. Appl. Polym. Sci.*, 2007, **106**, 730.

114. T. Wu and C. Wu, *Polym. Degrad. Stabil.*, 2006, **91**, 2198.

115. K. Fukushima, D. Tabuani, C. Abbate, M. Arena and L. Ferreri, *Polym. Degrad. Stabil.*, 2010, **95**, 2049.

116. Y. B. Luo, X. L. Wang and Y. Z. Wang, *Polym. Degrad. Stabil.*, 2012, **97**, 721.

117. Y. S. Nam, J. J. Yoon and J. G. Lee, *J. Biomater. Sci. Polym. Ed.*, 1999, **10**, 1145.

118. R. M. Ho, W. H. Tseng, H. W. Fan, Y. W. Chiang, C. C. Lin, B. T. Ko and B. H. Huang, *Polymer*, 2005, **46**, 9361.

119. J. Rzayev and M. A. Hillmyer, *J. Am. Chem. Soc.*, 2005, **127**, 13373.

120. S. Kamei, Y. Inoue, H. Okada, M. Yamada, Y. Ogawa and H. Toguchi, *Biomaterials*, 1992, **13**, 953.

121. L. Avérous, *Monomers, Polymers and Composites from Renewable Resources*, ed. M. Belgacem and A. Gandini, Elsevier, Amsterdam, 2008, pp. 433–450.

122. J. W. Leenslag, A. J. Pennings, R. R. M. Bos, F. R. Rozema and G. Boering, *Polymer*, 1987, **8**, 311.

123. T. Karjalainen, M. Hiljanen-Vainio, M. Malin and J. Seppälä, *J. Appl. Polym. Sci.*, 1996, **59**, 1299.

124. T. I. Croll, A. J. O'Connor, G. W. Stevens and J. J. Cooper-White, *Biomacromolecules*, 2004, **5**, 463.

125. H. Tsuji, M. Nishikawa, Y. Osanai and S. Matsumura, *Macromol. Rapid Comm.*, 2007, **28**, 1651.

126. H. Tsuji, M. Nishikawa, Y. Sakamoto and S. Itsuno, *Biomacromolecules*, 2007, **8**, 1730.

Industrial and Legislative Issues

CHAPTER 13

Industrial Uses of PLA

STEFANO FIORI

Condensia Química, S.A. C/Junqueras, 16, Barcelona, Spain
Email: s.fiori@condensiaquimica.es

13.1   Introduction

Poly(lactic acid) (PLA) is a thermoplastic polyester characterized by mechanical and optical properties similar to polystyrene (PS) and polyethylene terephthalate (PET). It is obtained from natural sources, completely biodegradable and compostable in controlled conditions as already stated in previous chapters.[1] PLA offers some key points with respect to classic synthetic polymers, since it is a bioresource and renewable, while raw materials are cheap and abundant compared to oil. From a commercial point of view, a non-secondary approach, it can embellish with the word "green" so fashioned for the major stream consumers. Legislation can also help the commercial diffusion of biopolymers. As an example, a decisive leap has been made with the control of non-biodegradable shopping bags distribution in the European Commission and many of its member states. In addition, PLA has received some interest from the industrial sectors because of its relatively low price and commercial availability compared with other bioplastics. This is the very key point for any successful polymer application. In fact, the current price of commercial PLA falls between 1.5 and 2 € kg^{-1}, which is sufficiently close to other polymers like polyolefins, polyesters or poly(vinyl chloride) (PVC). Clearly, the PLA market is still in its infancy, but it is expected that the decrease in the production costs and the improvement in product performance will result in a clear acceleration in the industrial interest for PLA uses. It is estimated that PLA consumption should reach

RSC Polymer Chemistry Series No. 12
Poly(lactic acid) Science and Technology: Processing, Properties, Additives and Applications
Edited by Alfonso Jiménez, Mercedes Peltzer and Roxana Ruseckaite
© The Royal Society of Chemistry 2015
Published by the Royal Society of Chemistry, www.rsc.org

500 000 tons by 2015, and it is expected to exceed more than one million tons in 2020.[2] In addition, according to the proven oil reserves, and relative to current consumption, it is a common prediction that global oil resources are only sustainable for about 30 years. For this reason, it is expected that the market price of oil and derivatives will abruptly increase, and consequently renewable materials, such as PLA, should become consolidated products in the marketplace. This scenario gives PLA a not-to-be-missed market opportunity and strong consumption potential. According to statistics, the world's plastics consumption was about 280 million tons in 2011, and it is realistic to consider a potential worldwide demand for PLA easily reaching about 20 million tons per year.[3] From the industrial point of view, good mechanical, thermal and barrier properties offered by PLA as well as the easy processability have brought attention to this material. Historically, PLA was primarily synthesized during the mid 1930s[4] and the firm DuPont patented its polymerization process during the 1950s.[5] At the very beginning, just applications in the biomedical field were pursued since manufacturing costs were still too high for low-cost applications. In fact, its high cost/benefit ratio reduced this polymer's uses to high performance materials, such as biocompatible sutures, implants, controlled release devices *etc.* In the 1970s, the company Purac produced clinic materials from PLA in low volume scale[6] using ring-opening polymerization (ROP).

That PLA grade was an ultra-pure low-molecular-weight polymer and it was extremely expensive. An important milestone was laid at the beginning of the 1990s when the company Cargill obtained high-molecular-weight PLA starting from the ROP of lactides.[7,8] This was the first step of a commercially viable form of PLA, permitting this biopolymer to be elevated from a niche material to a large-scale production polymer. During the last two decades, a vast number of articles and patents arose from the R&D departments of companies like Cargill, Mitsui, Purac, Teijin, *etc.* NatureWorks LLC maintains the legacy of that pioneering project and it is currently the major producer of PLA with a yearly production capacity over 140 000 tons (see NatureWorks LLC Ingeo™).[9] Nevertheless, although an intensive work on R&D about PLA science and technology has been carried out and large-scale production is a reality, the industrial application of the polymer is still in progress and only niche applications have been considered so far. For instance, flexible film packaging, cold drink cups, cutlery, apparel, staple fibres, bottles, injection moulding products, coatings, textiles *etc.* PLA can be also used in the paper industry.[10,11] This chapter offers an overview of the frontiers of the industrial uses of PLA.

13.2   PLA Mechanical and Thermal Properties

PLA is thermoplastic polyester characterized by good mechanical, thermal and optical properties (see Table 13.1). Neat PLA is generally characterized by good mechanical properties, depending on the grade and stereochemistry, as indicated in Chapters 1 and 2 of this volume, with elastic

Table 13.1   Properties of PLA.

Properties of PLA	
Crystallinity (%)	10–40%
Glass transition temperature (°C)	55–65
Melt-index range (g 10 min^{-1})	2–20
Melting temperature (°C)	130–170
Molecular weight M_w (Daltons)	100 000 to 300 000
Specific gravity	1.25

modulus 3000–4000 MPa and tensile strength in the range 50–70 MPa. Only elongation at break is poor (2–10%) and this restricts the pure PLA properties for some applications, such as film manufacturing. From the thermal point of view, PLA has a glass transition temperature (T_g) between 55 and 65 °C and it is a brittle material at room temperature.[12,13] The impact resistance of PLA is about 13–20 kJ m^{-2} (unnotched, Izod) for low crystallinity and 18–35 kJ m^{-2} for high crystalline grades. This value changes according to different authors.[14,15] Summarizing these values, PLA can be considered a material with poor or low impact resistance and the use of additives to improve this characteristic is advisable. Another point to be considered is the melt rheology and stability of PLA, since it is distinct from other common polymers. In fact, the chemical stability of PLA can be adversely affected at processing temperatures typically reached by semicrystalline PLA, normally above its melting temperature, since both hydrolysis and pyrolysis can become non-negligible. Fortunately, commercially available PLAs are typically stabilized against thermal degradation. Moreover, an important factor in PLA processing is the presence of water, since a large amount of hydrolysis may occur when PLA with high moisture is utilized as indicated in Chapter 12. Finally, the PLA appearance is affected by the crystalline content. In fact, amorphous or low-crystalline PLA is a clear material with high gloss while highly crystalline PLA is an opaque white material.

13.3   Stereochemistry

As has been extensively discussed in Chapter 2 of this volume, PLA is a stereopolymer, since its monomer, lactic acid, has a stereocentre. This characteristic offers the basis for several developments in PLA technologies and industrial uses. In fact, just changing the L- and D- enantiomers ratio imparts distinctive crystalline properties to the final polymer thus resulting in different grades of the material tailored for distinctive applications. Poly(D-lactic acid) (PDLA) can potentially act as a high performance polymer with T_m about 230 °C, which is clearly higher than the 180 °C for typical Poly(L-lactic acid) PLLA.[16] The degree of stereo-complexation is affected by the synthesis condition and chain structure composition but above all by the molecular weight of the PDLA/PLLA polymers. In high-molecular-weight PLA, the stereo-complexation is hindered by homochiral crystal forms while

Table 13.2 Dependence of thermal properties on D/L ratio.

Stereo-block PLDA/PLLA	T_m	HDT[a]
0	180	
20	214	180
30	214	197
50	216	198
100	230	

[a]Heat deflection temperature (°C).

a complete stereo-complexation can be attained in low-molecular-weight polymers.[17] For a long time the selection of the most adequate polymer degree has been a challenge but nowadays, and thanks to the availability of pure D-lactide from Purac,[18] it is quite easy to obtain stereopolymers. In the near future, "stereo-PLA" will be available at low prices with thermal characteristics sufficiently good for high-temperature applications. The heat deflection temperature (HDT) parameter is only slightly dependent on the molecular weight, but it is extremely dependent on the polymer stereochemistry. In Table 13.2, the thermal characteristics of PDLA/PLLA are reported, presenting T_g, T_m and crystallinity of PLA, which slightly decreases when lowering the amount of PLLA.[19,20]

13.4   Use of Additives

The thermal and mechanical properties of PLA, such as thermal stability and impact resistance, are lower than those for conventional synthetic polymers used for thermoplastic applications. For this reason, neat PLA (regardless of the grade and stereo-complexes ratio) is not ideally suitable for competing against oil derivatives polymers without a substantial modification in the sense of copolymers, blends or additives. PLA needs to be modified to improve its processing or material properties. These additives are available into the market for several applications in liquid or granule form. For example, the inherent PLA brittleness can be improved using ethylene copolymers (*e.g.* DuPont's BIOMAX[21]) to give PLA better toughness for rigid applications, such as casting or injection moulding. These additives when used in a concentration range between 1 and 5 wt% have no effects on film transparency. Carbonates can be also added to improve PLA impact toughness (*e.g.* EMForceBio by Specialty Minerals[22]). In fact at loadings between 20 and 30 wt% the filler acts as a nucleating agent promoting the development of crystallinity. In this way, the impact energy can be absorbed and partially dissipated avoiding the crack propagation. In addition, PLA has low melt strength compared with some classic polymers, such as PET or PP. This drawback tends to create difficulties in processing, such as sagging of the melt, necking, collapsing of the parison *etc.*, and above all when blow film extrusion, sheet extrusion or foaming operations are involved. Acrylic-based additives for PLA (*e.g.* Dow's Paraloid,[23] Arkema's Biostrength *etc.*[24]) are available for enhancing the melt strength of the PLA mass during processing. These additives, used in the 2–5 wt% range, create a 3D physical network able to resist the melt breakage.

Another possible approach is to use chain extenders. In fact, these additives can chemically bind to PLA chains forming a chemical network and improving the melt strength properties (as an example, BASF's Joncryl[25]). Some caution should be applied when similar additives are used. In fact a large extent of these additives will improve processing and use performance, particularly melt strength, but compromising the intrinsic biodegradation characteristics of PLA, the result being that a maximum loading around 1 wt% is advisable.

Surface stickiness of PLA films or artefacts can lead to a major issue when non-treated or using additives. In fact, PLA has an inherent superficial friction coefficient largely due to its high density of polar ester linkages and consequently important electrostatic forces can interact. From a practical point of view, PLA pieces and films can stick mutually (*i.e.* blocking) or stick to the production machinery (*e.g.* rollers, moulds, guides, bobbins *etc.*). In order to avoid or overcome these production drawbacks and issues, anti-blocking or slipping agents can be added to the PLA-based formulations. Fatty acid amides are commonly used because they are not compatible with PLA. They have a high tendency to migrate to the material's surface, reducing the coefficient of friction by intercalating between the adjacent layers (*e.g.* PolyOne's OnCap[26] or Clariant's Cesa-block[27]). In all cases, some care has to be taken using these materials because they can heavily affect the biodegradation behaviour of the final artefact.

13.4.1 Plasticizers

Plasticizer additives deserve a more detailed presentation. Plasticizers play an important role in the processing of PLA since they can completely affect the final mechanical properties of the material. A classic approach describes plasticizers as substances incorporated into a material to modify its mechanical properties. A plasticizer may reduce the melt viscosity, the glass transition and the elastic modulus of the material. Plasticizers are inert organic substances with low vapour pressures, predominantly esters, which interact physically with high polymers to form a homogeneous material.[28] In general terms, plasticizers are usually esters offering low volatilities, low migration, high flash point and good thermal stability. Low toxicity and high biodegradability are a *plus* for this class of additives. In the case of PLA, several products have been successfully utilized, particularly low-molecular-weight polyethylene glycols,[29] acetyl tributyl citrate, glycerol triacetine,[30] bis(2-ethylhexyl)adipate,[31] polyadipates[32] (Condensia Quimica's Glyplast 206/3NL) and low-molecular-weight PLA (Condensia Quimica's Glyplast OLAs).[33] Plasticizers also affect the PLA glass transition and crystallization temperatures, as a function of the additive's quantity used (Figure 13.1). The decrease in T_g leads to increases in chain mobility and this effect in turn increases the crystallization rate.

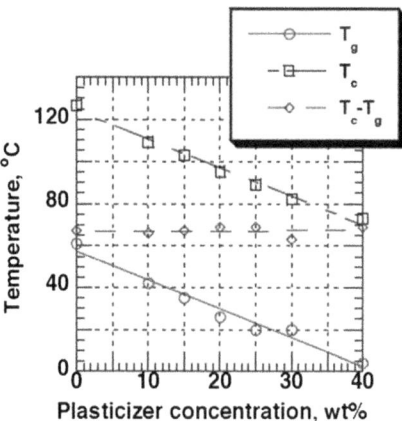

Figure 13.1 T_g and T_c for plasticized PLA as a function of typical plasticizer concentrations.

13.5 PLA Processing

13.5.1 Extrusion

Extrusion is one of the most important processing techniques to produce PLA-based artefacts *i.e.* injection moulding, blow moulding, film blowing and melt spinning. Commercial PLA grades can be processed using conventional extruder machines, like those utilized for PET processing. The process temperature is set up to 40–50 °C above the PLA melting temperature (*i.e. ca.* 210 °C) to ensure good homogeneity and optimal melt viscosity. Temperatures exceeding this limit can lead to thermal degradation with the formation of low-molecular-weight species, while a severe drop in molecular weight can occur to the resin (*i.e.* depolymerization). These by-products, such as low-molecular-weight lactic polymers or acetaldehyde, have to be avoided when food grade materials are produced. In fact, they can impart off-flavours to food and stay over the permitted migration level allowed by international food regulations. Furthermore, thermal degradation can reduce the PLA melt viscosity, elasticity and mechanical properties, and last but not least fuming and fouling of the machinery can reduce the throughput of the production plant.

13.5.2 Film and Sheet Casting

In cast film extrusion, the molten PLA is extruded through a dye and successively rolled. The production of PLA films or sheets by this processing method is very similar, with the only difference being the thickness of the resulting material. Typically films are in the order of less than 70 µm whereas sheets have thicknesses higher than 0.25 mm. By using the same technique, PLA is often co-extruded along with other polymers to form multilayer films to enhance its properties, for example in coupled films for fried food or sweets. In this case it is possible to couple metalized film to tune the barrier properties of the material.

13.5.3 Stretch Blow Moulding

Injection stretch blow moulding (ISBM) is often used for the production of PLA bottles. This process induces biaxial orientation of the polymer, which results in improved mechanical and barrier properties with respect to typical injection moulded bottles. ISBM is a two-step process: the first step is the injection moulding of the pre-form (Figure 13.2), the second is the inflation of the heated pre-form in a blow moulding machine. Since the stretching behaviour of PLA is similar to PET, the conversion of conventional production lines to the production of PLA-based materials is feasible although design modifications are necessary to achieve an optimized bottle wall thick homogeneity. A typical bottle of water made of PLA can be seen in Figure 13.3.

13.5.4 Blown Film Extrusion

Blown film extrusion is a fundamental technique for high-throughput production and it also permits the manufacturing of bi-axially oriented films with enhanced properties. In this process the molten PLA is extruded to form a cylinder of melt polymer by using an annular dye. By blowing air or gas through the dye head, the formed tube is inflated in the shape of a thin tubular bubble and rapidly cooled. The resulting tube is subsequently clamped at the top of the machine and flattened using a series of rollers to be reduced to a thin film and wound up in bobbins.

Although extrusion blow film moulding is an efficient process, the polymer needs to have pretty high melt strength in order to stabilize the bubble and impede its collapse and/or break. For this reason, melt strength additives are often used and moreover these additives help to improve

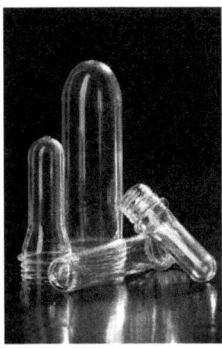

Figure 13.2 Pre-forms for stretching blow moulding applications. © Shutterstock.

Figure 13.3 Typical bottle of water made of PLA obtained by blow stretch moulding (Courtesy of Acqua Sant'Anna SpA).

the elongation at break of the final material. For example, the oligomeric form of lactic acid has been used with excellent results (*e.g.* Condensia Quimica's Glyplast OLA[34]). These highly compatible oligomeric plasticizers resulted in PLA formulations with elongation at break over 300%, maintaining ageing and biodegradation characteristics for a long time. It has been demonstrated that the plasticizer action decreases the crystalline domains and impedes successive annealing of these domains. Anti-slipping agents can also be used to avoid film stick and the consequent bobbin block.

13.5.5 Thermoforming

Thermoforming is a commonly used technique for the production of packaging containers, such as disposable cups, single-use food trays, lids and blisters (Figure 13.4). In this process, a PLA sheet is heated above T_g and by application of vacuum or pressure it is forced to accommodate into a mould, allowed to cool, removed and trimmed to obtain the final object. The processing conditions have to be optimized to obtain homogeneous wall thickness. In fact, cooling time, pressure, mould geometry and product design have to be chosen with care. The thermoformed parts exhibit improved impact and heat resistance compared to PLA manufactured by other production techniques due to the crystalline reorientation imparted by the combination of high temperatures and pressures during processing.

Figure 13.4 Typical PLA thermoformed trays.
© Shutterstock.

13.5.6 Spinning and Electrospinning

In spinning techniques, molten polymers or solutions are extruded from a fine hole and they are further elongated by applying an external tensile force. A subsequent cooling of the extruded polymer permits the formation of a solid filament. Post-treatments permit to use this filament for textile applications.[35] In the case of PLA, its spinning processability is equivalent to that of petroleum-based synthetic materials, where PLA polymer uses conventional polyester type fibre melt spinning processes. PLA fibres use conventional spinning machinery and similarly PLA fabrics use conventional dyeing and finishing machinery.

One special procedure of spinning is electrospinning, as detailed in Chapter 6 of this volume. It is an innovative technique for preparing smaller diameter PLA fibres with respect to conventional dying spinning techniques. Due to their small diameters these PLA fibres have large surface area, making them ideal materials for high-performance applications (*e.g.* medical devices, highly porous filters, technical fabrics, *etc.*).[36] Nevertheless, a major issue for this technique is the use of solvents, and consequently important developments have still to be carried out to apply the use of electrospinning in PLA production.

13.5.7 Foaming

Plastic foams can be defined as materials with cellular structure. These cells are normally filled with a gas generated by a foaming agent. The foam properties are largely affected by density and cell geometry. Generally speaking, cellular plastics are classified as open or closed cells. In closed cells structure, thin walls permit the gas to be retained into small bubbles, separating all cells. Instead, in the open cell structure, all voids are inter-connected in some way permitting the gas to move all around the structure. These two types of foams are very suitable for acoustic insulation; protective cases and cushioning while closed cell materials are more adequate for thermal insulation. The gas filler plays an important role in the final application of these materials. In fact, depending on their thermodynamic characteristics they can be good or bad thermal or pressure carriers. At the same time, a correct geometry of the foam cells can improve their mechanical properties to withstand heavy loads.

13.6 PLA Commercial Applications

As has been highlighted throughout this chapter and generally in the whole volume, PLA shows interesting characteristics for composites formulation to be used in multiple applications. In fact, PLA is a high-strength and high-modulus polymer with acceptable barrier properties and easy processing with conventional machinery. However, PLA applications are still limited due to its higher cost with respect to conventional materials, but also for its brittleness and low thermal stability.[37] Fibrous reinforcements have been introduced to decrease the overall cost and to improve mechanical properties, while powdered fillers have been also used. Recent interest has grown in using nanoscale fillers. In fact, a dramatic improvement in mechanical properties can be achieved using a very small quantity of these materials. Thermal properties, for example, can be steadily improved by introducing natural fillers like cellulosic fibres. Heat deflection temperature (HDT) is a key parameter for the utilization of objects at relatively high temperature, and it has been clearly improved in PLA formulations by using natural fibre reinforcements. For instance, the HDT for neat PLA is 64 °C raising up to 170 °C for PLA/Kenaf composites.[38] Several applications of these materials are currently under investigation,[39] in particular in the automotive industry. For example, Toyota and Ford have used PLA/fibre composites for the interior parts of their cars *i.e.* canvas roof, carpet mats *etc.*[40] Toyota has been using PLA resins since 2003 in its Raum and Prius car models (Figure 13.5).

13.6.1 Food Contact Materials

The packaging industry, particularly food packaging, plays a fundamental role in the use of non-renewable and non-biodegradable materials, resulting in a very important source of disposal waste. Therefore, the replacement of the current commodities by biopolymers is desirable, particularly PLA. The use of PLA in contact with food is considered safe with no health risks for consumers, particularly when used at temperatures under T_g as stated by international regulations launched by the European Commission and the US Food and Drug Administration, as is detailed in Chapter 14. In fact the degradation pathway of PLA leads to the formation of lactic acid, which is

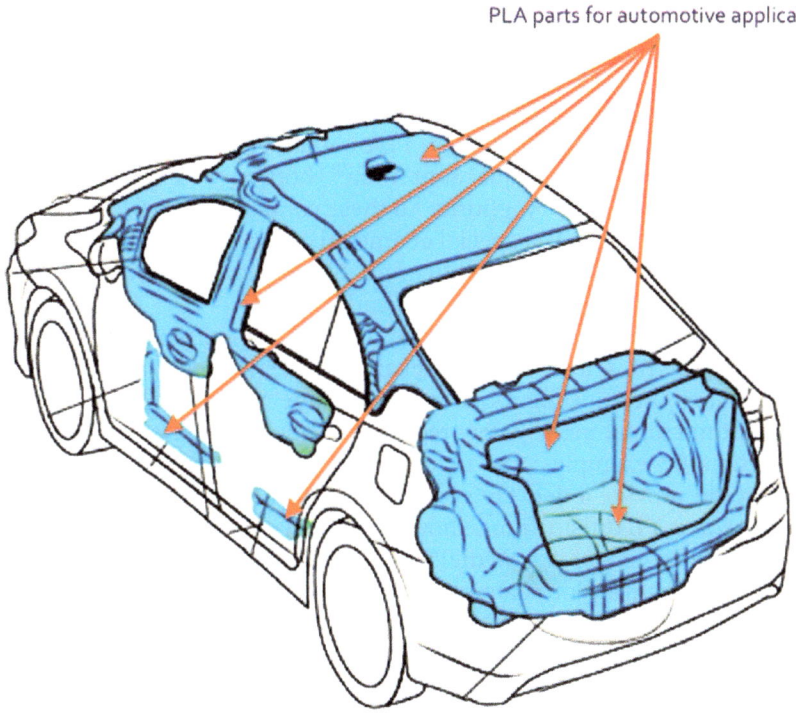

PLA parts for automotive applications.

Figure 13.5 Applications of biodegradable materials in automotives.

considered safe in many applications. But food packaging requires the food organoleptic and nutritional properties to maintain unaltered, and consequently materials should be chosen with outstanding characteristics in terms of barrier properties against radiation, water vapour, atmospheric gases and organic compounds, preventing food degradation and oxidation while preserving aromas and flavours.[41] In food packaging applications a key point is represented by light and UV absorption and transmission, particularly when light-sensitive foods are involved. It is important to remark than UV radiation is increasingly utilized in food sanitization, as an alternative method to chemical preservatives and consequently food packaging materials should not be decomposed by radiation.

On the contrary, sensitive components of food like flavours, vitamins or pigments may undergo degradation when exposed to radiation and oxygen. For these reasons, the plastic packaging plays a key role in preventing food photodegradation. From this point of view, PLA shows better behaviour than classical materials like polystyrene, cellophane, PET or LDPE.[42] Another important aspect for PLA is the light coloration imparted by the polymer. In fact, PLA is a highly transparent material, falling in the range over 95% with a transparency degree similar to PS, LDPE or PET. The only drawback in this sense is its natural yellowish coloration and this aesthetic aspect can create a consumer perception of "old stuff". For this reason, during the last years colorizing agents have been introduced into the polymer, changing the visual aspect of the final articles. It is remarkable that there have been recent developments in Europe and North America involving the use of PLA-based packaging for supermarket products. PLA containers have been used for packaging foods, such as Acqua Sant'Anna® bottled water, Noble® PLA bottled juices and Danone® yoghurts. All these containers meet EU and USA food grade requirements. The special characteristics of PLA, such as "GRAS" (Generally Recognized As Safe) status and biodegradability put PLA in a unique position for food-packaging applications.

Another important issue deals with antimicrobial food packaging, as detailed in Chapter 10. Antimicrobial packaging systems based on PLA would be superior to other antimicrobial materials due to its comparable cost, effective antimicrobial activity, few regulatory concerns and environmental friendliness. For example, the incorporation of bacteriocins into PLA matrices could provide a delivery system to improve their efficiency in food applications, *e.g.* PLA/nisin films show this behaviour and they have been proposed for commercial use. Actually, the PLA/nisin polymers can be used in bottle manufacturing or coated onto the bottle surface for use in liquid food packaging, as well as into films or coated on the surface of films for use in solid food packaging.

13.6.2 Medical and Biomedical Applications

Since the beginning of the commercial applications of PLA, it has been used in medical applications such as biodegradable scaffold, catheter, tissue regeneration and many other devices where biodegradability and biocompatibility are important issues. PLA has been extensively modified by copolymerization for medical devices. For example, in biomedical applications a surface modification is desired and plasma discharge, corona discharge or surface chemical attack have been successfully used.[43] In these cases, tissue compatibility is improved, cells can easily grow through the PLA-based templates and the resulting implants are more stable. Good results have been obtained also in regulation of angiogenic factors or gene transfer in growing tissues.[44] In addition, biodegradation in regeneration tissue has been extensively investigated[45] demonstrating the non-toxic pathway of the polymer decomposition. Lately, PLA microspheres have been reported for use in controlled drug delivery systems with both hydrophobic and hydrophilic behaviour. In fact, depending on the capsule structure and shape it is possible to insert different therapeutic agents targeted to organs and tissues. The controlled degradation of capsules permits a prolonged lifetime in the bloodstream and increased pharmacokinetics. Peculiarly, amorphous PLA (PDLLA) shows a rapid degradation pathway, being an excellent candidate for drug delivery vehicles and low-strength scaffolding material for tissue regeneration.

Figure 13.6 Typical example of mulch films.
© Shutterstock.

13.6.3   Agriculture

Agricultural mulch films are used for the production of vegetables and fruits (see Figure 13.6). Thin plastic films (typically polyethylene) are spread along the rows of plants at the beginning of the growing season. Holes are made in the films to plant the seeds and to permit aeration and sunlight contact. The use of mulch films speeds the ripening of crops, conserves moisture and fertilizer and inhibits weed growth, fungus infection and insect infestation. At the end of the growing season (generally 3–5 months) the film, contaminated with dirt and vegetation, must be collected, transported and disposed of. As the use of mulching films becomes more and more common, the disposal of the polyethylene films becomes a significant waste problem. The development of a truly degradable mulching film will reduce labour and disposal costs for agricultural producers. From these bases, the biodegradability of PLA can begin to be exploited in agricultural applications such as sandbags, weed prevention nets, vegetation nets and pots *etc*.

13.6.4   Electrical Appliances

In this field some uses of PLA have been proposed by leading industries. For example, Sony adopted PLA for making the external shell of its famous Walkman in 2001 by using the injection-moulding technique (Figure 13.7). Samsung and Toshiba used PLA for remote control and phone chassis (Figure 13.8). Sanyo uses PLA for the production of CD/DVD discs and cases (Figure 13.9). Fuji-Xerox used PLA in the interior of several copying machines and Fujitsu and PEGA successfully used PLA for the hard shell case of notebooks.

Figure 13.7   One of the first applications of PLA in electronics.

Figure 13.8   PLA cell phone frame.

13.6.5   Textiles

The use of biopolymers for textile applications is currently increasing and the technical substitution potential, which can be derived from the set of

Figure 13.9   CD/DVD case made of PLA.
© Shutterstock.

material properties, is estimated to be 33% of the total current polymer production. PLA for its ready-to-market presence, low cost and properties could be considered one of the most promising polymers for textile applications. For example, PLA fabrics can be found in knitted fabrics, non-woven textiles, headrest covers, vehicle seats, tarpaulins *etc.* (see for example Trevira's fibres made from PLA non-woven applications). For instance in Japan, JR trains use headrests made of PLA spunbond fabrics. In addition, thanks to the intrinsic antifungal/microbial properties, PLA can be used for air and water filters. Body towels of made PLA have a good reputation for not generating bad odours thanks to the above-mentioned antimicrobial properties of PLA. For these characteristics, potential application can be found in non-woven wipes, car and household wipes, personal care towels, diapers, napkins *etc.*

13.6.6   Home Furnishing

Typical applications of PLA in the home/office furnishing sector are carpets, tiles and rugs, draperies, curtains, pillows *etc.* PLA's low flammability and antimicrobial characteristics when loaded with the adequate additives permit its use in pillows, cushions, mattresses, sofas *etc.* A further application can be seen in PaperMate's Eco element, an innovative series of pens and pencils with the barrel made of PLA.

References

1. R. Auras, L. T. Lim and S. E. Selke, *Poly(lactic acid): Synthesis, Structures, Properties*, Wiley, Hoboken, USA, 2010.

2. D. Platt, *Biodegradable Polymers (market report)*, Rapra, Shawbury, UK, 2006.

3. Plastics Europe Association, *Plastics – the Facts 2012*, Brussels, 2012.

4. H. Benniga, *A History of Lactic Acid Making*, Springer, New York, USA, 1990.

5. W. H. Carothers, G. L. Dorough and F. J. Natta, *J. Am. Chem. Soc.*, 1932, **54**, 761.

6. J. Nieuwenhuis, *Clin. Mater*, 1992, **10**, 59.

7. P. R. Gruber, E. S. Hall, J. J. Kolstad, M. L. Iwen, R. D. Benson and R. L. Borchardt, *US Pat.*, 6,326,458, 2001.

8. P. R. Gruber, E. S. Hall, J. J. Kolstad, M. L. Iwen, R. D. Benson and R. L. Borchardt, *US Pat.*, 5,357,035, 1994.

9. www.natureworksllc.com. Accessed December 2013.

10. J. C. Bogaert and P. Coszach, *Macromol. Symp.*, 2000, **153**, 287.

11. R. Auras, B. Harte and S. Selke, *Macromol. Biosci.*, 2004, **4**, 835.

12. A. Södergård and M. Stolt, *Progr. Polym. Sci.*, 2002, **27**, 1123.

13. H. Urayama, T. Kanamori and Y. Kimura, *Macromol. Mater. Eng.*, 2001, **286**, 705.

14. D. W. Grijpma, H. Altperter, M. J. Bevis and J. Feijen, *Polym. Int.*, 2002, **51**, 845.

15. K. S. Anderson, K. M. Shreck and M. A. Hillmyer, *Polym. Rev.*, 2008, **48**, 85.

16. Y. K. Ikada, *Macromolecules*, 1987, **20**, 904.

17. H. Tsuji, S. H. Hyon and S. H. Ikada, *Macromolecules*, 1991, **24**, 5651.

18. www.purac.com. Accessed December 2013.

19. H. Tsuji and Y. Ikada, *Macromol. Chem. Phys.*, 1996, **197**, 3483.

20. L. T. Lim, R. Auras and M. Rubino, *Progr. Polym. Sci.*, 2008, **33**, 820.

21. www.dupont.com. Accessed December 2013.

22. www.mineralstech.com. Accessed December 2013.

23. www.dow.com. Accessed December 2013.

24. www.arkema.com. Accessed December 2013.

25. www.basf.com. Accessed December 2013.

26. www.polyone.com. Accessed December 2013.

27. www.clariant.com. Accessed December 2013.

28. J. K. Sears and J. R. Darby, *The Technology of Plasticizers*, Wiley, New York, USA, 1982.

29. E. Hiltunen, J. H. Selin and M. Skog, *US Pat.*, 6,117,928, 2000.

30. A. Masanobu and S. Ikado, *US Pat.*, 5,763,513, 1998.

31. V. P. Martino, A. Jiménez and R. A. Ruseckaite, *J. Appl. Polym. Sci.*, 2009, **112**, 2010.

32. V. P. Martino, R. A. Ruseckaite and A. Jiménez, *Polym. Int.*, 2009, **58**, 437.

33. N. Burgos, D. Tolaguera, S. Fiori and A. Jiménez, *J. Polym. Environ.*, 2014, **22**, 227.

34. www.condensiaquimica.es. Accessed December 2013.

35. O. Avinc and A. Khoddami, *Fibre Chem.*, 2010, **42**, 68.

36. Z. M. Huang, Y. H. Zhang and M. Kotaki, *Compos. Sci. Tech.*, 2003, **63**, 2223.

37. M. A. Huneaul and H. Li, *Polymer*, 2007, **48**, 270.

38. T. Nishino, K. Hirao, M. Kotera, K. Nakamae and H. Inagaki, *Compos. Sci. Tech.*, 2003, **63**, 1281.

39. biostructproject.eu. Accessed December 2013.

40. S. Serizawa, K. Inoue and M. Iji, *J. Appl. Polym. Sci.*, 2006, **100**, 618.

41. N. S. Oliveira, J. Oliveira, T. Gomes, A. Ferreira, J. Dorgan and I. M. Marrucho, *Fluid Phase Equil.*, 2004, **222**, 317.

42. J. Lunt, in *Biodegradable and Sustainable Fibres*, Woodhead Publishing Ltd, Cambridge, UK, 2005.

43. J. H. Lee, G. Khang, J. W. Lee and H. B. Lee, *J. Colloids Interface Sci.*, 1998, **205**, 323.

44. M. H. Sheridan, L. D. Shea, M. C. Peters and D. J. Mooney, *J. Contr. Release*, 2000, **64**, 91.

45. S. Li and M. Vert, *Polym. Degrad. Stabil.*, 2001, **71**, 61.

CHAPTER 14

Legislation Related to PLA

MERCEDES A. PELTZER* AND ANA BELTRÁN-SANAHUJA

Analytical Chemistry, Nutrition & Food Sciences Department,
University of Alicante, PO Box 99, 03080, Alicante, Spain
*Email: mercedes.peltzer@ua.es

14.1   Food Packaging and Legislation

Food is packaged to prevent physical, chemical, physiological and microbial spoilage during transport, distribution, handling and storage. The main roles of food packaging materials are to protect food products from external influences and damage, to contain the food and to provide and to give information to the consumer about ingredients and nutritional data about the product.[1] All these functions are important to control and modify their mechanical and barrier properties that consequently depend on the structure of the polymeric packaging material. In addition, it is important to study the changes that can occur in the packaging materials characteristics during their interaction with the packaged food.

The food packaging sector has been subjected in recent years to important evolution, since new plastics and designs have been recently developed to improve their performance in protecting food and improving shelf-life and organoleptic and nutritional properties. The use of bio-based and bio-degradable polymers for food packaging applications is rising in the food industry to reduce CO_2 emissions and the dependency on fossil resources. Among them, poly(lactic acid) (PLA) is suitable in the manufacture of cups, bottles, films and containers. The main applications include rigid thermo-forms such as trays and lids, bottles for water, candy and flower wraps, disposable salad cups and cold drink cups.[2]

RSC Polymer Chemistry Series No. 12
Poly(lactic acid) Science and Technology: Processing, Properties, Additives and Applications
Edited by Alfonso Jiménez, Mercedes Peltzer and Roxana Ruseckaite
© The Royal Society of Chemistry 2015
Published by the Royal Society of Chemistry, www.rsc.org

When food is in contact with the packaging material, a mass transfer of components from the packaging towards the food and *vice versa* begins. This food/packaging interaction includes the so-called "chemical migration", a process characterized by high complexity, diffusion being the main process responsible for this type of migration, *i.e.* the macroscopic movement of molecular structures from higher to lower concentrations, while adsorption can be another way of leading to this type of migration.[3]

The Framework Regulation (EC) No 1935/2004[4] covers all food contact materials (FCM). It is based on the two general principles of inertness and safety of the material and states that no components of the materials and articles that constitute risk for the health of the consumers or can lead to unacceptable change in the composition of the foodstuffs or deteriorate their organoleptic properties should migrate into foods. In the EU, Framework Regulation (EC) No 1935/2004 together with the EC Regulation No 10/2011[5] are the basic European Community legislative tools that define food contact materials to all articles that may be brought into contact with food, including packaging, dishes, cutlery and table surfaces in food preparation areas. This legislation set exceptions for active and intelligent materials and for materials that are destined to be consumed with food.

Regulation No 1935/2004 sets three requirements in order to ensure safe and good quality of the food. Firstly, the FCM should not transfer their components into the food in quantities that could endanger human health; secondly, FCM should not change the composition of the food in an un-acceptable way; and the last requirement is that FCM should not cause de-terioration in the taste, odour or texture of the food. The first requirement deals with the migration and food safety, while the other two are related to food quality. Regulation No 1935/2004 does not include the details for the implementation of these principles, but Regulation No 10/2011 does. This regulation was created to ensure that all the plastic packaging systems comply with Article 3 of Regulation 1935/2004, and establishes the specific rules for plastic materials and articles to be applied for their safe use. In addition, this regulation repeals Directive 2002/72/EC on plastic materials and articles intended to come into contact with foodstuff. Annex I of the current regulation lists all the authorized substances, such as monomers and other starting substances, macromolecules obtained from microbial fermentation, additives and polymer production aids with the specific mi-gration limit. Annex II presents the restrictions on materials and articles, indicating those substances that materials should not release. Annex III describes the food simulants to be used in migration testing and checking of compliance.

Related to food packaging materials, it is important to highlight the Commission Regulations (EU) No 1282/2011[6] and No 1183/2012[7] amending and correcting Regulation (EU) No 10/2011 on plastic materials and articles intended to come into contact with food, in which the European Food Safety Authority (EFSA) issued favourable scientific evaluations for additional substances that should be added to the current list in Regulation (EU) No 10/2011.

Until now, only ceramics, regenerated cellulose, plastics and active and intelligent materials have been covered by specific legislation. Due to the importance of plastics, they were the first to be covered by the Community harmonization. Regulation (EU) No 10/2011 covers plastic monolayers, multilayers and multi-material multilayers. With this last term, legislation covers all materials and articles that are not constituted exclusively by plastics. This was introduced by the new Regulation (EU) 10/2011, indicating that evaluation of overall migration and specific migration is not required for multi-material multilayers. The compliance shall be evaluated in terms of chemical composition of plastic layers involved (plastic layers applied on metallic substrates, for example). In this regulation, a plastic layer which is not in direct contact with food and is separated from the food by a functional barrier need not to comply with the restrictions and specifications, except for the vinyl chloride monomer.

As was mentioned before, according to Regulation 1935/2004, the release of substances from FCM should not bring any change in the food composition. It is feasible to manufacture plastics that do not release more than 10 mg of substance per 1 dm^2 of plastic surface area. The overall migration limit of 10 mg per 1 dm^2 results for cubic packaging containing 1 kg of food to a migration of 60 mg per kg food. In the case that any substance does not have a lower limit, this level should be applied as a generic limit for the inertness of a plastic material. This limit should be demonstrated by the overall migration tests, following standardized test conditions of temperature, time and test medium (food simulant), representing the worst-case scenario of use of these materials.

The introduction of new materials and technologies has changed the view of food packaging and the use of biopolymers is one of the latest technological trends with the aim to improve the current materials and/or produce environmentally friendly packaging.[8] In the absence of specific regulations, these materials are treated in exactly the same way as conventional plastics and lie on the scope of European regulations and good manufacturing practice guidelines for the compliance of any FCM, following the above-detailed legislative framework. However, due to differences in origin and properties between conventional and bio-based materials, including PLA, some differences have been noticed.[8,9]

Those biopolymers derived from biomass, such as starch, pectin or casein, are referred to the EU positive list of monomers and additives but others, such as chitosan, gelatine, keratin, gluten, zein and soybean, are authorized as food additives (they may fulfil the required legislation for food) but they are not specified in the FCM legislation. As a consequence, polysaccharide- or protein-based materials are submitted to the overall migration limit set by the plastic regulation where no specific limit is specified. The difficulty in this situation for use in hydrolyzable biopolymers, such as PLA, is that migration tests were designed for water-resistant materials. Those materials that present low water resistant properties would be used for dry food. In these cases, the use of the new simulant E Tenax® (poly(2,6-diphenyl-pphenylene oxide)) is recommended.[8]

Very few studies are referenced about the safety assessment of biopolymer-based packaging materials. PLA is the most studied biopolymer and very limited migration was found in food packed with PLA, within the fixed overall migration limit in all cases.[10,11] Conn *et al.*[12] evaluated PLA to prove its safety to be used as FCM by considering the identity and toxicological properties and quantities of substances that migrate into food during the intended use. The potential migrants from PLA were lactic acid, lactoyllactic acid (linear dimer of lactic acid), other small oligomers of PLA and the lactide dimer. All these compounds that migrate from PLA contacting food are known to hydrolyze in aqueous media into lactic acid, which is already authorized as a monomer and additive (Regulation No. 10/2011), with no restrictions or specifications.[8,9] The main component that might migrate into food is lactic acid or its oligomers, which will hydrolyze in aqueous systems and will produce the monomer. Lactic acid is a common ingredient in food which has been shown to be safe at levels far above those migrating from PLA in contact with food simulants.

14.2 Biodegradation and Compostability Legislation

Besides the multiple advantages of plastic materials, their increased use has created a serious problem in municipal solid waste generation. Over the last few decades (between 1980 and 2010), municipal solid waste (MSW) generation has increased from 3.66 to 4.43 pounds per person per day. Although recycling also increased as an average from 10% of MSW in 1980 to 34% in 2010[13] and disposal of waste residues to landfills decreased from 89% in 1980 to about 54% of MSW in 2010, the absolute numbers of residues are continuously increasing. Plastics constitute around 12% of total MSW generation in 2010.

The slow degradation and the rapid rate of production of petroleum-based plastics create an immense imbalance on their life cycle. Therefore, bio-based and biodegradable plastics have received increasing attention from researchers and many organizations around the world, since they can reduce and/or replace petroleum-based plastics in many applications, in particular food packaging.[13,14] Growing environmental awareness imposes both user-friendly and eco-friendly attributes onto packaging films and processes. As a consequence biodegradability is not only a functional requirement but also an important environmental attribute. Regarding biodegradation issues, biopolymers break down to produce environmentally friendly products, such as carbon dioxide, water and quality compost.[15]

Many biodegradable plastics were not very well labelled due to some misapprehension about the difference between a degradable and a bio-degradable plastic. Due to this fact, many international standards were developed in order to clarify this misconception. Among the associations dealing with these products are the American Society for Testing and

Materials (ASTM), European Committee for Standardization (CEN), German Institute of Standardization, Japanese Industrial Standards and the International Organization for Standardization (ISO).

The ASTM D883-12 standard[16] defines a degradable plastic as a plastic that undergoes a significant change in its chemical structure under specific environmental conditions resulting in a loss of some properties. The degradation of the plastic could be induced chemically, biologically, through heat or by ultraviolet light and, depending on the process of degradation, the material is classified as photodegradable, when degradation is by UV light, or biodegradable when degradation is performed by micro-organisms.[17] This international standard defines biodegradable plastics as plastics in which their degradation results from the action of naturally occurring micro-organisms such as bacteria, fungi and algae.

Plastics biodegradation could be aerobic or anaerobic, depending upon the environment in which they are degrading, and biodegradation is measured by the amount of carbon by-products that are produced by the biodegradation process in which the sum of all the carbon by-products is equal to the total amount of carbon in the non-degraded material.

The main form of aerobic biodegradation is through composting. The compostability attribute is very important for biopolymer materials because while recycling is energy expensive, composting allows disposal of the packages in the soil producing only water, carbon dioxide and inorganic compounds without toxic residues. In this sense, composting is considered to be nature's way of recycling and could be prepared either at industrial scale or by consumers at a residential level.[18]

The European Directive 94/62/EC,[19] amended by 2004/12/EC,[20] indicates that composting of packaging waste is a form of recycling, since the original product, "the package", is transformed into a new product, "the compost". Following this Directive the European Standard EN 13432[21] was introduced. This European Standard defines the characteristics that a material must accomplish to be defined as "compostable":[22]

a) Biodegradability, which is determined by measuring the actual metabolic conversion of the compostable materials into carbon dioxide. This property is quantitatively measured by using the standard test method, EN14046:2003[23] (which is also published as ISO 14855: biodegradability under controlled composting conditions). This method is based on the measurement of the actual metabolic microbial conversion, under composting conditions, of the packaging into water, carbon dioxide and new cell biomass. The acceptance level is 90% of carbon dioxide, which must be reached in less than 6 months.

b) Disintegrability is the fragmentation and loss of visibility in the final compost (absence of visual contamination). This is measured with a composting test (EN14045).[24] The packaging sample is mixed with organic waste and maintained for 3 months. After this time no more than 10% of material fragments are allowed to be larger than 2 mm.

c) Absence of negative effects on the composting process, checked with a composting test.

d) Low level of heavy metals (Potentially Toxic Elements) and no adverse effect on the quality of the produced compost.

The terms "biodegradation", "biodegradable polymer" and "compostability" are frequently misused and they are a source of misunderstanding: solubility in water is considered as a synonym of biodegradability, and biodegradability as a synonym of compostability. The term biodegradable by itself is not useful, since biodegradation is not always ensured. If the environment is not favourable for the degradation to occur, the material will not degrade in a short time, even with the presence of enzymes that could speed up the breaking rate of the chemical bonds. Therefore, it is important to couple the term biodegradable with the specification of the particular environment where the degradation is expected to happen, during the time scale of the process.[25]

According to these standard norms, a compostable plastic demonstrates a satisfactory rate of biodegradation when achieving 90% of the organic carbon in the whole item or for each organic constituent, which is present in the material at a concentration of more than 1 wt% (by dry mass), and is converted to carbon dioxide by the end of the test period when compared to a positive control.

14.3 Nanomaterials in Food Packaging Legislation

Nanotechnology involves the characterization, fabrication and/or manipulation of nanomaterials. The Commission Recommendation of 18 October 2011 defines a "nanomaterial" as "a natural, incidental or manufactured material containing particles, in an unbound state or as an aggregate or as an agglomerate and where, for 50% or more of the particles in the number size distribution, one or more external dimensions is in the size range 1 nm–100 nm".[26] When particle size is reduced below this threshold, the resulting material exhibits physical and chemical properties that are significantly different from those properties of the common materials.

Research in nanotechnology has seen a huge improvement over the last decade, and already there are numerous companies specializing in the fabrication of new forms of nanosized materials for different applications including medical therapeutics and diagnostics, energy production, molecular computing and structural materials.[27]

In food contact materials, polymer nanocomposites-based packaging materials are developed by the inclusion of nanoscale fillers resulting in clear improvement of food quality and extension of the shelf-life through minimizing microbial growth. They can serve not only as barriers to moisture, water vapour, gases and solutes, but also as carriers of some active substances.[28] The enhancement of the polymer barrier properties is the most obvious application of nanocomposites in the food industry.

However, these materials present additional characteristics such as stronger mechanical properties (*e.g.* reinforcement of polymer matrices), flame resistance and better thermal properties (higher melting point, degradation and glass transition temperature) than common polymers with no nanofillers. In addition, surface wettability and hydrophobicity are improved, resulting in surfaces more prone to bonding with other polymers and additives.[29]

Nano-biocomposites use biodegradable matrices and incorporate nanofillers to minimize the disadvantages of not using traditional packaging materials.[30] These materials, besides protecting food from its surroundings and prolonging its shelf-life, are more environmentally friendly.[31]

The use of nanotechnologies in the food industry may present potential risks to both human health and the environment due to the use of novel materials in novel ways, and risk assessments must be carried out. Three different ways of entrance penetration of nanoparticles in the organism are possible: inhalation, skin penetration and ingestion. Free nanoparticles can cross cellular barriers and that exposure may lead to oxidative damage and inflammatory reactions. In the case of nanomaterials for food packaging, many people fear risk of indirect exposure due to potential migration of nanoparticles from packaging materials, in particular nano-biocomposites.

All applications of this new technology must be assessed for safety use. In the EU, the Directorate General of Health and Consumer Protection (SCENIHR)[32] provides opinions on emerging or newly identified health and environmental risks and other issues that require a comprehensive assessment of risks to consumer safety or public health. SCENIHR focuses on nanotechnologies' risk assessment with particular interest in establishing recognized terminology in the field so that research can be integrated to a certain extent. Since the applications of nanotechnologies are increasing very fast, SCENIHR needs to identify what could be considered a nanomaterial by clear descriptions. The ISO/TS 27687:2008 standard is related to nanotechnologies terminology and definitions for nano-objects, nano-particle, nanofibre and nanoplate.[33] The document lists various terms and definitions related to particles in the area of nanotechnologies.

It is important to note that there may be risks to the consumer in the form of nanoparticles migration from packaging materials into food. The results of this exposure have not been fully determined and the lack of such data is an obstacle to the assessment of risk posed by the consumption of foodstuff in contact with nanopackaging materials. The exposure of nanomaterials and consumers is evaluated by migration tests, and predicted safe levels of exposure are based on animal exposure trials. In this sense, for the final consumers of food packaged with nanomaterials the first concern is to verify the extent of migration of nanoparticles from the package into food and then, if this migration happens, the effect of the ingestion of these nanoparticles inside the body from the mouth to the final gastrointestinal tract.

The case of PLA nano-biocomposites used in food packaging raises the issue of environmental contamination as the result of nanomaterials being released from the biopolymer bulk following polymer degradation. Eco-toxicity tests are required to determine the impact and risks of the nano-materials to human health and the environment. In fact, the Novel Food Regulation EC No 258/97 concerns the commercial distribution and intro-duction to the market within the European Union of novel foods or novel food ingredients. Under this regulation, foods deemed to be novel must undergo a safety assessment prior to being placed on the market. This could be the piece of legislation where nano-biocomposites based on PLA could be associated.[34]

Nanotechnology has also been applied to the field of active packaging already treated in Chapter 10 of this volume. Metal nanoparticles, metal oxide nanomaterials and carbon nanotubes are the most used nanoparticles to develop antimicrobial active packaging nanocomposites. In this sense, the regulation entitled Active and Intelligent Materials and Articles intended to come into contact with Food (EC) No 450/2009 states that if the legislation determines a certain amount of a substance in the food, the total amount should not exceed the indicated limits regardless of the source, even in-cluded in the foodstuff or released from the packaging material. The ap-proach for nanofillers and nanomaterials is the same as in all the other additives able to be used in food, which is that these substances should undergo a safety assessment by EFSA and an authorization from the Euro-pean Commission. Once this substance is authorized, it could be added to the positive list and used within specific constraints.

On the other hand, some countries have different approaches for nano-biocomposites in food packaging applications. For instance, Taiwan has introduced the Nano-Mark System: this is a quality-like symbol of assurance to consumers which certifies that a product uses genuine nanotechnology.[35] The so-called "nanoingredients" are included as a category to which this symbol is assigned. In Australia, nanotechnologies are considered by the National Industrial Chemicals Notification and Assessment Scheme (NIC-NAS), which regulates chemicals for the protection of human health and the environment. New administrative processes were introduced to address nanotechnology.[36,37] Other countries, such as the United States of America, are currently working on a federal regulatory framework to provide con-sistent and comprehensive screening and protection for consumers.

14.4   Life Cycle Assessment (LCA)

Products obtained from renewable materials are considered to be environ-mentally beneficial, since they save fossil resources and are considered po-tentially biodegradable. But an important issue to be considered is how big the influence on the environment is, since they usually involve the use of agricultural products for non-food uses and sometimes the evolution of gases like carbon dioxide after biodegradation/compostability could be harmful for the atmosphere. Therefore, it is necessary to assess the overall impact of the use of PLA in many applications and the Life Cycle Assessment (LCA) methodology should be applied.

LCA is a technique to assess the potential environmental aspects associated with products or services. It is a compilation of an inventory of relevant inputs and outputs, evaluating the potential environmental impacts associated with them and interpreting the results of the inventory and impact phases in relation to the objectives of the study. The term "life cycle" indicates that all stages in a product's life, from resource extraction to ultimate disposal, should be taken into account.

In other words, LCA is a quantitative analysis of resource depletion and production of pollutants from the production systems under study, but it is possible to find in these studies a qualitative analysis for those issues difficult to quantify, for example: biodiversity. The assessment of the environmental impact and sustainability is necessary to evaluate the life-cycle of the product, from the raw materials extraction through production protocols used in the manufacture, the use of the product, recycling and/or ultimate disposal of some of its constituents. Such a complete life-cycle is also often named "cradle to grave" and should also be performed on PLA-based products. This evaluation incorporates manufacturing practices, energy input/output and overall material flows. Transportation, storage, retail and other activities between the life-cycle stages are included where relevant. This life cycle of a product is hence identical to the complete supply-chain of the product and end-of-life treatment.[38] In the LCA analysis, the use of resources, raw materials, parts and products, energy carriers, electricity, *etc.* are considered the "inputs". While the emission to air, water and land as well as waste and by-products are recorded as the "output".

The ISO standard on Environmental Management Systems concerns the Life Cycle Assessment of products and processes. ISO 14040[39] is the general framework for the specific standards ISO 14041,[40] ISO 14042[41] and ISO 14043.[42] These ISO standards are integrated, since 2006, by the general standard ISO 14044.[43] This standard indicates the next phases to perform a LCA analysis:

- Stage 1. Defining the goal and scope of LCA: **ISO 14041**. Identification of materials, processes and products to be considered and how broadly they will be defined.
- Stage 2. Life cycle inventory analysis: **ISO 14042**. Uses quantitative data to establish levels and types of energy and materials used/released from the processes.
- Stage 3. Impact Analysis: **ISO 14043**. Relating outputs of system to the impacts on the external world into which the outputs flow.
- Stage 4. Interpretation: **ISO 14044**. Uses findings from previous stages to draw conclusions and make recommendations for reducing environmental impact.

The life cycle of PLA starts with corn production. All free energy used for such production is captured by photosynthesis[44] by following the basic stoichiometric equation:

$$H_2O + CO_2 \xrightarrow{\text{light}} (CH_2O) + O_2$$

In this equation, CH_2O represents carbohydrate, such as sucrose or starch. Therefore all carbon, hydrogen and oxygen found in the starch molecule and in the final polylactide molecule is originated from water and carbon dioxide. Several steps are performed to obtain PLA: 1) Corn production and transport; 2) Corn processing and conversion of starch into dextrose; 3) Conversion of dextrose into lactic acid; 4) Conversion of lactic acid into lactide; 5) Polymerization of lactide to PLA.

The lactic acid production presents the inputs of the corn production and all the elements required, such as fertilizers, limestone, electricity and fuels (natural gas, diesel, propane and gasoline) used in the farm, the atmospheric carbon dioxide utilization through photosynthesis, the irrigation water applied to the cornfield and the production of the herbicides and insecticides used to grow the corn. On the output side, emissions including nitrous oxide, other nitrogen oxides, nitrates and phosphates should be considered. The inputs used in the PLA production from lactic acid are natural gas, electricity, steam, potable and cooling water, nitrogen and chemicals. The outputs are air emissions, water emission, solid waste and co-products.

In general terms, polymers obtained from renewable resources can be significantly lower in greenhouse gas emissions and fossil energy use as compared with conventional petrochemical- based polymers. Over the long term, LCA demonstrates that PLA production processes can become both fossil-energy free and a source of carbon credits.[45]

14.5 Conclusions

PLA is a sustainable alternative to petrochemical-derived products and it is approved by the Food and Drug Administration (FDA) for use as a food contact material. These PLA-based materials are treated in the same manner as conventional plastics regarding food contact material legislation. Although the interest in nanotechnology is growing, uncertainties for risk assessment and exposure assessment of nanomaterials arise due to the limit information about toxicology, behaviour and bioaccumulation. Biobased polymers can influence to a lesser extent greenhouse gas emissions and fossil energy use today as compared with conventional petrochemicalbased polymers. Over the longer term, LCA demonstrates that PLA production processes can become both fossil-energy free and a source of carbon credits.

References

1. R. Coles, in *Food Packaging Technology*, ed. R. Coles, D. McDowell and M. J. Kiran, Blackwell Publishing, CRC Press, London, United Kingdom, 2003, pp. 1–31.
2. Y. Bor, J. Alin and M. Hakkarainen, *Packag. Tech. Sci.*, 2012, **25**, 427.
3. I. S. Arvanitoyannis and K. V. Kotsanopoulos, *Food Bioprocess Technol.*, 2014, **7**, 21.
4. Regulation No 1935/2004 of the European Parliament and of the Council of 27 October 2004 on materials and articles intended to come in contact with food and repealing Directives 80/590/EEC and 89/109/EEC, *Official Journal of the European Union*, 08/07/2009.
5. Commission Regulation (EU) No 10/2011 of 14 January 2011 on plastic materials and articles intended to come into contact with food, *Official Journal of the European Union*, 21/01/2011.
6. Commission Regulation (EU) No 1282/2011 of 28 November 2011 amending and correcting Commission Regulation (EU) No 10/2011 on plastic materials and articles intended to come into contact with food, *Official Journal of the European Union*, 10/12/2011.
7. Commission Regulation (EU) No 1183/2012 of 30 November 2012 amending and correcting Regulation (EU) No 10/2011 on plastic materials and articles intended to come into contact with food, *Official Journal of the European Union*, 12/12/2011.
8. N. Gontard, H. Angellier-Coussy, P. Chalier, E. Gastaldi, V. Guillard, C. Guillaume and S. Peyron, in *Biopolymers. New Materials for Sustainable Films*, ed. D. Plackett, John Wiley & Sons Ltd, United Kingdom, 2011, pp. 57–70.
9. T. Sipiläinen-Malm, U. Thoden van Velzen and A. Leufvén, in *Biobased Packaging Materials for the Food Industry*, ed. C. J. Weber, KVL Department of Dairy and Food Science, Denmark, 2000, pp. 85–104.
10. M. Mutsuga, Y. Kawamura and K. Tanamoto, *Food Addit. Contam.*, 2008, **25**, 1283.
11. E. Fortunati, M. Peltzer, I. Armentano, L. Torre, A. Jiménez and J. M. Kenny, *Carbohydr. Polym.*, 2012, **90**, 948.
12. R. E. Conn, J. J. Kolstad, J. F. Borzelleca, D. S. Dixler, L. J. Filer Jr, B. N. LaDu Jr and M. W. Pariza, *Food Chem. Toxicol.*, 1995, **33**, 273.
13. K. Dagnon, M. Pickens, V. Vaidyanathan and N. D'Souza, *J. Polym. Environ.*, 2013, DOI: 10.1007/s10924-013-0596-9.
14. M. Alboofetileh, M. Rezaei, H. Hosseini and M. Abdollah, *J. Food Process. Preserv.*, 2013, DOI: 10.1111/jfpp.12124.
15. R. N. Tharanathan, *Trends Food Sci. Tech.*, 2003, **14**, 71.
16. ASTM D883-12 Standard terminologies relating to plastics, 2012.
17. R. Jayasekara and I. Harding, *J. Polym. Environ.*, 2005, **13**, 231.
18. T. Kijchavengkul and R. Auras, *Polym. Int.*, 2008, **57**, 793.

19. Directive 94/62/EC of the European Parliament and of the Council of 20 December 1994 on packaging and packaging waste, *Official Journal of the European Union*, 31/12/1994.
20. Directive 2004/12/EC of the European Parliament and of the Council of 11 February 2004 amending Directive 94/62/EC on packaging and packaging waste, *Official Journal of the European Union*, 18/02/2004.
21. EN 13432:2000 Packaging. Requirements for packaging recoverable through composting and biodegradation. Test scheme and evaluation criteria for the final acceptance of packaging.
22. http://www.ilex-envirosciences.com/leaflets/EN13432.pdf. Accessed October 2013.
23. EN 14046:2003. Packaging. Evaluation of the ultimate aerobic bio-degradability and disintegration of packaging materials under controlled composting conditions. Method by analysis of released carbon dioxide.
24. EN 14045:2003. Packaging. Evaluation of the disintegration of packaging materials in practical oriented tests under defined composting conditions.
25. F. Degli-Innocenti, *Biobased Packaging Materials for the Food Industry*, ed. C. J. Weber, A European Concerted Action, 2000, pp. 107–113.
26. Commission Recommendation of 18 October 2011 on the definition of nanomaterial (2011/696/EU), *Official Journal of the European Union*, 20/10/2011.
27. T. Duncan, *J. Colloid Interface Sci.*, 2011, **363**, 1.
28. A. M. Youssef, *Polym. Plast. Tech. Eng.*, 2013, **52**, 635.
29. Q. Zhou, K. P. Pramoda, J. M. Lee, K. Wang and L. S. Loo, *J. Colloid Interface Sci.*, 2011, **355**, 222.
30. A. Sorrentino, G. Gorrasi and V. Vittoria, *Trends Food Sci. Tech.*, 2007, **18**, 84.
31. N. Sozer and J. L. Kokini, *Trends Biotechnol.*, 2009, **27**, 82.
32. SCENIHR, Scientific Committee on Emerging and Newly-Identified Health Risk, 2010. Available at http://ec.europa.eu/health/scientific_committees/emerging/index_en.htm#. Accessed October 2013.
33. Scientific Basis for the Definition of the Term "nanomaterial" available in http://ec.europa.eu/health/scientific_committees/emerging/docs/scenihr_o_032.pdf. Accessed October 2013.
34. M. Cushen, J. Kerry, M. Morris, M. Cruz-Romero and E. Cummins, *Trends Food Sci. Tech.*, 2012, **24**, 30.
35. C. F. Chau, S. H. Wu and G. C. Yen, *Trends Food Sci. Tech.*, 2007, **18**, 269.
36. NICNAS 2010, National Industrial Chemicals Notification and Assessment Scheme, Australian Government, Gazette.
37. K. Lyons and J. Whelan, *NanoEthics*, 2010, **4**, 53.
38. G. Finnveden, M. Z. Hauschild, T. Ekvall, J. Guinée, R. Heijungs, S. Hellweg, A. Koehler, D. Pennington and S. Suh, *J. Environ. Manag.*, 2009, **91**, 1.

39. ISO 14040:2006. Environmental management. Life cycle assessment. Principles and framework.

40. ISO 14041:1998. Environmental management. Life cycle assessment. Goal and scope definition and inventory analysis.

41. ISO 14042:2000. Environmental management. Life cycle assessment. Life cycle impact assessment.

42. ISO 14043:2000. Environmental management. Life cycle assessment. Life cycle interpretation.

43. ISO 14044:2006. Environmental management. Life cycle assessment. Requirements and guidelines.

44. E. T. H. Vink, D. A. Glassner, J. J. Kolstad, R. J. Wooley and R. P. O'Connor, *Ind. Biotechnol.*, 2007, **3**, 58.

45. E. T. H. Vink, K. R. Rabago, D. A. Glassner and P. R. Gruber, *Polym. Degrad. Stabil.*, 2003, **80**, 403.

Subject Index

Active packaging, 245–262
    antimicrobial active packaging,
      249–257
    antioxidant active packaging,
      257–262
Adult stem cells (ASCs), 268
Aggregated morphology, 219
Agricultural mulch films, 329
Antimicrobial active packaging,
   249–257
    antimicrobial agents, 250–251
    PLA-based packaging
      developments, 251–257
    polymer matrix, incorporation
      of agent, 251
Antimicrobial PLA-based packaging
   developments, 251–257
    coating, PLA films, 251
    electrospinning methodology,
      256
    extrusion, 255–256
    other procedures, 256–257
    polymeric coatings and film
      casting, 253–255
Antioxidant active packaging, 257–262
    antioxidant agents, 258
    antioxidant PLA-based
      packaging developments,
      258
    extrusion, 261–262
    other packaging applications,
      262
    polymeric coatings and film
      casting, 258–260

polymer matrix, incorporation
   of agent, 258
Antiplasticization, 126
Article 3 of regulation (EC) N° 450/
   2009, 246
Association Internationale pour
   l'Etude des Argiles (AIPEA), 216
Avrami parameters, 82, 83, 234

Back-biting mechanism, 8
3S-(Benzyloxymethyl)-6S-methyl-1,4-
   dioxane-2,5-dione (BMD), 23–24
Biodegradable polymers, 196–197
Biomass, 160–163
Block copolymers, 84–92
Butyl lactate, 6

Carbon nanotubes (CNT), 182–183
CED. *See* Cohesive energy density
Cellulose nanocrystals (CNCs),
   227–228
Cellulose nanofibrils (CNFs),
   226–227
Chemical structure, hydrolytic
   degradation rate, 296–297
    branching levels, 297
    crosslinking, 297
    tacticity, 296–297
Clay, definition of, 216
Clay Minerals Society (CMS), 216
CMC. *See* Critical micellar
   concentration
CNCs. *See* Cellulose nanocrystals
CNFs. *See* Cellulose nanofibrils

Coaxial electrospinning
  (co-electrospinning), 174
Cohen–Turnbull–Fujita model, 126
Cohesive energy density (CED), 127
Coordination-insertion mechanism,
  105
Copolymerization, 22–25, 195–196
Coupling reactions, 107–113
    epoxy-based, 110–113
    isocyanate-based, 108–110
Crazing process, 206
Critical micellar concentration
  (CMC), 51–53

Differential scanning calorimetry
  (DSC) thermograms, 41, 43
Differentiation induction factors,
  269
Drug delivery, 188–191
Drug delivery systems (DDS), 50
Dynamic light scattering (DLS), 24

Elastomers, 198–203
Electrical appliances, 329–330
Electrospinning process, 171–191,
  256, 325
Electrospun-PLA fibres
    applications of, 183–191
    drug delivery, 188–191
    tissue engineering, 184–187
    wound dressing, 187–188
Embryonic stem cells (ESCs), 268
Encapsulation efficiency (EE), 53, 55
Enhanced permeation retention
  effect (EPR effect), 51
Epoxidized palm oil (EPO), 163
Epoxidized soybean oil (ESO), 162
European Directive 94/62/EC, 338
Exfoliated morphology, 219

Factors controlling hydrolytic
  degradation
    molecular weight, 296–305
    pH, 292–294
    shape and geometry, 295–296
    temperature, 294–295

Film and sheet casting, 322–323
Flory–Huggins interaction, 127
Foaming process, 325–326
Food contact materials, 326–328
Framework Regulation (EC) No
  1935/2004, 335
Free-radical grafting
    of acrylated poly(ethylene
      glycol) (Acryl-PEG), 116–117
    of maleic anhydride (MA), 115
Free Volume Theory, 126

Glass transition temperature, 68–73,
  135–146
Glyceric acid synthesis, 24
Graphene, 183

Half times, crystallization, 234
Halloysite nanotubes (HNT),
  180–181, 218
Hildebrand solubility theory, 127
HNT. *See* Halloysite nanotubes
Home furnishing, 331
Hydrolytic degradation
    benefits of, 305–308
    degradation medium
      conditions, 292–295
    factors controlling, 292–305
    mechanisms of, 290–292
    structural modification,
      306–308
    surface hydrophilicity level
      modifications, 306
Hydroxyapatite (HA), 181–182
2-Hydroxypropanoic acid. *See* Lactic
  acids

Induced pluripotent stem cells
  (iPSCs), 268–269
Injection stretch blow moulding
  (ISBM). *See* Stretch blow
  moulding
Intercalated morphology, 219
Interfacial compatibilization
  reaction, 113
ISO/TS 27687:2008 standard, 340

Joncryl®, 112

Kaolinites, 218

Lactic acids, 4
    chemical synthesis of, 7
    fermentation, bacteria, 5
    isolation of, 5–6
    melt/solid polycondensation,
      27–29
    polycondensation of, 25–30
    purification of, 5–6
    solution polycondensation, 27
    stereo-block polycondensation,
      29–30
    stereoisomers of, 4–5
    structures of, 4–5
    synthesis of, 4–7
D-Lactic acids. *See* Lactic acids
DL-Lactic acids. *See* Lactic acids
L-Lactic acids. *See* Lactic acids
Lactides
    back-biting mechanism, 8
    cationic catalysts, 15
    copolymerization, 22–25
    metal catalysts, 11–14
    organic catalysts, 15–18
    polymerization of, 8–25
    polymerization, reversible
      activation of triazolylidene
      carbenes, 19
    purification of, 7–8
    ring-opening polymerization
      (ROP), 8, 10
    stereo-controlled
      polymerization, 18–21
    stereoisomers of, 7
    structures of, 7
    synthesis of, 7–8
    thermal properties of, 7
    thermodynamics of
      polymerization, 9–10
Lactobacillus helvetics, 5, 7
Lactonitrile, hydrolysis of, 7
Langmuir film balance technique,
  47

Lauritzen and Hoffman (LH) theory,
  76–81
Layer-by-layer (LbL) assembly, 61
Legislation
    biodegradation and
      compostability, 337–339
    food packaging and, 334–337
    nanomaterials in food
      packaging, 339–341
Life Cycle Assessment (LCA),
  341–343
Loading content, micelles, 53, 55
Long-term stability, 155–164
Long-term stability, plasticizers
    chemical stability and
      degradation, 159–160
    physicochemical studies of,
      155–159

Material thickness, hydrolytic
  degradation mechanism and, 295
Melt electrospun, 173–174
Melting temperature, 68–73
Melt/solid polycondensation, 27–29
Micelles, 50–56
Microfibrilated cellulose (MFC). *See*
  Cellulose nanofibrils (CNFs)
Microparticles, 49–50
Mixed plasticizers, 153
Modulus changes, 57–59
Molecular weight, hydrolytic
  degradation rate, 296–305
    additives/(nano)particles,
      blending with, 302–305
    blending with other polymers,
      301–302
    chemical structure, 296–297
    copolymerization, 299–301
    orientation of polymer chains,
      297–298
Monomeric plasticizers, 134–147,
  164–165
Montmorillonite (MMT), 180,
  217–218, 220, 222
Multi-walled carbon nanotubes
  (MWCNTs), 183

Nanocellulose
    surface modification of,
        228–230
    types of, 226–228
Nanofibre scaffolds, 48–49
Nanofibrilated cellulose (NFC). *See*
    Cellulose nanofibrils (CNFs)
Nanoingredients, 341
Nano-Mark System, 341
Nanostructured PLA, 270–279
Nanostructured poly(lactic acid)
        blending approach, 271–274
        nanocomposites, 276–277
        nanoparticles (NPs), 274–275
        nanoshell particles (NSs), 276
        surface modification of,
            277–279
Natural rubber, 203–208
Non-biodegradable petroleum-based
    polymers, 197–198
Novel Food Regulation EC No 258/
    97, 341

Oilgolactide, 8
Oligomeric plasticizers, 147–152,
    165

Phase morphology, 199
pH-sensitive stereo-complex
    nanoparticles, 61
Phyllosilicates, 216–218
Physical ageing, 155–158
PLA blending
    with biodegradable polymers,
        196–197
    with elastomers, 198–203
    with natural rubber,
        203–208
    with non-biodegradable
        petroleum-based polymers,
        197–198
PLA electrospun nanocomposite
    fibres, 179–183
        with carbon nanotubes,
            182–183
        with graphene, 183

with halloysite nanotubes,
    180–181
with hydroxyapatite, 181–182
with montmorillonites, 180
PLA electrospun nanofibres
    crystallinity, 177
    diameter, morphology and
        orientation, 175–177
    mechanical properties,
        177–179
PLA fibres production, 171–174
PLA homopolymers, synthesis of,
    38–39
PLA-nanocellulose biocomposites
    barrier properties, 238
    crystallization, 232–233
    mechanical properties,
        233–238
    nanocellulose, surface
        modification of, 228–230
    nanocellulose types, 226–228
    processing/mixing strategies,
        230–231
    properties, 232–238
    thermal properties, 233–238
PLA plasticizers
    ageing effects of, 155–158
    from biomass, 160–163
    glass transition temperature,
        135–146
    long-term stability of, 155–164
    mechanical properties, impact
        on, 134–153
    mixed, 153
    monomeric, 134–147
    oligomeric, 147–152
    as processing aids, 153–154
    stress-strain plots, 151–152
PLA stereo-complex
    biomedical applications of,
        47–61
    by co-crystallization, 39–40
    enzymatic degradation of,
        44–47
    micelles, 50–56
    microparticles, 49–50

nanofibre scaffolds, 48–49
properties of, 40–44
stereo-complex hydrogels,
  56–60
structure-properties of, 38–47
synthesis of, 38–47
PLA stereo-copolymers, synthesis of,
  38–39
Plasticization
  Free Volume Theory for, 126
  inverse effect of, 126
  of poly(lactide), 124–165
  principles of, 125–129
Plasticizers
  additives, 321–322
  interaction with contact media,
    129–134
  migration, 129–134
  miscibility of, 129, 130–132
  permanence, 129–134
  solubility parameters of, 130
Poly(butylene adipate-co-
  terephthalate) (PBAT), 196
Poly(butylene succinate) (PBS), 196
Poly(ε-caprolactone) (PCL), 273
Poly(ethylene succinate) (PES), 196
Poly(hydroxyalkanoates) (PHAs), 273
Poly(lactic acid) (PLA)
  abiotic-hydrolytic degradation
    of, 289–309
  and active packaging, 245–262
  additives, use of, 320–322
  agricultural mulch films, 329
  benefits of hydrolytic
    degradation, 305–308
  block copolymers, 84–92
  blown film extrusion, 323–324
  commercial applications,
    326–331
  copolymers *vs.* blends, PLLA
    and PCL, 87–89
  crystallization kinetics, 74–84
  crystallization of, 66–93
  crystal structure and single
    crystals, 67–69
  crystal unit parameters, 69

electrical appliances, 329–330
electrospinning, 325
extrusion, 322
factors controlling hydrolytic
  degradation of, 292–305
film and sheet casting, 322–323
foaming, 325–326
food contact materials,
  326–328
glass transition temperatures,
  68–73
half-crystallization time,
  inverse of, 80, 82
home furnishing, 331
industrial uses of, 317–331
Lauritzen and Hoffman (LH)
  theory, 76–81
legislation related to, 334–343
mechanical properties,
  318–319
medical and biomedical
  applications, 328
melting temperature, 68–73
molecular hydrolytic
  degradation mechanisms of,
  290–292
PE-*b*-PLA block copolymers,
  89–92
PLLA–PEO copolymers, 84–86
processing techniques,
  322–326
spherulite growth kinetics,
  74–76
spinning techniques, 325
stereochemistry, 319–320
stretch blow moulding, 323
superstructural morphology,
  73–74
textile applications, 330–331
thermal properties, 318–319
thermoforming, 324–325
Poly(lactic-co-glycolic acid) (PLGA),
  274
Poly(lactide) (PLA)
  free-radical grafting reactions
    of, 113–117

Poly(lactide) (*continued*)
    mechanical properties of, 125
    modification, blending with
        elastomers, 195–209
    plasticization of, 115–116,
        124–165
    reactive extrusion (REX)
        processing, 102–105
    structural diversities of, 8–9
Polyanhydrides, 272–273
Polycarbodiimide, 110
Polyethylene glycol acrylate (PEGA),
    grafting of, 116, 117
Polylactide (PLA)/clay
    nano-biocomposites
        nanoclays for, 216–218
        processing of, 219–220
        properties of, 220–222
Polyolefins, 248
Polyurethane elastomer (PU),
    198–203
Porosity, 295–296

QSAR. *See* Quantitative Structure-
    Activity Relationship (QSAR)
Quantitative Structure-Activity
    Relationship (QSAR), 127

Racemic DL-lactic acid, synthesis
    of, 7
Reactive extrusion (REX) processing,
    102–105
    coupling reactions, 107–113
    free-radical grafting reactions,
        113–117
    ring-opening polymerization
        (ROP), 105–107
    transesterification (exchange)
        reactions, 117–118
Regulation (EU) No 10/2011, 336
REX. *See* Reactive extrusion (REX)
    processing
Ring-opening polymerization (ROP),
    102
    aluminium catalysts, 19–20
    cationic catalysts, 15

    of DL-lactide, 39
    of lactide, 105–107
    of L-lactide, 39
    mechanism for, 17–18
    metal catalysts, 11–14
    organic catalysts, 15–18
    two-step, 21–22
ROP. *See* Ring-opening
    polymerization

Sepiolite, 218
Single-walled carbon nanotubes
    (SWCNTs), 182–183
Solution polycondensation, 27
Solvent-based electrospinning, 173
Spherulite growth kinetics,
    74–76
Spinning techniques, 325. *See also*
    Electrospinning process
Sporolactobacillus, 5
Stem cells, 267–269
    adult stem cells (ASCs), 268
    differentiation induction
        factors, 269
    embryonic stem cells (ESCs),
        268
    induced pluripotent stem cells
        (iPSCs), 268–269
Stereo-block polycondensation,
    29–30
Stereo-complex hydrogels, 56–60
Stereo-controlled polymerization,
    18–21
Stress-strain curves
    PLA/NR blend, 205
    PLA/PU blend, 201
Stretch blow moulding, 323
Structural modification, hydrolytic
    degradation, 306–308
    crystallization and thermal
        properties, variations in,
        306–308
    grafting, surface reactive sites
        for, 308
    mechanical properties,
        improvements of, 308

Superstructural morphology,
73–74
Surface area, 295–296
SWCNTs. *See* Single-walled carbon
nanotubes

Taylor cone, 172
Textile applications, 330–331
Thermoforming, 324–325
Thermoforming process,
324–325
Tissue engineering (TE),
184–187, 266–267
  biomaterials and
    nanotechnology,
    269–270

nanostructured PLA and,
270–279
stem cells, 267–269
Transesterification (exchange)
reactions, 117–118
T-shaped REX technology, 110

Universal Functional Activity
coefficient for Polymers (UNIFAP),
127

Viscoelastic curves, 200

Wide angle X-ray diffraction (WAXD)
spectrum, 42
Wound dressing, 187–188